NONLINEAR REGRESSION MODELING FOR ENGINEERING APPLICATIONS

Wiley-ASME Press Series List

Introduction to Dynamics and Control of Mechanical Engineering Systems	To	March 2016
Fundamentals of Mechanical Vibrations	Cai	May 2016
Nonlinear Regression Modeling for Engineering Applications	Rhinehart	August 2016
Stress in ASME Pressure Vessels	Jawad	November 2016
Bioprocessing Piping and Equipment Design	Huitt	November 2016
Combined Cooling, Heating, and Power Systems	Shi	January 2017

NONLINEAR REGRESSION MODELING FOR ENGINEERING APPLICATIONS

MODELING, MODEL VALIDATION, AND ENABLING DESIGN OF EXPERIMENTS

R. Russell Rhinehart

WILEY

This edition first published 2016

First Edition published in 2016

Registered office
John Wiley & Sons Ltd, The Atrium, Southern Gate, Chichester, West Sussex, PO19 8SQ, United Kingdom

For details of our global editorial offices, for customer services and for information about how to apply for permission to reuse the copyright material in this book please see our website at www.wiley.com.

Library of Congress Cataloging-in-Publication Data:

Names: Rhinehart, R. Russell, 1946- author.
Title: Nonlinear regression modeling for engineering applications : modeling, model validation, and enabling design of experiments / R. Russell Rhinehart.
Description: Chichester, UK ; Hoboken, NJ : John Wiley & Sons, 2016. | Includes bibliographical references and index.
Identifiers: LCCN 2016012932 (print) | LCCN 2016020558 (ebook) | ISBN 9781118597965 (cloth) | ISBN 9781118597934 (pdf) | ISBN 9781118597958 (epub)
Subjects: LCSH: Regression analysis–Mathematical models. | Engineering–Mathematical models.
Classification: LCC TA342 .R495 2016 (print) | LCC TA342 (ebook) | DDC 620.001/519536–dc23
LC record available at https://lccn.loc.gov/2016012932

A catalogue record for this book is available from the British Library.

Set in 10/12pt, TimesLTStd by SPi Global, Chennai, India.
Printed and bound in Malaysia by Vivar Printing Sdn Bhd

1 2016

Contents

Part III REGRESSION, VALIDATION, DESIGN

Series Preface

The Wiley-ASME Press Series in Mechanical Engineering brings together two established leaders in mechanical engineering publishing to deliver high-quality, peer-reviewed books covering topics of current interest to engineers and researchers worldwide.

The series publishes across the breadth of mechanical engineering, comprising research, design and development, and manufacturing. It includes monographs, references and course texts.

Prospective topics include emerging and advanced technologies in Engineering Design; Computer-Aided Design; Energy Conversion & Resources; Heat Transfer; Manufacturing & Processing; Systems & Devices; Renewable Energy; Robotics; and Biotechnology.

Preface

Utility

Mathematical models are important.

Engineers use mathematical models to describe the natural world and then rearrange the model equations to answer the question, "How do I create an environment that makes Nature behave the way I want it to?" The answer to the mathematical rearrangement of the model equations reveals how to design processes, products, and procedures. It also reveals how to operate, use, monitor, and control them. Modeling is a critical underpinning for engineering analysis, design, control, and system optimization.

Further, since mathematical models express our understanding of how Nature behaves, we use them to validate our understanding of the fundamentals about processes and products. We postulate a mechanism and then derive a model grounded in that mechanistic understanding. If the model does not fit the data, our understanding of the mechanism was wrong or incomplete. Alternately, if the model fits the data we can claim our understanding may be correct. Models help us develop knowledge.

These models usually have coefficients representing some property of Nature, which has an unknown value (e.g., the diffusivity of a new molecule in a new medium, drag coefficient on a new shape, curing time of a new concrete mix, a catalyst effective surface area per unit mass, a heat transfer fouling factor). Model coefficient values must be adjusted to make the model match the experimentally obtained data, and obtaining the value of the coefficient adds to knowledge.

The procedure for finding the model coefficient values that makes a model best fit the data is called regression.

Although regression is ages old, there seem to be many opportunities for improvements related to finding a global optimum; finding a universal, effective, simple, and single stopping criterion for nonlinear regression; validating the model; balancing model simplicity and sufficiency with perfection and complexity; discriminating between competing models; and distinguishing functional sufficiency from prediction accuracy.

I developed and used process and product models throughout my 13-year industrial career. However, my college preparation for the engineering career did not teach me what I needed to know about how to create and evaluate models. I recognized that my fellow engineers, regardless of their *alma mater*, were also underprepared. We had to self-learn as to what was needed. Recognizing the centrality of modeling to engineering analysis, I have continued to explore model development and use during my subsequent academic career.

This textbook addresses nonlinear regression from a perspective that balances engineering utility with scientific perfection, a view that is often missing in the classroom, wherein the focus is often on the mathematical analysis, which pretends that there are simple, first-attempt solutions. Mathematical analysis is intellectually stimulating and satisfying, and sometimes useful for the practitioner. Where I think it adds value, I included analysis in this book. However, development of a model, choosing appropriate regression features, and designing experiments to generate useful data are iterative procedures that are guided by insight from progressive experience. It would be a rare event to jump to the right answers on the first try. Accordingly, balancing theoretical analysis, this book provides guides for procedure improvement.

This work is a collection of what I consider to be best practices in nonlinear regression modeling, which necessarily includes guides to design experiments to generate the data and guides to interpret the models. Undoubtedly, my view of best has been shaped with my particular uses for the models within the context of process and product modeling. Accordingly, this textbook has a focus on models with continuous-valued variables (either deterministic, discretized, or probabilities) as opposed to rank or classification, nonlinear as opposed to linear, constrained as opposed to not, and of a modest number of variables as opposed to Big Data.

This textbook includes the material I wish I had known when starting my engineering career and now what I would like my students to know. I hope it is useful for you.

The examples and discussion presume basic understanding of engineering models, regression, statistics, optimization, and calculus. This textbook provides enough details, explicit equation derivations, and examples to be useful as an introductory learning device for an upper-level undergraduate or graduate. I have used much of this material in the undergraduate unit operations lab course, in my explorations of model-based control on pilot-scale units, and in modeling of diverse processes (including the financial aspects of my retirement and the use of academic performance in the first two college years to project upper-level success). A person with an engineering degree and some experience with regression should be able to follow the concepts, analysis, and discussion.

My objective is to help you answer these questions:

- How to choose model inputs (variables, delays)?
- How to choose model form (linear, quadratic, or higher order, or equivalent model structures or architectures such as dimension or number of neurons)?
- How to design experiments to obtain adequate data (in number, precision, and placement) for determining model coefficient values?
- What to use for the regression objective (vertical least squares, total least squares, or maximum likelihood)?
- How to define goodness of model (r-square, fitness for use, utility, simplicity, data-based validation, confidence interval for prediction)?
- How to choose the right model between two different models?
- What optimization algorithm should be used for the regression to be able to handle the confounding issues of hard or soft constraints, discontinuities, discrete and continuous variables, multiple optima, and so on?
- What convergence criteria should be used to stop the optimizer (to recognize when it is close enough to optimum)?
- Should you linearize and use linear regression or use nonlinear regression?

- How to recognize outliers?
- How can you claim that a model properly captures some natural phenomena?

The underlying techniques needed for the answers include propagation of uncertainty, probability and statistics, optimization, and experience and heuristics. The initial chapters review/develop the basics. Subsequent chapters provide the application techniques, description of the algorithms, and guides for application.

Access to Computer Code

Those interested can visit the author's web site, www.r3eda.com, for open access to Excel VBA macros to many of the procedures in this book.

Years back our college decided to standardize with Visual Basic for Applications (VBA) for the undergraduate computer programming course. As a result, routines supporting this text are written in VBA, which is convenient to me, and also a widely accessible platform. However, VBA is not the fastest, and some readers may not be familiar with that language. Therefore, this text also provides a VBA primer and access to the code so that a reader may convert the VBA code to some other personally preferred platform. If you understand any structured text procedures, you can understand the VBA code here.

Preview of the Recommendations

Some of the recommendations in this book are counter to traditional practice in regression and design of experiments (DoE), which seem to be substantially grounded in linear regression. As a preview, opinions offered in this textbook are:

1. If the equation is nonlinear in the coefficients, use nonlinear regression. Even if the equation can be log-transformed into a linear form, do not do it. Linearizing transformations distort the relative importance of data points within the data set. Unless data variance is relatively low and/or there are many data points, linearizing can cause significant error in the model coefficient values.
2. Use data pre-processing and post-processing to eliminate outliers.
3. Use direct search optimizers for nonlinear regression rather than gradient-based optimizers. Although gradient-based algorithms converge rapidly in the vicinity of the optimum, direct search optimizers are more robust to surface aberrations, can cope with hard constraints, and are faster for difficult problems. Leapfrogging is offered as a good optimizer choice.
4. Nonlinear regression may have multiple minima. No optimizer can guarantee finding the global minimum on a first trial. Therefore, run the optimizer for N trials, starting from random locations, and take the best of the N trials. N can be calculated to meet the user desire for the probability of finding an optimum within a user-defined best fraction. The equation is shown.
5. Pay as much attention to how constraints are defined and included in the optimization application as you do to deriving the model and objective function (OF) statement. Constraints can have a substantial influence on the regression solution.

6. The choice of stopping criteria is also influential to the solution. Conventional stopping criteria are based on thresholds on the adjustable model coefficient values (decision variables, DVs), and/or the regression target (usually the sum of squared deviations) that we are seeking to optimize (OF). Since the right choice for the thresholds requires *a priori* knowledge, is scale-dependent, and requires threshold values on each regression coefficient (DV) and/or optimization target (OF), determining right threshold values requires substantial user experience with the specific application. This work recommends using steady-state identification to declare convergence. It is a single criterion (only looking at one index – statistical improvement in OF relative to data variability from the model), which is not scale-dependent.
7. Design the experimental plan (sequence, range, input variables) to generate data that are useful for testing the validity of the nonlinear model. Do not follow conventional statistical DoE methods, which were devised for alternate outcomes – to minimize uncertainty on the coefficients in nonmechanistic models, in linear regression, within idealized conditions.
8. Design the experimental methods of gathering data (measurement protocol, number and location of data sets) so that uncertainty on the experimental measurements has a minimal impact on model coefficient values.
9. Use of the conventional least-squares measure of model quality, $\sum(y_{data} - y_{model})^2$, is acceptable for most purposes. It can be defended by idealizing maximum likelihood conditions. Maximum likelihood is more compatible with reality and can provide better model coefficient values, but it presumes knowledge of the variance on both experimental inputs and output, and requires a nested optimization. Maximum likelihood can be justified where scientific precision is paramount, but adds complexity to the optimization.
10. Akaho's method is a computationally simple improvement for the total east-squares approximation to maximum likelihood.
11. Establish nonlinear model validity with statistical tests for bias and either autocorrelation or runs. Do not use *r*-square or ANOVA techniques, which were devised for linear regression under idealized conditions.
12. Eliminate redundant coefficients, inconsequential model terms, and inconsequential input variables.
13. Perform both logic-based *and* data-based tests to establish model validity.
14. Model utility (fitness for use) and model validity (representation of the truth about Nature) are different. Useful models often do not need to be true. Balance perfection with sufficiency, complexity with simplicity, rigor with utility.

Philosophy

I am writing to you, the reader, in a first-person personal voice, a contrast to most technical works. There are several aspects that led me to do so, but all are grounded in the view that humans will be implementing the material.

I am a believer in the Scientific Method. The outcomes claimed by a person should be verifiable by any investigator. The methodology and analysis that led to the outcomes should be grounded in the widely accepted best practices. In addition, the claims should be tempered and accepted by the body of experts. However, the Scientific Method wants decisions to be purely rational, logical, and fact based. There should be no personal opinion, human emotion, or human bias infecting decisions and acceptances about the truth of Nature. To preserve the image of no human involvement, most technical writing is in the third person. However, an author's choice of idealizations, acceptances, permissions, assumptions, givens, basis, considerations, suppositions, and such, are necessary to permit mathematical exactness, proofs, and the consequential absolute statements. However, the truth offered is implicitly infected by the human choices. If a human is thinking it, or if a human accepts it, it cannot be devoid of that human's perspective and values. I am not pretending that this book is separate from my experiences and interpretations so I am writing in the first person.

Additionally, consider the individuals applying techniques. They are not investigating a mathematical analysis underlying the technique, but need to use the technique to get an answer for some alternate purpose. Accordingly, utility with the techniques is probably as important as understanding the procedure basis. Further, the application situation is not an idealized simplification. Nature confounds simplicity with complexity. Therefore, as well as proficiency in use, a user must understand and interpret the situation and choose the right techniques. The human applies it and the human must choose the appropriate technique. Accordingly, to make a user functional, it is important for a textbook to understand the limits and appropriateness of techniques. The individual is the agent and primary target, the tool is just the tool. The technique is not the truth, so I am writing to the user.

It is also essential that a user truly understands the basis of a tool, to use it properly. Accordingly, in addition to discussing the application situations, this text develops the equations behind the methods, includes mathematical analysis, and reveals nuances through examples. The book also includes exercises so the user can develop skills and understanding.

In the 1950s Benjamin Bloom chaired a committee of educators that subsequently published a taxonomy of Learning Objectives, which has come to be known as Bloom's Taxonomy. One of the domains is termed the Cognitive, related to thinking/knowing. There are six levels in the Taxonomy. Here is my interpretation for engineering (Table 1).

Notably most of classroom instruction has the student working in the lower three levels, where there are no user-choices. There is only one way to spell "cat," only one right answer to the calculation of the required orifice diameter using the ideal orifice equation and givens in the word problem, and so on. In school, the instructor analyzes the situation, synthesizes the exercise, and judges the correctness of the answer. By contrast, competency and success in professional and personal life requires the individual to mentally work in the upper levels where the situation must be interpreted, where the approach must be synthesized, and where the propriety of the approach and answer must be evaluated. When instruction prevents the student from working in the upper cognitive levels, it misrepresents the post-graduation environment, which does a disservice to the student and employers who have to redirect the graduate's perspective. Accordingly, my aim is to facilitate the reader's mental activity in the upper levels where human choices have to be made. I am therefore writing to the human, not just about the technology.

A final perspective, on the philosophy behind the style and contents of this book is grounded in a list of desired engineering attributes. The members of the Industrial Advisory Committee for our School helped the faculty develop a list of desired engineering attributes, which we use to shape what we teach and shape the student's perspectives. Engineering is an activity, not a body of knowledge. Engineering is performed by humans within a human environment; it is not the intellectual exercise about isolated mathematical analysis. There are opposing ideals in judging engineering and the list of Desired Engineering Attributes reveals them. The opposing ideals are highlighted in bold (Table 2).

Table 1 Bloom's taxonomy

Level	Name	Function – person does	Examples
6	Evaluation (E)	Judge goodness, sufficiency, and completeness of something, choose the best among options, know when to stop improving. Must consider all aspects	Decide that a design, report, research project, or event planning is finished when considering all issues (technical completeness, needs of all stakeholders, ethical standards, safety, economics, impact, etc.)
5	Synthesis (S)	Create something new: purposefully integrate parts or concepts to design something new that meets a function	Design a device to meet all stakeholders' approvals within constraints. Create a new homework problem integrating all relevant technology, design a procedure to meet multiple objectives, create a model, create a written report, design experiments to generate useful data
4	Analysis (An)	Two aspects related to context *One*. Separate into parts or stages, define and classify the mechanistic relationships of something within the whole	*One*. Describe and model the sequence of cause-and-effect mechanisms: tray-to-tray model that relates vapor boil-up to distillate purity, impact of transformer start-up on the entire grid, impact of an infection on the entire body and person health
		Two. Critique, assess goodness, determine functionality of something within the whole	*Two*. Define and compute metrics that quantify measures of utility or goodness
3	Application (Ap)	Independently apply skills to fulfill a purpose within a structured set of "givens"	Properly follow procedures to calculate bubble point, size equipment, use the Excel features to properly present data, solve classroom "word problems"
2	Understanding/ comprehension (U/C)	Understand the relation of facts and connection of abstract to concrete	Find the diameter of a 1-inch diameter pipe, convert units, qualitatively describe staged equilibrium separation phenomena, explain the equations that describe an RC circuit, understand what Excel cell equations do
1	Knowledge (K)	Memorize facts and categorization	Spell words, recite equations, name parts of a valve, read resistance from color code, recite the six Bloom levels

Table 2 Desired engineering attributes

Engineering is an activity that delivers solutions that work for all stakeholders. Desirably engineering:

- Seeks **simplicity** in analysis and solutions, while being **comprehensive** in scope.
- Is **careful**, correct, self-critical, and defensible; yet is performed with a **sense of urgency**.
- Analyzes **individual mechanisms** and integrates stages to **understand the whole**.
- Uses state-of-the-art **science** and **heuristics**.
- Balances **sufficiency** with **perfection.**
- Develops **sustainable solutions** – profitable and accepted **today**, without burdening **future stakeholders**.
- Tempers **personal gain** with **benefit to others**.
- Is **creative**, yet **follows codes**, regulations, and standard practices.
- Balances probable **loss** with probable **gain** but not at the expense of EHS&LP – **manages risk**.
- Is a collaborative, **partnership activity**, energized by **individuals**.
- Is an **intellectual analysis** that leads to **implementation and fruition**.
- Is **scientifically valid**, yet **effectively communicated** for all stakeholders.
- Generates **concrete** recommendations that honestly reveal **uncertainty**.
- Is grounded in **technical fundamentals** and the **human context** (societal, economic, and political).
- Is grounded in **allegiance to the bottom line of the company** and to **ethical standards of technical and personal conduct**.
- Supports **enterprise harmony** while seeking to **cause beneficent change**.

Engineering is not just about technical competence. State-of-the-art commercial software beats novice humans in speed and completeness with technical calculations. Engineering is a decision-making process about technology within human enterprises, value systems, and aspirations, and I believe this list addresses a fundamental aspect of the essence of engineering. As a complement to fundamental knowledge and skill of the core science and technical topics, instructors need to understand the opposing ideals, the practice of application, so that they can integrate the issues into the student's experience and so that student exercises have students practice right perspectives as they train for technical competency.

A straight line is very long. Maybe the line goes between pure science on one end and pure follow-the-recipe and accept-the-computer-output on the other end. No matter where one stands, the line disappears into the horizons to the left and to the right. No matter where one stands, it feels like the middle, the point of right balance between the extremes. However, the person way to the left also thinks they are in the middle. If Higher Education is to prepare graduates for industrial careers, instructors need to understand the issues surrounding Desired Engineering Attributes from an industrial perspective, not their academic/science perspective. Therefore, I am writing to the human about how to balance those opposing ideals when using nonlinear regression techniques for applications.

Acknowledgments

My initial interest in modeling processes and products arose from my engineering experience within industry, and most of the material presented here benefited from the investigations of my graduate students as they explored the applicability of these tools, guidance from industrial advisors as they provided input on graduate projects and undergraduate education outcomes, and a few key mentors who helped me see these connections. Thank you all for revealing issues, providing guidance, and participating in my investigations.

Some of the techniques in this text are direct outcomes of the research performed by Gaurav Aurora, R. Paul Bray, Phoebe Brown, Songling Cao, Chitan Chandak, Sandeep Chandran, Solomon Gebreyohannes, Anand Govindrajanan, Mahesh S. Iyer, Suresh Jayaraman, Junyi Li, Upasana Manimegalai-Sridhar, Siva Natarajan, Jing Ou, Venkat Padmanabhan, Anirudh (Andy) Patrachari, Neha Shrowti, Anthony Skach, Ming Su, John Szella, Kedar Vilankar, and Judson Wooters.

A special thanks goes to Robert M. Bethea (Bob) who invited me to coauthor the text *Applied Engineering Statistics*, which was a big step toward my understanding of the interaction between regression, modeling, experimental design, and data analysis. Another special thank you to Richard M. Felder, always a mentor in understanding and disseminating engineering science and technology.

As a professor, funding is essential to enable research, investigation, discovery, and the pursuit of creativity. I am grateful to both the Edward E. and Helen Turner Bartlett Foundation and the Amoco Foundation (now BP) for funding endowments for academic chairs. I have been fortunate to be the chair holder for one or the other, which means that I was permitted to use some proceeds from the endowment to attract and support graduate students who could pursue ideas that did not have traditional research support. This book presents many of the techniques explored, developed, or tested by the graduate students.

Similarly, I am grateful for a number of industrial sponsors of my graduate program who recognized the importance of applied research and its role in workforce development.

Most of all, career accomplishments of any one person are the result of the many people who nurtured and developed the person. I am of course grateful to my parents, teachers, and friends, but mostly to Donna, who for the past 26 years has been everything I need.

Nomenclature

Accept	Not reject. There is not statistically sufficient evidence to confidently claim that the null hypothesis is not true. There is not a big enough difference. This is equivalent to the not guilty verdict, when the accused might have done it, but the evidence is not beyond reasonable doubt. Not guilty does not mean innocent. Accept means cannot confidently reject and does not mean correct.
Accuracy	Closeness to the true value, bias, average deviation. In contrast to precision.
AIC	Akiake Information Criterion, a method for assessing the balance of model complexity to fit to data.
A priori	Latin origin for "without prior knowledge."
Architecture	The functional form of the mathematical model.
ARL	Average run length, the average number of samples to report a confident result.
Autocorrelation	One value of a variable that changes in time is related to prior values of that variable.
Autoregressive	A mathematical description that one value of a variable that changes in time is related to prior values of that variable; the cause would be some fluctuating input that has a persisting influence.
Batch regression	The process of regression operates on all of the data in one operation.
Best-of-N	Start the optimizer N times with independent initializations and take the best of the N trials as the answer.
Bias	A systematic error, a consistent shift in level, an average deviation from true.
Bimodal	A pattern in the residuals that indicates there are two separate distributions, suggesting two separate treatments affected the data.
Bootstrapping	A numerical, Monte Carlo, technique for estimating the uncertainty in a model-predicted value from the

	inherent variability in the data used to regress model coefficient values.
Cardinal	Integers, counting numbers, a quantification of the number of items.
Cauchy's technique	An optimization approach of successive searches along the line of local steepest descent.
CDF	The cumulative distribution function, the probability of obtaining an equal or smaller value.
Chauvenet's criterion	A method for selecting data that could be rejected as an outlier.
Class	The variable that contains the name of a classification – nominal, name, category.
Coefficient correlation	When the optimizer does not find a unique solution, perhaps many identical or nearly identical OF values for different DV values, a plot of one DV value w.r.t. another reveals that one coefficient is correlated to the other. Often termed parameter correlation.
Coefficient or model coefficient	A symbol in a model that has a fixed value from the model use perspective. Model constants or parameters. Some values are fundamental such as Pi or the 2 in square root. Other values for the coefficients are determined by fitting model to data. Such coefficient values will change as new data is added.
Confidence	The probability that a statement is true.
Constraints	Boundaries that cannot be violated, often rational limits for regression coefficients.
Convergence	The optimizer trial solution has found the proximity of the optimum within desired precision.
Convergence criterion	The metric used to test for convergence – could be based on the change in DVs, change in OF, and so on.
Correlation	Two variables are related to each other. If one rises, the other rises. The relation might be confounded by noise and variation, and represent a general, not exact relation. The relation does not have to be linear.
Cross correlation	Two separate variables are related to each other. Contrast to autocorrelation in which values of one variable are related to prior values.
Cumulative sum	CUSUM, cumulative sum of deviations scaled by the standard deviation in the data.
CUSUM	Cumulative sum of deviations scaled by the standard deviation in the data.
Cyclic heuristic	CH, an optimizer technique that makes incremental changes in one DV at a time, taking each in turn. If the OF is improved, that new DV value is retained and the next increment for that DV will be larger. Otherwise, the

	old DV value is retained and the next increment for that DV will be both smaller and in the opposite direction.
Data	As a singular data point (set of conditions) or as the plural set of all data points.
Data-based validation	The comparison of model to data to judge if the model properly captures the underlying phenomena.
Data model	The calculation procedure used to take experimental measurements to generate data for the regression modeling, the method to calculate y and x experimental from sensor measurements.
Data reconciliation	A method for correcting a set of measurements in light of a model that should make the measurements redundant.
Decision variables	DVs are what you adjust to minimize the objective function (OF). In regression, the DVs are the model coefficients that are adjusted to make the model best fit the data.
Dependent variable	The output variable, output from model, result, impact, prediction, outcome, modeled value.
Design	Devising a procedure to achieve desired results.
Design of experiments	DoE, the procedure/protocol/sequence/methodology of executing experiments to generate data.
Deterministic	The model returns one value representing an average, or parameter value, or probability.
Deviation	A variable that indicates deviation from a reference point (as opposed to absolute value).
Direct search	An optimization procedure that uses heuristic rules based on function evaluations, not derivatives. Examples include Hooke–Jeeves, leapfrogging, and particle swarm.
Discrete	A variable that has discrete (as opposed to continuum) values – integers, the last decimal value.
Discrimination	Using validation to select one model over another.
Distribution	The description of the diversity of values that might result from natural processes (particle size), simulations (stochastic process, Monte Carlo simulation), or an event probability.
DoE	Design of experiments.
DV	Decision variable.
Dynamic	The process states are changing in time in response to an input, often termed transient.
EC	Equal concern – a scaling factor to balance the impact of several measures of undesirability in a single objective function. Essentially, the reciprocal of the Lagrange multiplier.

Empirical	The model has a generic mathematical functional relation (power series, neural network, wavelets, orthogonal polynomials, etc.) with coefficients chosen to best shape the functionalities to match the experimentally obtained data.
Ensemble	A model that uses several independent equations or procedures to arrive at predictions, then some sort of selection to choose the average or representative value.
Equal concern factor	The degree of violation of one desire that raises the same level of concern as a specified violation of another desire, weighting factors in a penalty that are applied as divisors as opposed to Lagrange multipliers.
Equality constraints	A constraint that relates variables in an equality relation, useful in reducing the number of DVs.
EWMA	Exponentially weighted moving average, a first-order filtered value of a variable.
EWMV	Exponentially weighted moving variance, a first-order filtered value of a variance.
Experiment	A procedure for obtaining data or results. The experiment might be physical or simulated.
Exponentially weighted moving average	EWMA, a first-order filtered value of a variable.
Exponentially weighted moving variance	EWMV, a first-order filtered value of a variance.
Final prediction error	FPE, Ljung's take on Akaike's approach to balancing model complexity with reduction in SSD. Concepts are similar in Mallows' Cp and Akaike's information criterion.
First principles	An approach that uses a fundamental mechanistic approach to develop an elementary model. A phenomenological model, but not representing an attempt to be rigorous or complete.
First-order filter	FOF – an equation for tempering noise by averaging, an exponentially weighted moving average, the solution to a first-order differential equation, the result of an RC circuit for tempering noise on a voltage measurement.
FL	Fuzzy logic – models that use human linguistic descriptions, such as: "Its cold outside so wear a jacket." This is not as mathematically precise as, "The temperature is 38 °F, so use a cover with an insulation *R*-value of 12," but fully adequate to take action.
FOF	First-order filter.
FPE	Final prediction error, which is Ljung's take on Akaike's approach to balancing model complexity with reduction in SSD. Concepts are similar in Mallows' Cp and Akaike's information criterion.

Fuzzy logic — FL – models that use human linguistic descriptions, such as: "Its cold outside so wear a jacket." This is not as mathematically precise as, "The temperature is 38 °F, so use a cover with an insulation *R*-Vvalue of 12," but fully adequate to take action.

Gaussian distribution — The bell-shaped or normal distribution.

Generalized reduced gradient — GRG, a gradient-based optimization approach that reduces the number of DVs when a constraint is encountered by replacing the inequality with an equality constraint as long as the constraint is active.

Global optimum — The extreme lowest minima or highest maxima of a function.

Gradient — The vector of first derivatives of the OF w.r.t. each DV, the direction of steepest descent. Gradient-based optimizers include Cauchy's sequential line search, Newton–Raphson, Levenberg–Marquardt, and GRG.

Gradient-based optimization — Optimization approaches that use the gradient, the direction of steepest descent. Gradient-based optimizers include Cauchy's sequential line search, Newton–Raphson, Levenberg–Marquardt, and GRG.

GRG — Generalized reduced gradient, a gradient-based optimization approach that reduces the number of DVs when a constraint is encountered by replacing the inequality with an equality constraint as long as the constraint is active.

Hard constraint — May not be violated, because it leads to an operation that is impossible to execute (square root of a negative, divide by zero) or violates some physical law (the sum of all compositions must be less than or equal to 100%).

Histogram — A bar graph representing the likelihood (probability, frequency) of obtaining values within numerical intervals.

HJ — Hooke–Jeeves, an optimization procedure that searches a minimal pattern of local OF values to determine where to incrementally move the pattern, moves the pattern center, and repeats.

Homoscedasity — Constant variance throughout a range.

Hooke–Jeeves — HJ, an optimization procedure that searches a minimal pattern of local OF values to determine where to incrementally move the pattern, moves the pattern center, and repeats.

Imputation — The act of creating missing data values from correlations to available data.

Incremental regression — The model coefficients are incrementally adjusted at each sampling, so that the model evolves with the changing process that generates the data.

Incremental steepest descent	ISD, an optimization technique that makes incremental steps in the steepest descent direction, re-evaluating the direction of steepest descent after each incremental TS move.
Independent variable	Input variable, input to the model, cause, influence.
Inequality constraints	A constraint that related variables in an inequality relation, a less than or greater than relation. Could be treated as either a hard or soft constraint.
Input variable	An influence, cause, source, input, forcing function, independent value to the model.
Inverse	The model is used "backward" to answer the question, "What inputs are required to provide a desired output?"
ISD	Incremental steepest descent, an optimization technique that makes incremental steps in the steepest descent direction, re-evaluating the direction of steepest descent after each incremental TS move.
Lag	In statistics it refers to the time interval between time-discretized data. A lag of 5 means a delay of five samples. In dynamic modeling it refers to a first-order, asymptotic dynamic response to a final value. Both definitions are used in this book.
Lagrange multipliers	Weighting factors in a penalty that are applied as multipliers as opposed to equal concern factors.
Leapfrogging	LF, an optimization technique that scatters players throughout DV space and then leaps the worst over the best, to converge on the optimum.
Levenberg–Marquardt	LM, an optimization technique that blends incremental steepest descent and Newton–Raphson.
LHS	Left-hand side, the terms on the left-hand side of an equation (either equality or inequality).
Likelihood	A measure of the probability that a model could have generated the experimental data.
Linear	The relation between two variables is a straight line.
Linearizing transforms	Mathematical operations that linearize an OF, providing the convenience of using linear regression solution methods. Be cautious about the weighting distortion that results.
LF	Leapfrogging, an optimization technique that scatters players throughout DV space, then leaps the worst over the best, to converge on the optimum.
LM	Levenberg–Marquardt, an optimization technique that blends incremental steepest descent and Newton–Raphson.
Local optimum	One of several minima or maxima of a function, but not the extreme.

Logic-based validation	Comparison of model functionality to rational, logical expectations.
MA	Moving average, the average of the chronologically most recent *N* data values in a time series.
Maximum	Highest or largest value.
Maximum error	An estimate of the uncertainty in a calculated value, based on all sources of uncertainty providing their maximum perturbation and influencing the outcome in the same direction.
Mean	The expected average in a list of data.
Median	The middle value in a list of data. To find it, repeatedly exclude the high and low values. If an odd-numbered list, the one that remains is the median. If an even-numbered list, average the two that remain.
Minimum	Lowest or smallest value.
Model	A mathematical representation of the human's understanding of Nature's response to the influence.
Model architecture	The mathematical structure, the functional relations within a model.
Moving average	MA, the average of the chronologically most recent *N* data values in a time series.
Nature	A respectful anthropomorphic representation of the mystery of the processes that generate data and which tortures us with complexity and variation.
Nelder–Mead	NM, a direct search optimization technique that uses the Simplex geometry and moves the worst local trial solution through the centroid of the others.
Neural network	NN – a modeling approach that was intended to mimic how brain neurons "calculate."
Newton–Raphson	NR, an optimization method that uses the local OF derivatives and a second-order series approximation of the OF to predict the optimum, jumps there and repeats.
$NID(\mu, \sigma)$	Normally (Gaussian) and independently distributed with a mean of μ and a standard deviation of σ.
NM	Nelder–Mead, a direct search optimization technique that uses the Simplex geometry and moves the worst local trial solution through the centroid of the others.
NN	Neural network – a modeling approach that was intended to mimic how brain neurons "calculate."
Noise	Random, independent perturbations to a conceptually deterministic value.
Nominal	Latin origin for name, a class/category/string variable.
Nonlinear	Means not linear. This could be any not linear relation.
Nonparametric test	The category of statistical tests that do not presume a normal model of the residuals.

Normal equations	The set of linear equations that arise in linear regression when the analytical derivatives of the OF w.r.t. each coefficient are set to zero.
Normal SSD	The sum of squared differences between the model and data in both the x and y axes, normal to the model, perpendicular to the model, often called total SSD.
NR	Newton–Raphson, an optimization method that uses the local OF derivatives and a second-order series approximation of the OF to predict the optimum, jumps there and repeats.
Not reject	There is not statistically sufficient evidence to confidently claim that the null hypothesis is not true. There is not a big enough difference. This is equivalent to the not guilty verdict, when the accused might have done it, but the evidence is not beyond a reasonable doubt. Not guilty does not mean innocent. Not reject does not mean correct. We use accept, instead of not reject.
Null hypothesis	The supposition that two treatments are equal, that differences are due to experimental vagaries, not mechanistic cause-and-effect relations.
Objective function	The procedure or equation that provides a measure of badness to be minimized (or goodness to be maximized). Usually the OF is the sum of squared deviations of modeled output to data.
Objective function value	The value of the OF.
OF	Objective function.
Optimization	A procedure for determining best values. In regression it is applied to determine model coefficient values that make model best fit to the data.
Optimum	Either lowest or highest value, either minimum or maximum.
Ordinal	Sequence or rank from ordering things relative to some attribute.
Outlier	A data point that is expected to be part of the cluster, but deviates too far from the expected location to be accepted as "all is well."
Output variable	A response, effect, output, consequent, prediction, state variable, modeled value.
Overfitting	When there are too many adjustable model coefficients and the model begins to fit noise in the data, not just the underlying trend. In the neural network community this is termed memorization.
Parametric analysis	The exploration of trends in model output variables as coefficient values change. Used in logical validation and sensitivity analysis.

Parametric test	The category of statistical tests that presumes a normal model of the residuals.
Parity plot	A graph of the model predicted values w.r.t. the actual data. Ideally, points fall on the 1:1 line. Patterns in deviations reveal model issues.
Particle swarm optimization	PSO, an optimization technique that scatters individuals (particles, players) throughout DV space and then each explores the local area with random perturbations while being drawn to both their personal best spot and the global best of all players.
Partitioned model	A model that has an IF-THEN structure that directs which equation or procedure should be used. For instance, if laminar flow, use the Hagan–Poiseuille relation, but if turbulent flow, use the Fanning–Moody–Darcy relation.
PDF	Probability distribution function, a mathematical equation representing the normalized histogram.
Penalty	A value added to the OF representing the degree of violation of a soft constraint.
Phenomenological	The mathematical model was derived from a conceptual understanding of the behavior of Nature. Other terms include first principles, mechanistic, fundamental, theoretical, rigorous, physical, and scientific.
Post-processing	Adjustment or culling of data after data work-up, in the light of modeling outcomes.
Precision	A measure of reproducibility of replicated results, perhaps the standard deviation. This is in contrast to accuracy, which is a measure of the average deviation from true.
Pre-processing	Adjustment or culling of data prior to use in modeling.
Probable error	An estimate of the uncertainty in a calculated value, based on the multiple sources of uncertainty providing independent perturbation magnitudes and signs.
Probability model	A model that predicts the probability distribution, either PDF or CDF.
Process/product model	The nominal input–output, mechanistic, influence–response, composition–property model of the process or product.
PSO	Particle swarm optimization, a multiplayer direct search approach.
PV	Process variable, a measured value, typically from a continuously operating process.
r-lag-1	A ratio of variances used to indicate autocorrelation of the residuals.
r-square	A ratio of variances to indicate the amount of variance removed by the model.

r-statistic	A ratio of variances used in steady- and transient-state identification.
Random	Independent values in a sequence. The distribution may or may not be Gaussian.
Rank	The variable that contains the order, precedence, placement, preference.
Rational	A variable with continuous-valued values (as opposed to integer or discrete values); the value preserves the ratio of some quantification, a real number.
Real	A variable with continuous-valued values (as opposed to integer or discrete values) – rational, double precision, scientific, rational.
Realization	A particular result of a stochastic outcome, it represents one possible result, not the average.
Recursive	The identical procedure is repeated, but using the most recent outcomes. The values are iteratively updated.
Regression	The procedure of adjusting model coefficient values to minimize the SSD based on the "y" distance (or some alternate measure of closeness) between data and model.
Regressors	The name for input variables to the model. The user chooses these. They need to be the complete set, represent the right delays, and not contain extraneous variables.
Reject	There is statistically sufficient evidence to confidently claim that the null hypothesis is not true, that there is a difference, that the process is not a steady state, that a variable is not zero.
Replicate	Repeated, with the intent to exactly reproduce experimental conditions.
Right	The model properly captures the natural phenomena that it seeks to represent, has fidelity, is true, correct. In possible contrast to being useful.
RHS	Right-hand side, the terms on the right-hand side of an equation (either equality or inequality).
rms	Root-mean square, the square root of the average of squared deviations. It would be the standard deviation if it were normalized by $N-1$ terms in the sum.
Run	A sequence of residuals with the same sign; the pattern in the runs will change when the residuals are ordered by different variables.
Runs test	A test to see if there are too few or too many runs in the data.
Scaled variables	Values of the variables are normalized by their range. The variables are dimensionless and often scaled to be

	in the range of 0–1. The scaling might be the actual data range or an expected maximum range.
Self-tuning filter	STF, a first-order filter that adapts filter factor to the data variability.
Semi-empirical	The model is a blend of phenomenological and empirical.
Significant digits	Those digits in a measurement or calculation representing confidently known values, not uncertain values.
Simulation	Solving a mathematical procedure to mimic what Nature would do.
Soft constraint	Should not be violated, but a penalty is added to the OF proportional to the degree of violation.
SPC	Statistical process control, techniques for monitoring processes and triggering action only after a change is statistically significant.
SQ	Successive quadratic, an optimization approach that uses a surrogate model, a quadratic functionality, of the local OF response to DVs.
SS	Steady state.
SSD	Sum of squared deviations between modeled and actual values.
SSID	Steady state identification.
Static	Something that does not change in time.
Statistic	A value that represents some property of a distribution of values (average, variance, median, runs, etc.)
Statistical process control	SPC, techniques for monitoring processes and triggering action only after a change is statistically significant.
Steady state	The process has settled to a state in which it does not change in time, but sequential measurements will not have exactly reproducible values because of noise. The process may have come to thermodynamic equilibrium, but it might be off equilibrium due to persistent influences.
STF	Self-tuning filter, a first-order filter that adapts the filter factor to the data variability.
Stochastic	A process that does not return a unique value after each trial, but returns a distribution of values. Rolling a die returns a 1, 2, 3, 4, 5, or 6, a uniform distribution with six possible values. Measuring heights of people results in a normal distribution with continuous values.
Successive quadratic	SQ, an optimization technique that uses local OF values to generate a quadratic surrogate model of the function, jumps to the optimum of the surrogate model, and then repeats.

Systematic error	A bias, a consistent average deviation from true.
Total SSD	The sum of squared differences between the model and data in both the x and y axes, normal to the model, perpendicular to the model.
Transient	The process states are changing in time in response to an input, often termed dynamic.
Trial	The run of an experiment or the run of an optimization that produces results.
Trial solution	An optimizer guess for the DV values that move toward the optimum.
Truth about Nature	The impossible-to-know mechanisms within a real process. Often, it seems so simple, but each refinement of investigation takes the scientist toward greater complexity. Often idealized concepts and models are fully adequate for engineering.
TS	Transient state.
Type I error	The null hypothesis is true (there is no difference), but the data provided unusual extreme values to lead to rejecting the null hypothesis – true but rejected.
Type II error	The null hypothesis is false (there is a difference), but the data provided values that were not definitively different, that could not justify rejecting the null hypothesis – false but accepted.
Uncertainty	A measure of the possible range or a value. The value might be experimental and range could be estimated from replicate measurements. The value might be from a model, and uncertainty can be mathematically propagated through the calculations.
Useful	The model balances perfection with sufficiency, that is, provides a good-enough representation, and that it is functional (convenient, reliable, sufficiently accurate) in use. In possible contrast to being right.
Validation	A procedure that determines whether the model can be rejected by the data or by logical considerations.
Variance	The square of the standard deviation in replicate data.
Verification	A procedure that determines that the model execution has fidelity to the model concepts.
Vertical SSD	The sum of squared deviations in the model prediction, parallel to the y axis.
Voting	Taking the middle measurement of three independent sensors as the data that represents the process.
w.r.t.	With respect to, the equivalent of "against" or "versus" when describing a response variable graphed with respect to an influence variable.

Symbols

$a, b, c, d, \ldots$ = model coefficients
$\alpha, \beta, \gamma, \delta, \ldots$ = model coefficients
α = level of significance, Type I error probability, distance factor in steepest descent
B = measure of badness, typically the amount of a constraint violation
β = Type II error probability
c = confidence in an event, number of columns
CDF = cumulative distribution function
d = deviation between data and model, disturbance
D = equal concern deviation
DV = decision variable value
δ = infinitesimal change, measure of variance
Δ = small change
∇ = gradient operator
E = expected count, desired half-range
ε = small deviation, error, correction
f = function, fraction
F = F-statistic for variance ratio
g = function derivative
H = Hessian matrix of second derivatives
H_0 = null hypothesis
I = identity matrix
J = objective function, Jacobean vector of first derivatives
λ = weighting factor, Lagrange multiplier, filter factor
m = number of model coefficients
M = number of players in a multiplayer optimization, number of decision variables, number of function evaluations
μ = mean, true value
n = noise, random and independent perturbation to a measurement or condition, delay counter, value of an index variable, number of occurrences
N = number of data sets, number of iterations, number of trials

O	=	observed count
OF	=	objective function value
p	=	parameter, coefficient, probability of an event
P	=	penalty for a constraint violation, probability of a compound event
PDF	=	probability distribution function
q	=	probability of not-an-event $= 1 - p$
r	=	residual, UID[0,1] random variable, number of rows
r_1	=	r-lag-1 autocorrelation statistic
r^2	=	variance reduction statistic
R	=	range of a variable (high minus low values), ratio statistic
rms	=	root-mean square $= \sqrt{SSD/N}$
s	=	estimate of the standard deviation, scaling parameter in the logistic model
S	=	distance along a line
SSD	=	sum of squared deviations
σ	=	standard deviation
t	=	time, t-statistic
τ	=	time constant
u	=	input variable to a dynamic process, the forcing function
υ	=	degrees of freedom, measure of variance
w	=	weighting coefficient
x	=	input variable (influence, independent, given)
X	=	process variable (either influence or response)
χ^2	=	chi-square statistic
$\tilde{y}$	=	modeled response variable
y	=	response variable (output, dependent, effect, outcome)
ψ	=	true value of y

Subscripts, Superscripts, and Marks

i	=	counter for data set number or time or iteration
$'$	=	scaled variable, deviation variable
SS	=	steady state
*	=	optimal value

Part I

Introduction

1

Introductory Concepts

1.1 Illustrative Example – Traditional Linear Least-Squares Regression

Consider this objective: find the best quadratic model, as described by the following equation:

$$y = a + bx + cx^2 \tag{1.1}$$

which matches the data in Figure 1.1.

Here, "x" represents the independent variable and "y" the dependent variable. Often, x and y are respectively termed *cause and effect, input and output, influence and response, property and condition*, and y is termed a function of x. Equation 1.1 is a human's mathematical description of how y responds to x and it is likely that the relation will not exactly match how nature actually works. In regression, in fitting a model to data, the values of the model coefficients (a, b, and c) will be adjusted to create the best model.

Conventionally, the best model is the one that minimizes the sum of squared distances from data point to model curve, where distance is that for the dependent variable (parallel to the vertical y axis). One data-to-model deviation is indicated as "*d*" on Figure 1.1. The sum of squared deviations (SSD) is defined as

$$SSD = \sum_{i=1}^{N} [y_i - \tilde{y}(x_i)]^2 \tag{1.2}$$

where N indicates the number of data points on Figure 1.1 and "*i*" the number of a particular data point within the set of N. The number associated with a data point does not necessarily correspond with either x or y values. More likely the data point number corresponds to the chronological order of experimental trials that implemented the x value and measured the y response, as the sequential trial number is indicated on Figure 1.1. The data set might appear as illustrated in Table 1.1.

Continuing the explanation of Equation 1.2, y_i represents the ith measured y value, the data value, from Table 1.1, and $\tilde{y}(x_i)$ indicates the model-calculated y value from Equation 1.1 using the ith x value from Table 1.1. The tilde accents on the symbols $\tilde{y}$ and $\tilde{y}(x_i)$ are both explicit indications that $\tilde{y}(x_i)$ represents the modeled y value. Redundancy in symbols is often not used, and here the $\tilde{y}(x_i)$ term will be represented by either $\tilde{y}_i$ or $\tilde{y}(x_i)$.

Nonlinear Regression Modeling for Engineering Applications: Modeling, Model Validation, and Enabling Design of Experiments, First Edition. R. Russell Rhinehart.

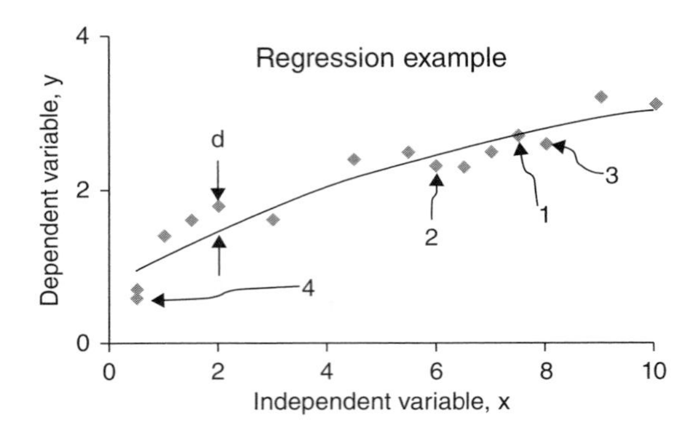

Figure 1.1 Illustration of regression concepts

Table 1.1 Illustration of data for Figure 1.1

Trial number	*X*, Input variable value	*Y*, Response variable value
1	7.5	2.7
2	6	2.3
3	8	2.6
4	0.5	0.7
.	.	.
.	.	.
.	.	.

The objective, find values for coefficients a, b, and c that minimize the SSD, defines an optimization procedure. Conventionally, the optimization application is stated by:

$$\underset{\{a,b,c\}}{Min} \; J = \sum_{i=1}^{N} [y_i - \tilde{y}(x_i)]^2 \tag{1.3}$$

In the jargon of optimization, Equation 1.3 reads, "The objective is to *Min*(imize) the *Objective Function J* (equal to the SSD) by adjustment of values of the decision variables (DVs) a, b, and c." This fully describes the regression "problem." DVs are what you adjust to minimize the objective function (OF) value. The DVs are the model coefficients that are adjusted to make the model best fit the data.

As a model of the y response to x, Equation 1.1 is nonlinear. *Nonlinear* means not linear, but does not indicate what the nonlinearity is (quadratic, cubic, reciprocal, exponential, etc.). If the cx^2 term was not in Equation 1.2 then the model would describe a linear y–x relation. However, in regression we adjust coefficient values, not the x or y values, and in Equation 1.1 each coefficient appears linearly (holding all else constant, the value of y is a linear response to the value of either a or b or c). The exponent for each coefficient is +1 and none of the coefficients are imbedded within a functionality that would make it have a nonlinear impact

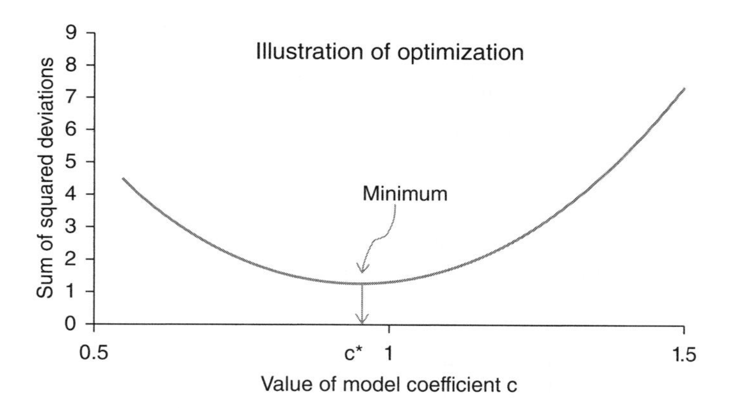

Figure 1.2 Illustration of regression coefficient optimization

on *y*. A formal definition of linearity is given later. Linearity simplifies determination of the optimum values for coefficients.

In linear regression, the model coefficients have a linear impact on the model prediction even if the model has a nonlinear $\tilde{y}(x_i)$ relation.

At the optimum value of each coefficient, SSD is a minimum. This means that any change in the coefficient value (either larger or smaller) makes the SSD larger. This is illustrated in Figure 1.2 for coefficient *c*. The concept is the same for each coefficient. The figure also illustrates that the SSD w.r.t. the coefficient graph has a slope (derivative) with a value of zero at the optimum value, c^*, of the coefficient. This property provides the classic method to determine values of the coefficients.

First, expand Equation 1.2 to explicitly reveal the model coefficients:

$$SSD = \sum_{i=1}^{N} [y_i - a - bx_i - cx_i^2]^2 \tag{1.4}$$

where x_i and y_i represent the values from Table 1.1 for the *i*th trial number.

Second, take the derivative of SSD with respect to each of the three coefficients and set the derivative (the slope) of each to zero (the value of the slope at the optimum value for each DV). This yields

$$\begin{aligned}
\frac{\partial SSD}{\partial a} &= 0 = -2\sum_{i=1}^{N} [y_i - a - bx_i - cx_i^2] \\
\frac{\partial SSD}{\partial b} &= 0 = -2\sum_{i=1}^{N} [y_i - a - bx_i - cx_i^2]x_i \\
\frac{\partial SSD}{\partial c} &= 0 = -2\sum_{i=1}^{N} [y_i - a - bx_i - cx_i^2]x_i^2
\end{aligned} \tag{1.5}$$

Divide each of the three equations by the value of −2 and rearrange to produce the equation set 1.6, often termed the "normal equations." Note that the three model coefficients appear linearly:

$$
\begin{aligned}
aN + b\Sigma x_i + c\Sigma x_i^2 &= \Sigma y_i \\
a\Sigma x_i + b\Sigma x_i^2 + c\Sigma x_i^3 &= \Sigma x_i y_i \\
a\Sigma x_i^2 + b\Sigma x_i^3 + c\Sigma x_i^4 &= \Sigma x_i^2 y_i
\end{aligned} \tag{1.6}
$$

Third, solve the three linear equations for the three unknowns, a, b, and c. There are many linear algebra methods for the solution of equation set 1.6, one being Gaussian elimination.

Equation set 1.6 is often presented in matrix-vector, linear algebra notation:

$$
\begin{bmatrix} N & \Sigma x_i & \Sigma x_i^2 \\ \Sigma x_i & \Sigma x_i^2 & \Sigma x_i^3 \\ \Sigma x_i^2 & \Sigma x_i^3 & \Sigma x_i^4 \end{bmatrix} \begin{bmatrix} a \\ b \\ c \end{bmatrix} = \begin{bmatrix} \Sigma y_i \\ \Sigma x_i y_i \\ \Sigma x_i^2 y_i \end{bmatrix} \tag{1.7}
$$

or as

$$
\underline{\underline{M}}\, \underline{c} = \underline{RHS} \tag{1.8}
$$

If the reader is seeking to understand classic linear regression, the reader should create a simple problem and perform those three steps.

There are several efficiencies associated with linear regression. One is the direct extension to more complicated models. For example, if the model represented by Equation 1.1 was a cubic relation between y and x, then there would be an additional x^3 term, with a fourth coefficient, d. In that case, following the same procedure of minimizing SSD by taking the derivative of SSD w.r.t. each coefficient and setting the derivatives to zero, there would be four normal equations with four unknowns. If, for another example, there were two independent variables, x_1 and x_2, in a model with M number of linear coefficients there would be M normal equations. In linear regression there are M normal equations, each a linear relation in the M model coefficients, and each representing the derivative-equal-zero for each of the M coefficients. Regardless of the number of linear coefficients, the representation in Equation 1.8 remains the same and the linear algebra method for solving Equation 1.8 remains the same. This is one convenience of traditional linear regression.

The second convenience is, as long as the determinate of the M matrix is not zero, there is a unique solution that can be obtained from classic linear algebra solution methods.

Linear regression provides one universally applicable method that is guaranteed to return a unique solution.

In regression, the unknowns are the values of the model coefficients. By contrast, when using the model, model coefficients appear as the known values in Equation 1.1; but, in equation set 1.6 their roles are reversed. The a, b, and c values are the unknowns, and since the experimental values for N and each y_i and x_i are known, the values of the multipliers of the model coefficients are the known values. Since the unknowns, a, b, and c, appear linearly within the three equations of equation set 1.6, linear algebra procedures will solve for the coefficient values. This is predicated on the three equations being linearly independent, which requires relatively straightforward planning of the experimental procedure that generated the x–y data. Two heuristics that guide experimental design are to (i) have three or more independent data points for each coefficient to be evaluated and (ii) choose conditions such the data represent the entire x and y range.

The preceding example represents a *batch* data process in which all data are available and the objective is to find the best model to match the batch of data. However, in many applications related to modeling dynamic processes (those that change in time) the long-past data are irrelevant and model coefficients are incrementally updated using the new data at each sampling in an *incremental* or *recursive* manner.

The coefficients appear linearly in Equation 1.1. Although the model output, y, is a nonlinear (quadratic) function of the model input, x, the model output is linearly dependent on each coefficient. If you were to fix the value of x and plot the value of y w.r.t. any of the model coefficients, the graph would be a straight line. Contrasting this in nonlinear regression the sensitivity of model output to model coefficient is not linear.

1.2 How Models Are Used

Mathematical models represent a human conceptual understanding of how Nature works. They relate inputs to outputs, which are alternately termed causes to effects, influences to outcomes, stimulus to responses. Mathematical models are statements of relations such as: at a particular temperature the reaction rate will be such-and-such. When the load exceeds xxx the beam will fail or when the speed is xxx the fuel consumption will be yyy.

We use models in two prediction manners. Once we have the model, we use it to predict, forecast, or anticipate what Nature will do. In developing the model, we start with concepts and then convert the concepts into mathematical statements that are equivalent to our linguistic sentences. If the mathematical model of the human concept is a correct representation of Nature, then the model will match the data. If the concept is wrong the model will not match the data. Therefore, we also use models to test our concepts of how Nature works.

We also use the inverse of the model. Once we have a mathematical "sentence" of how Nature works, we can reverse the sentence to answer the question "If we want a particular outcome, what has to be done?" For instance: "What temperature is required to make the reaction go at a desired rate?" "If I want the beam to support xxx weight, how thick must it be?" In these applications we determine the input that gives the desired output, which is termed the inverse calculation.

1.3 Nonlinear Regression

Often, useful models are not linear in their coefficient influence on the model output and this leads to unique functionalities that do not have a generic method of solution. Consider a simple power law model

$$y = ax^b \tag{1.9}$$

This has two model coefficients, a and b. Following the classic analytical procedure for minimization of SSD, first take the derivatives of SSD w.r.t. each coefficient:

$$\frac{\partial SSD}{\partial a} = -2\sum (y_i - a{x_i}^b){x_i}^b \tag{1.10}$$

$$\frac{\partial SSD}{\partial b} = -2\sum (y_i - a{x_i}^b)a{x_i}^b \ln(x_i) \tag{1.11}$$

and set each to zero:

$$0 = \sum y_i x_i^{\,b} - \sum a x_i^{\,2b} \tag{1.12}$$

$$0 = \sum y_i a x_i^{\,b} \ln(x_i) - \sum a^2 x_i^{\,2b} \ln(x_i) \tag{1.13}$$

This leads to two equations in two unknowns, and in general to M equations in M unknowns, but there does not appear to be a rearrangement that permits a linear algebra solution for the unknowns a and b. The a and b coefficients are neither independent (a and b are in the same term) nor linear (neither are to the first power).

Further, there is not a unique resulting functionality for the derivative relations. Each new model functionality results in new relations in the derivative equations. Following the procedure for another simple, two-parameter model,

$$y = \frac{a}{x - b} \tag{1.14}$$

leads to

$$0 = \sum \frac{y_i}{x_i - b} - a \sum \frac{1}{(x_i - b)^2} \tag{1.12}$$

$$0 = \sum \frac{y_i}{(x_i - b)^2} - a^2 \sum \frac{1}{(x_i - b)^3} \tag{1.13}$$

Again, this results in two equations in two unknowns, but they are nonlinear and not separable.

There are techniques for solving for the roots of a set of nonlinear equations (roots are the values of the coefficients that make the function equal zero), such as Newton's method, and it would appear that after the derivative equations are determined (in these examples either Equations 1.10 and 1.11 or 1.12 and 1.13), one could apply a standard procedure to solve for the unknowns. However, nonlinear equations can provide relationships that make root finding algorithms unstable and can generate multiple sets of roots. We need a better method.

Further, often in nonlinear regression there are constraints on variables (delays must be greater than zero, for example) and root finding algorithms cannot cope with constraints.

Finally, more often than not, one cannot analytically determine the derivative function. The equivalent of Equations 1.10 and 1.11 or 1.12 and 1.13 may not be available.

However, we desire a universal method for nonlinear regression that does not depend on the user defining analytical derivatives, that is, stable, and that can handle constraints. This book offers numerical optimization procedures as the answer. We will use numerical optimization to determine the numerical values of the coefficients in Equation 1.3.

1.4 Variable Types

Proper choices of model, optimization procedure, and objective function are dependent on the types of variables that are relevant. This section describes variable types.

There are a variety of naming conventions for numbers representing origins in mathematics, philosophy, statistics, and computer languages. However, the concepts are aligned. My explanation of number types comes from an engineering and computer language view. It somewhat

Table 1.2 Nomenclature for variable types

Here	Stephens' names	Other names
Class	Nominal	Classification, Category, Text, String
Rank	Ordinal	Opinion, Scale, Rating
Discrete	Cardinal	Integer
Continuum	Rational	Continuous-Valued, Scientific, Double Precision, Real
Deviation	Interval	Gage, Reference, Relative
Scaled	—	Dimensionless, Normalized

follows the taxonomy proposed by Stephens in 1946, a statistician, which still does not have total agreement within the statistics community. He proposed that variables are of the following types (Table 1.2).

Class represents classification and *Nominal* is derived from Latin for "name," but the name has no quantitative relation to the item. In a rock collection, there is the first rock you found, then the second, third, and so on. However, the numbers are just names and do not represent any quantity related to properties of the rock such as weight, size, value, thermal conductivity, surface roughness, or beauty. The names could have well been "A," "B," "C," or "alpha," "beta," "gamma," or "un," "deux," and "trois," We name trial runs sequentially in chronological order, even though trials should be run in a manner that randomizes the relation to any input variable. The "1" in the name "Trial 1" has no relevance to the flow rate, the impact, the difficulty, the benefit, and so on.

Nominal labels represent class, or category, or type. These could be city names, brand names, people names, social security numbers, colors, process types (Kraft or Solvay), pump types (centrifugal, turbine, positive displacement), separation types (distillation, absorption, crystallization), modeling types (neural network, finite-impulse), and so on. The names refer to items or procedures that have characteristic properties, but there is no correspondence of name spelling to the properties. You cannot mathematically relate the name to the value. In computer programming languages these are called string, text, or class variables.

Class or category can be related to dichotomous (two-valued) values (good/bad, on/off, fail/success, 1/0, tumor/not-tumor), and models are often used to predict whether something will be categorized as 1 or 0.

Rank means placement and *Ordinal* means sequence or order. A runner comes in first place, or second, or third, and so on, in a race. Here the number is a measure of placement, or goodness, or desirability, and relates to some property of the event. Alternately, it could relate to the property of an item (largest, second largest, … , or brightest, second brightest, …). Ordinal numbers reveal order, but they do not indicate a relative or proportional quantification of the property. For example, in a race, first and second may be nearly tied, and far ahead of third. Alternately, first may be far ahead of second and third. Ranking, "It came in second," indicates a relative relationship, but does not represent a ratio of quantity or goodness. Second place may be nearly as desirable as first, or far less desirable than first.

Similarly, opinion ranking scales (rate your opinion of the movie from best 5-star to worst 1-star), often called Likert scales, can place items in order, but the betterness of the 1 (outstanding) over the 2 (very good) may be much greater than the betterness of the 2 over the 3 (average). In school the standard quality points associated with letter grades of "A," "B," "C," "D," and "F" are 4, 3, 2, 1, and 0. These are used to calculate a grade point average (GPA) representing a student's average performance. However, if the normal distribution is used to assign grades, the "A" could represent the top 10 percentile, a "B" the 30 to 10 percentile (a 20% interval) and the "C" a 30% interval. An average of rankings may be a continuous-valued statistic, not just the integers associated with one ranking, but it is likely to be a nonlinear representation of "goodness."

Cardinal numbers, integers, are the counting numbers, the whole numbers, the indications of the number of whole items. They can only have integer values. They can only have values that exist at intervals of unity. The integer value is in direct proportion to the quantity it represents. Twelve is twice as much as six and six is twice three. Further, 10 is 2 more than 8 and 3 is 2 more than 1. The difference in values also represents the same counted quantity. The relation to quantity is preserved with addition, subtraction, multiplication, or division.

Integers are a type of *discrete* number, but the discretization interval is unity. However, the interval could be half, or quarters, or 16ths. The discretization does not have to be unity. The Westminster Clock chimes on the quarter hour. Its interval is $^1/_4$ of an hour. Or is it an interval of 15 minutes? Either way the time interval is not unity. In bit representation of computer storage, an 8-bit storage location might be filled with the binary sequence 00101101. The discrete interval is 1 bit. However, representing a display range of 0–100% the 2^8 possible numbers in an 8-bit storage have a 1-bit discretization interval that represents a $(100\% - 0\%)/2^8 = 0.390625\%$ interval. Observing such a number display, you would see all reported values as multiples of the 0.390625% interval. This minimum quantity represents discretization error, or discretization uncertainty. Look at a table of viscosity or table of t-statistic critical values and you will find that the table does not report infinite numbers for each entry. It may report one decimal digit, in which case the discretization interval is 0.1, or four decimal digits, in which case the discretization interval is 0.0001.

Whether the discrete numbers have unity or some other discretization interval value, they have several properties. One is that their value is linearly proportional to the quantity they represent. As a consequence differences between any two also relate to the quantity whether the numbers are high or low.

By contrast, a rank (or ordinal) number representing value scale, opinion scale, or ranking does not necessarily have the linearly proportional value. If the highest GPA is a 4.00 and the lowest a 0.00, then the difference between a 4.00 and a 3.50 is not the same as the difference between a 2.50 and a 2.00. The person with a 4.00 did not do twice as much work as one with a 2.00. They both may have completed all assignments. The 4.00 person did not get twice as many answers right as the 2.00 person. The 4.00 person might have got 93% correct on all tests, and the 2.00 person got 77% correct on all tests, representing a 1.2:1 performance advantage.

Continuum, *Real*, or *Rational* numbers refer to a continuum and are proportional to the quantity. We often imagine that properties of length, time, mass, and temperature are continuous and that the property can be divided into infinitesimal intervals. Continuum numbers permit having infinite decimal places. The ratio of 1 to 3 is 0.3333333 … , but on an atomic view, mass is not continuous. If you want to increase the amount of water in a glass, the smallest increment you can add is one molecule of water. If you tried to add half of an H_2O molecule, it would not be a molecule of water. Effectively, on an engineering scale the continuum view seems valid and we consider the measurements of properties to be real numbers or continuum, with the discretization interval effectively zero. In one community these continuous-valued numbers that are proportional to the quantity they represent are termed rational numbers. In computer programming they are called real numbers or single or double precision numbers.

There are other common definitions for the terms rational and real. From a mathematical view, real numbers are either rational (meaning ratio) if they can be represented by a ratio of integers or irrational if they cannot be represented by a ratio of integers. When the decimal part has a repeated pattern the real number is rational. If there is no repeated pattern (such as the square root of 2, the golden ratio, the base of the natural logarithm, or pi), the real number with infinite decimal digits is irrational. Further, in mathematics, real numbers are distinguished from imaginary or complex numbers that contain $\sqrt{-1}$ as an element. However, this text will use the term real to mean a continuous-valued number that is linearly proportional to the quantity it represents.

Next in the table comes a *deviation* variable, Stephens' category of *interval*. The zero of the interval or deviation scale does not represent that the property or characteristic of the item is zero. Zero degrees Fahrenheit does not mean zero thermal energy. You can still remove thermal energy and cool it to minus 17 °F. Zero gage pressure does not mean zero pressure. "Sea level" does not mean zero height. You cannot use deviation variables or interval values in many scientific calculations. Reaction kinetics are based on absolute temperature and pressure. Gravitational pull is based on distance from the center of the Earth. However, deviation variables preserve relative order and relative changes. The number of molecules needed to go from 5 to 6 psig is the same as that needed to go from 32 to 33 psig (ideally). The amount of thermal energy needed to raise the temperature from 38 to 39 °C is the same that is required to raise the temperature from 18 to 19 °C (ideally). In subtraction, deviation variables preserve relative difference, but they cannot be used in multiplication, addition, or division. Four degrees Celsius is not twice as hot as 2 °C. They do not preserve proportion or relative quantity.

Continuum numbers (rational or irrational) and discrete numbers (which includes integers) are directly related, linearly proportional, to the property of the event or item (time, mass, weight, intensity, value, etc.). As a result, real and discrete numbers have a ratio or proportional property. If the numerical representation of one event is twice in magnitude as the numerical representation of the other, then the one event has twice the magnitude, impact, value, of the other. Real and discrete numbers preserve this ratio or proportional relation, making them useful for modeling.

Finally, there is the category of *scaled* numbers. Usually these are either continuum or discrete numbers that are expressed as a ratio of full scale. A continuum example is "the glass is half full." An integer example is "we filled 90% of the seats." In those examples scaled variables are a fraction of the maximum possible value. Alternately, classic dimensionless groups, such as the Reynolds number or Fourier number, scale the extent of one mechanism with another. Here the scaled variable value is not bounded between 0 and 1. One can also scale deviation variables by their low to high value, and alternately in neural network modeling they are scaled from low to high on a −0.8 to +0.8 basis. Scaled variables have the same properties of ratio and relative proportion as their basis continuum, discrete, or deviation variables.

This textbook is mainly about the use of discrete (cardinal, integer) and continuum numbers (continuous valued, real, rational), both of which are proportional to the characteristic, within engineering models about the physical and chemical world. This is about how to generate the data, how to generate the models, and how to evaluate the models.

By contrast, there are methods related to the use of class or category or nominal variables as either the input or output (cause or effect, independent or dependent) variables in a model.

Although this text contains some discussion of such operations, the content of this book is not about that.

1.5 Simulation

The aspects of regression modeling can be simulated, as Chapter 5 more thoroughly directs, and doing so permits the reader to rapidly explore methods of this text.

With experimental data you do not know whether your regression modeling results are correct or not, because you do not know the truth about Nature. However, with simulation, you assume the truth about Nature (as represented with equations and coefficient values) and generate simulated experimental data. Then you can compare regression models to the simulated truth to see if the results of the procedures are correct. If the regression procedure consistently provides correct results over a range of critically devised tests, then you have a high confidence that the procedure is valid and will provide correct results on an unknown application.

A first step is to generate data that would simulate experimentally obtained data. The concept is that the experimenter would set the independent variables (experimental conditions), execute the experiment, and measure the response (dependent) variable. The simulation of this is illustrated in Figure 1.3, in which x is the independent variable and y is the response variable. Also indicated in the figure is that the measured response is corrupted by both noise (random perturbations) and a deviation, d_2 (systematic bias), and that the input to the process, P, is also corrupted by a deviation, d_1. In reality, the values of the deviations and noise are not knowable; however, for a simulation the experimenter can choose those values. The deviations would represent longer term calibration drifts, or the effects of input material variability, ambient conditions, device warm-up, operator training, operator fatigue, or equipment aging. These values would change slowly over time. By contrast, "noise" would represent sample-to-sample independent perturbations due to random events such as the vagaries of incomplete mixing, turbulence, sampling, mechanical vibrations, stray electronic field disturbances, and so on.

Figure 1.3 represents a traditional view with no uncertainty on x, and drift and noise linearly added to the otherwise truth.

To create a simulator using this model, first define the equation or procedure to obtain y_{true}, the truth about nature, from x:

$$y_{true} = f(x_{true}) \tag{1.14}$$

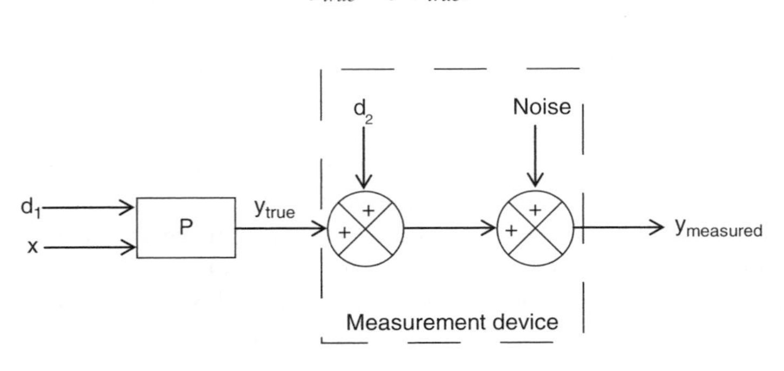

Figure 1.3 Simulation concept

Then perturb the values by the disturbances and noise:

$$y_{meas,i} = f(x_{nominal,i} + d_{1,i}) + d_{2,\,i} + n_{2,i} \tag{1.15}$$

You could add disturbance and noise to the input x in any of several appropriate ways. See Chapter 5 for a variety of options, and for models for noise and disturbance.

1.6 Issues

There are many types of models, but the best for engineering utility are often based on fundamental (alternately labeled scientific, first-principles, physical, theoretical, mechanistic, or phenomenological) derivations. In these, the coefficients that are adjusted to best fit the data often appear nonlinearly. Nonlinear regression is desired, but linear regression is easier.

Linear regression leads to the normal equations, equation set 1.6, that are deterministically solved, an aspect that makes it preferred over nonlinear regression. Often, to employ linear regression, we linearize functions by log-transforming power-law or exponential functions, or use other approaches that are mathematically correct when there is no uncertainty in either the x or y data. However, mathematically transformed data alters the relative locations and variance, which leads to biased coefficient values when there is experimental error in the data. By contrast, nonlinear regression can manage the equation in its nonlinear form. However, nonlinear optimization is an iterative procedure, which requires user-selected stopping criteria and which may become trapped in local minima. Normally, stopping criteria on either (or both) OF and DVs is scale-dependent and requires *a priori* user knowledge of the problem. Subsequent chapters show how to accommodate these aspects of nonlinear regression.

Should data be fitted with a linear, quadratic, cubic, or higher order model? Should it be a power law model? Does the model include the right input variables? There needs to be a method to determine the proper, justified, rational model structure. These are usually qualitative human choices and subsequent chapters provide a basis for making them.

You can fit a linear model to any data set. You can find a best model of any functional relation for any data set. Just because optimization finds the best coefficient values does not mean that the model is either *right* or *useful. Right* would imply that the model properly captures the natural phenomena that it seeks to represent. *Useful* would imply that the model balances perfection with sufficiency, that is provides a good-enough representation and that it is functional (convenient, reliable, sufficiently accurate) in use. Measures of *useful* are not normally included in lessons about the linear regression technique, and measures of *right* in linear regression lessons are usually based on r-square or other measures of removing variance. However, these statistical measures are independent of mechanistic reasoning. Subsequent sections show a technique for deciding whether the model is *right* for selecting the better from several equations and provide a set of criteria for evaluating *utility.*

Is there really no uncertainty on the x value? Perform any trial to generate data representing a y response to an x influence. Perhaps you wish to explore the influence of your kitchen oven temperature on the time it takes for the turkey thermometer to read 265 °F. You turn a knob or digitally specify the value of x, oven temperature. Is the oven really at that temperature when the power blinks on, then off? Or does the temperature control device lead you to believe it is at the specified temperature? In any experiment there is uncertainty on both the input and output. By contrast, conventional regression ascribes all of the uncertainty to the y response.

Further, the uncertainty may vary from one level to another. For example, differential pressure variability (noise) on orifice flow rate measurement increases with turbulence, which increases with flow rate. As another example of variability that changes with the y value, pH data may be highly sensitive to acid/base concentration variation near neutrality, but have minimal sensitivity in adjacent regions of nearly the same acid-to-base ratio. If one operating region produces data that is highly variable, the regression should reduce the weight of that data in setting model coefficients. Maximum likelihood approaches to the OF definition can accommodate these aspects. However, as a subsequent section reveals, they add complexity that is often not justified by utility aspects, by the improved functionality of the model due to the improved accuracy of the coefficient values.

Finally, my two favorite issues can be summed in the phrases "The model is wrong" and "Nature lies." *The model is wrong.* The mathematical model represents our human concept, an ideal version of the complexity of Nature. Regardless of how rigorous we try to make models, we always seem to find that our landmark breakthroughs still have an incompleteness. Nature is mysteriously complex, yet simple. In addition, we often simplify the models to obtain a mathematically tractable version. For example, we truncate the value of pi. We linearize. We truncate infinite series. We use numerical methods. Often, we intentionally make the model wrong, in a judgment that balances perfection with sufficiency.

Then, *Nature lies.* (Not to be disrespectful, Mother, but you seem to mask the reality about yourself.) Measurements are corrupted with uncertainty and calibration drifts, and experimental outcomes are corrupted with uncontrolled influences and variability of input materials and conditions. We call these noise or disturbances, random or systematic error.

Our objective in regression modeling is to use confounded data to determine coefficient values of an incomplete model, so that the model becomes functional.

Questions related to regression are:

- What should be used to determine goodness of the model fit to data?
- What model choices (model architecture and input variables) are best?
- Of competing model types, which is best?
- How many data points are best?
- Where should the data points be best located?
- What optimization approach is best?

To answer the questions, to use confounded data to create a functional model, we must understand:

- Models, model types, and measures of functionality.
- Systematic and random perturbations and how to quantify them.
- Nonlinear optimization, how to find the global optima, how to define convergence.
- How to propagate uncertainty in experimental conditions to data variability and from model coefficient variability to uncertainty of model prediction.
- How to test data for legitimacy and outliers.
- How to test model fit to the data.
- How to design experiments to generate data that provide defensible and useful models.

This book attempts to provide that knowledge, understanding, and tools.

1.7 Takeaway

Nature is usually nonlinear. Understand the variable types relevant to your application. Since the truth about Nature is unknown, simulations are a strong method to develop confidence and establish credibility.

Your model is wrong (not absolutely true) and the data are corrupted by fluctuations, noise, error, and so on. What you are attempting in regression is to fit a wrong model to data that misrepresents the truth. Quantifying the uncertainty in models is important to those who will want to use the models.

Exercises

1.1 Consider a linear $y(x)$ model and remove the term cx^2 from Equation 1.1. Follow the procedure to minimize SSD, obtain the two "normal equations" and solve for the a and b coefficient values.

1.2 Consider a cubic $y(x)$ model and add the term dx^3 to Equation 1.1. Show that minimizing SSD provides a linear set of equations in the form of Equation 1.8.

1.3 List some situations in which the modeled value is Class, Rank, Discrete, Real, or Deviation.

1.4 Provide several reasons for random error, noise on a measurement.

1.5 Provide several reasons for bias or systematic error on a measurement.

1.6 Provide several reasons for disturbance error, uncontrolled inputs, that affect experimental outcomes.

1.7 Consider that the objective is to find a linear regression model, $\tilde{y} = a + bx$, to make the average residual zero, $0 = \bar{r} = 1/N \sum_{i=1}^{N} r_i = 1/N \sum_{i=1}^{N} (y_i - \tilde{y}_i)$. Show that there are an infinite number of solutions.

2

Model Types

2.1 Model Terminology

The term *model* is used in multiple contexts. Sometimes you will find that model means a conceptual linguistic statement or textual description of how influence affects response. Alternately, model is used to refer to a small-scale physical entity, which can be used for laboratory tests.

For this textbook, *model* refers to the mathematical representation of the human concept. The model may be expressed as a series of algebraic, calculus, or differential equations. The model is a mathematical representation that guides the set of procedures needed to calculate the value of a result from an influence. For example, in Equation 1.1, the procedure to obtain a numerical value of y is to multiply the value of b times x, then c times x squared, and then add the values of the several terms. Alternately, the model equations may be converted to a computer procedural code, which specifies an automation of executable commands to create a simulator. Either way, the mathematical representation provides a method to calculate how the process, product, or procedure responds to influences.

Mathematical models range from theoretically derived, *phenomenological* (often termed fundamental, rigorous, mechanistic, scientific, theoretical, or first principles), to equations purely representing the data, *empirical* (with no grounding in process knowledge), to a blend of both. All are useful.

Mathematical models have *coefficients*, often termed *constants* or *parameters*. These are represented as the gas law constant, R, in Equation 2.1, the ideal gas law, and the additional a and b terms in Equation 2.2, the van der Waals equation of state:

$$PV = nRT \tag{2.1}$$

$$P = \frac{nRT}{V - nb} - a\left(\frac{n}{V}\right)^2 \tag{2.2}$$

Although the value of the gas law constant can be derived from first principles, our ever-naïve understanding of Nature yields imperfect values. The values of model coefficients are usually obtained by a best fit of the model to the data.

Nonlinear Regression Modeling for Engineering Applications: Modeling, Model Validation, and Enabling Design of Experiments, First Edition. R. Russell Rhinehart.

Models have *variables*. In contrast to coefficients, which conceptually have values that are constant in time or uniform in space, the variables are the terms that can change value in time or situation. Variables are represented as P, V, n, and T in Equations 2.1 and 2.2. They represent the cause-and-effect relations and are often called *independent and dependent*, *input and output*, *influence and response*, *cause and effect*, *givens and result*, or *state variables*. Which variables are interpreted as *cause* and which are *effect* is the result of the situation. For example, P could be the result of independent choices for a gas that is compressed with conditions V, n, and T. Accordingly, Equation 2.2 is set up to calculate the dependent variable, P, from independent "givens" V, n, and T. Alternately, the question may be what temperature is required to store a volume of gas below the pressure limits of a vessel. Then either Equation 2.1 or 2.2 can be rearranged to solve for the dependent variable T, given input variables V, n, and P.

2.2 A Classification of Mathematical Model Types

There are many model types and each creates particular categories of problems (challenges) for regression, which defines appropriate methods for regression. The diverse model types can be classified by opposing attributes:

1. *Steady-state versus transient.* Steady-state models are variously called static or equilibrium, as opposed to dynamic or seasonal models that reveal how a variable develops in time. Steady-state does not necessarily mean "at rest," "it stopped," or "no residual activity." For example, in vapor–liquid equilibrium, the rate that molecules leave the liquid and enter the vapor is matched by the rate that molecules leave the vapor and enter the liquid. There is constant molecular motion and exchange between phases. It never stops or becomes stationary. However, at equilibrium there is no net transfer of material, energy, or momentum and measures of system properties become steady in time. With steady-state models, there is a one-to-one correspondence of the influence and the result. By contrast, dynamic models describe how the system states evolve in time and are often expressed as a differential equation. Here an input change can initiate time-dependent evolution of the output – for one input value, the output will have a progression of values – and there is not a one-to-one correspondence of input to output. The mathematical procedures of solving steady-state and dynamic models are different.
2. *Phenomenological versus empirical.* Phenomenological is often termed first principles, mechanistic, or rigorous. Empirical means that there is no phenomenological basis for the model form. Common empirical model categories are time series, polynomial, and neural network. However, there is a spectrum within the extremes that blends both phenomenological and empirical techniques. These are variously termed hybrid, semi-empirical, truncated, appropriated, dimensional analysis, pseudo-component, and so on.
3. *First principles versus rigorous.* Both are phenomenological approaches. A first principles version uses the simple principles as they might be first taught in an introductory course, and may be termed reduced, simplified, or idealized models. A rigorous model also uses mechanistic representation, but attempts to include all of the complexity that Nature offers. Of course, the processes of Nature seem always to be one step more complicated than we understand. Therefore, even rigorous models are not complete and, of course, the balance between simplicity and completeness creates a range of modeling rigor between first principles and "complete."

4. *Deterministic versus stochastic*. Deterministic models return a single value such as a particle size, or material temperature, or average density. However, Nature is not uniform and there is a spatial distribution of particle sizes, molecular distribution of temperatures, or variation of density within a sample. Those are spatially distributed variations. There may also be a time-related, temporal, distribution of values such as wind speed at any one location over a time interval. Models that provide the distribution of values reveal the range of values as well as the average. Monte Carlo simulation or bootstrapping studies are of the stochastic type. They return probabilistic or stochastic results – a different value for each run. When all values are assembled, they provide a view of the distribution, from which you can obtain the average, variance, or other useful statistics.
5. *Deterministic versus distribution*. An alternate type of distribution model would mathematically calculate distribution properties such as expectation and variance, or probability, with methods like integrating probabilities (as one might develop in a reliability study) or propagating uncertainty. Here the values of the distribution coefficients would have deterministic values.
6. *Linear versus nonlinear*. If linear (straight line), the response changes linearly with the variable. The relation is $y = mx + b$. Nonlinear is not one particular form; it could be anything that is not linear (quadratic, reciprocal, exponential, etc.). Linear relations provide tractable mathematical analysis. Unfortunately, most about Nature is nonlinear. In regression modeling the linear or nonlinear relation has two meanings. After model coefficients are determined, in the use of the model, the term would describe the relation between design variables (operational variables, experimental inputs) and the response. However, in regression, the experimental input values have been implemented, they are fixed, and the objective is to adjust model coefficients to make the model best match the data. Here the linear relation is between the model coefficients and the model output. A simple reaction rate model is $r = ke^{-E'/T}c$, where k and E' are model coefficients and T and c are reaction temperature and concentration. The expression is linear in coefficient k, but nonlinear in coefficient E'. The relation is linear in variable c, but nonlinear in variable T. There are many special cases where some variables are linear and others not, termed bilinear or affine, which permit limited mathematical analysis.
7. *Continuous valued versus discrete*. Some variables are continuous valued, are infinitely divisible, in our human concept. Consider a glass of water. You can pour out half the water and the contents are still water, take out 83.59373441% and the remains are still water. The fluid appears as a continuum, until you are looking at the molecular level. When down to one molecule, you cannot remove half a molecule of water. For many purposes, however, the continuum concept is fully functional at our engineering macro scale. By contrast, in engineering analysis, many variables are integers, discontinuous, or discretized and can only have certain values. This discontinuity provides difficulty for some methods of determining model coefficients.
8. *Analytic versus nonanalytic*. An analytic model has continuous valued derivatives. There are no slope or level discontinuities. A nonanalytic model has slope or level discontinuities. Models that are in the category of partitioned, fuzzy, ensemble, or stochastic are nonanalytic. Regression on such values requires optimizer approaches that can cope with the nonanalytic objective function values.
9. *Single equation versus multiple (partitioned)*. Commonly, models have one equation. However, often there are several equations, a particular equation for one range of

variables and another for another range. For example, if the Reynolds number is greater than 2100 use the Fanning or Moody–Darcy relation for pressure drop; alternately use the Hagen–Poiseuille relation. Partitioned models are commonly used in gain scheduling, fluid dynamics, and fuzzy logic and its derivatives, such as Takagi–Sugeno–Kang (TSK) models.

10. *Single equation versus multiple (ensemble).* In several nonlinear modeling approaches we look at the average, median, or range of several models. A common example is weather forecasting. It is not uncommon to hear the forecaster say "Model A indicates storms here in 3 days, but Model B indicates it will be to our South." They use multiple forecasting models and when the models agree, they report the forecast with confidence. It is not uncommon to train neural networks (NNs) of diverse architectures, let each predict, and chose the intermediate value as the expectation and the range of values to indicate the uncertainty. When sensors are subject to failure, a common practice is to install three sensors and use the middle-of-three, the median value; the process instrumentation procedure is termed "voting."
11. *Explicit versus implicit.* In explicit models, the outcome variable can be isolated on one side of the equation and explicitly calculated from the other variables. This could be a sequence of equations. Solving the explicit equation to obtain a numerical value would be following the sequence of mathematical operations (+, −, ×, /, parenthesis, functions, etc.). This is often termed a computer procedure or structured text instruction, which is described by the assignment statement. By contrast, in implicit models, the result is needed to be able to calculate the result. Implicit models need to be "solved" with techniques like root-finding (successive substitution, Newton's, trial and error, interval halving, etc.) or optimization. Implicit models require an iterative procedure and a convergence criterion that goes beyond the simple execution of a sequence of assignment statements.
12. *Direct versus inverse.* Normally we consider the cause-to-effect, input-to-output, influence-to-response relations. We ask, "If I do this, how will Nature respond?" The response variables are variously termed state variables, outputs, or dependent variables. We use the model to solve for "y" given "x." However, in design and control we wish to determine the conditions that make a process behave in a desired manner. We ask, "What conditions are required to make Nature respond like I want?" In such a case we run the model inverse and we determine the input (influence, design condition, independent variable) that contributes to the specified output (desired behavior). We use the rearrangement of the model to solve for "x" given "y." If one is going to use the inverse of the model, then the regression structure should look at the "x" prediction from the given "y," the sum of squared deviations (SSD) on x, and not the conventional SSD on y.
13. *Single-valued versus multivalued.* A classic problem with inverse models is that there are multiple solutions. "What is 2-squared?" There is one answer, "4." "What number squared makes 4?" Now there are two answers to the inverse, "+2 and −2." The inverse of polynomial, ensemble, fuzzy, and neural network models can lead to multiple answers. One answer has to be chosen to match the data.
14. *Single versus multiple variables.* This refers to the number of variables in either the model input or output.
15. *Unconstrained versus constrained.* In unconstrained models there are no bounds on the variables. In constrained problems there are relations that cannot be violated. Constraints

might be on material and energy balances, non-negativity of variables, input limits of 0–100% throttle, rate of change constraints, and so on. In regression, it is common to recognize that model coefficients must be within bounded and feasible values (time constants and delays must be non-negative) or that coefficients values cannot lead to an unstable model.

16. *Primitive versus dimensionless variables*. When we derive a model from first principles, we are stating it in terms of primitive or natural variables. These might include viscosity, flow rate, density, and so on. Dimensional analysis is an approach to place all variables in the model into dimensionless groups, to nondimensionalize the model. Common dimensionless groups include Reynolds, Nusselt, Prandtl, Biot, and Fourier numbers. Each dimensionless group is composed of three or so primitive variables. The nondimensionalized model preserves the same functional relation between variables, but has the benefit of reducing the number of independent variables in the ratio of 3 (or so) to 1, which reduces the number of experiments needed to explore the model space to determine model coefficient values. Recognizing that the essential inputs to the model are the fewer number of dimensionless groups, some empirical modeling approaches (such as the Buckingham pi method) seek to determine the appropriate dimensionless groups and to use them in developing empirical models.
17. *Value versus probability*. Deterministic models determine a value, but often our analysis is related to the probability of finding a lower (or higher) value. "What is the probability of an accident when driving at this speed?" "How is the probability of getting an "A" in a course related to the number of homework problems completed?" "How does system reliability relate to the number of parallel devices?" A common empirical probability model is the logistic response $p = 1/(1 + e^{-s(x-c)})$, where p is the probability of an event, related to variable x and coefficients s and c represent the scale and center for x. The regression objective is to find model coefficients that best fit the model to probability data.
18. *Real-valued versus rank or category*. Most models predict a value, a real number. However, the procedure might be providing an analysis with a dichotomous outcome (stay/run, on/off, be watchful/take cover) or more categories (buy/sell/hold). It might be providing a rank. Here, a regression objective is to best match predicted categories to data and, rather than having a continuous-valued deviation to indicate the degree of model fit, one might have to count the number of right or wrong classifications to assess model goodness.
19. *Tractable versus intractable*. A tractable model leads to explicit mathematical statements. If there are no mathematical techniques to provide model solution or analysis, it is an intractable model. Intractable models are often analyzed through iterative procedures that include root-finding, optimization, and IF-THEN-ELSE conditionals.

As a complexity, the model types can be mixed. For instance a steady-state, first principles model might be stochastic with continuous valued variables, where another might be deterministic with integer values. Any one of the model opposing classifications can be associated with one of nearly any of the other 19 classifications.

The particular difficulties of the model and data types shape the choice of the most appropriate optimization method for regression, for data processing, for model validation, and for experimental design. The following sections on model types will point out such issues.

Those 19 classifications are based on mathematical methods, but the model type can also be differentiated by its function.

20. *Data model versus process model.* A data model is what is used to convert primitive sensor measurements into data that characterizes the process, into the data that would be graphed and fit to a process model. A process model would characterize the influence and response mechanism of a process or product. For instance, in calibrating an orifice flow rate device, one might time the collection of liquid in a bucket. The flow rate is then calculated by the quantity of the liquid divided by the collection time. Simultaneously the orifice device creates a pressure drop in the flowing line and a transducer transmits a signal (often an electrical current) that is a scaled representation of the pressure drop. The process model would relate milliamperes to flow rate and the regression will seek to find process model coefficient values that best fit the flow rate w.r.t. milliamperes data. However, flow rate is not measured; it is calculated by a data model. Depending on context, the process model might become a data model. Once the orifice device is calibrated, consider the exploration of a model to predict flow rate w.r.t. valve position. Here, the user sets a valve position and uses the orifice "measured" flow rate (actually calculated from the milliamperes signal and the orifice equation) for the data. What had been a process model has become a data model. The distinction is important when using a data model to propagate uncertainty in the data.

2.3 Steady-State and Dynamic Models

Processes take time to evolve. For example, place hot and cold items in contact with each other so that they can exchange heat. The temperatures of the two items do not instantly attain thermal equilibrium, but temperature progressively, spatially, changes in time, asymptotically approaching a steady-state temperature. *Dynamic models* describe how properties, attributes, or states evolve in time. These are often called *transient models*, and are described by differential equations, revealing a $^{d}/_{dt}$ term representing the rate of change of some inventory within some control volume. The inventory and control volumes might be, respectively, heat within a metal slab and temperature would be the state variable measure, molecules within a tank with concentration as the state variable, items within a warehouse, people within a city, or $ within an economy. If the inventory (hence the state variable) is spatially uniform within the control volume, then the model will only have time dependency and will be an ordinary differential equation (ODE).

Dynamic models indicate how a variable evolves over time. It does not predict one single value for the variable, but provides a value after one second, then two seconds, then three seconds, and so on. Accordingly, to test a dynamic model w.r.t. data, each point in the time-series forecast (or prediction or history) needs to be compared to the actual data.

By contrast to an ODE, if the inventory (state variable) changes within the control volume as well as time, then the model generates a partial differential equation (PDE). Consider how the temperature changes spatially and in time as a loaf of bread is cooking in an oven. First, throughout the dough, everything is at the room temperature and then the oven heats the outside of the material, which transmits heat to the inside. After a bit, the outside is oven-hot, but the inside is still cool and then the inside temperature progressively rises. To compare the PDE to the data you must compare the state variable at each point in time and do it for each point in space.

ODE dynamic models might describe how an average property, such as concentration or temperature, which is measured on the macro scale, changes in time. Alternately, dynamic models might describe how the probability distribution (as opposed to spatial distribution) of properties changes in time. For example, the distribution of particle size as crystallization proceeds within a unit operation changes in time. In any small internal section of the well-stirred crystallizer, in any sample taken, the particles will have a size distribution. In other examples, molecular weight distribution changes in time as a batch polymerization reaction proceeds, the spatial distribution of temperature changes in time as an item approaches thermal equilibrium, and the energy distribution population of neutrons changes in time during a power transition of a nuclear reactor.

By contrast to dynamic models *steady-state* or *static models* describe the average value or distribution coefficient values of the material property, rate of change, or the final or equilibrium state. For example, in making a batch of cookies the number that stick to the cookie sheet after baking depends on the amount of cooking grease. The rate of heat loss from a house depends on the thickness of insulation. The viscosity of a gas depends on temperature and molecular weight. The yield of products in a petroleum cracking operation depends on the crude composition and cracking temperature. Static models will not have the $^{d}/_{dt}$ terms, although they may have integrals with respect to time that express an average. Steady-state (static) models often are the constitutive relations within dynamic models.

2.3.1 Steady-State Models

As an example of a steady-state model, consider the Boltzmann ideal gas concept, which leads to the ideal gas law (Equation 2.1). As popular and useful as this model is as an equation of state, idealizations (spherical, point mass, elastic collisions, no interparticle forces) make it invalid for relatively dense (high pressure, low temperature) or polar gases. The value of the gas law constant R can be derived from fundamental physics, making it a phenomenological or first principles model.

Van der Waals conceptually improved the model by adding terms. One term with coefficient a represents the several types of molecular attraction forces between molecules; the other, with coefficient b, represents the volume occupied by a mole (quantity) of the molecules. A van der Waals representation was presented as Equation 2.2. That model represents an intuitive compensation, an appropriation, of the first principles model and since different molecules have different mutual attraction and volume, expectedly the a and b coefficient values will be dependent on the gas. Rather than determine coefficient a and b values from fundamental analysis, we determine the values from regression to experimental results. This is a semi-empirical model. However, as this equation has been used, and discrepancies to data noticed, researchers have devised a more complex conceptual understanding of Nature, which has led to more complex rigorous and semi-empirical models.

First principles models represent an elementary (introductory, initial, basic) conceptual understanding of Nature and are the tool of preference for many situations. By contrast, rigorous models attempt to include all confounding fundamental phenomena to become a complete description of Nature. However, even the most rigorous phenomenological models always seem shy of being perfect descriptions of Nature. The human conceptual understanding is always not quite complete, and we often use conceptual idealizations or mathematical conveniences (such as linearizing or truncating an infinite series) to obtain a model that balances sufficiency and utility with perfection.

One could consider that Equations 2.1 and 2.2 identify the key variables, but that the proper functionality escapes explicit analysis and that the relation can only be expressed conceptually as

$$P = f(n, T, V) \tag{2.3}$$

Then one might use a Taylor series expansion, truncated to quadratic terms, to provide an explicit relation that approximates Equation 2.3:

$$P = a + bn + cn^2 + dT + eT^2 + fV + gV^2 + hnT + inV + jTV \tag{2.4}$$

Equation 2.4 contains the relevant variables, but does not express their fundamental relationship. It is an empirical model and values for coefficients a through j would be determined by regression, fitting the model to data. Equation 2.4 is an empirical model.

Advantages of phenomenological models are (i) the coefficients have a physical interpretability because the coefficients are related to mechanistic concepts, (ii) the functional form of the models fairly well relates cause-and-effect responses over a broad range of values, (iii) there are relatively few coefficients whose values need to be obtained by a best fit to data, and (iv) data validation of the models validates our conceptual understanding and knowledge about Nature. In contrast to rigorous models, advantages of first principles phenomenological models are simplicity of understanding and deployment.

Equations 2.1 to 2.4 are steady-state models. Given values for n, T, and V, the models return a single value for P. Steady-state models can be expressed as the dependent variable, y, is a function of the independent variables, x, and the model coefficients, p:

$$y = f(x_1, x_2, x_3, \ldots, p_1, p_2, p_3, \ldots) = f(\underline{x}, \underline{p}) \tag{2.5}$$

Here, the underscore, the vector notation, implies that they are multiple variables or coefficients.

Experimentally, there would be separate trials or runs that would impose the conditions in x, wait until the process settles at a steady state, and measure the value of y. For a model with three independent variables, tabulated data and a modeled y-value based on particular coefficient values might appear as in Table 2.1.

Table 2.1 Steady-state data and model

Run number, data set, trial	Independent variable 1	Independent variable 2	Independent variable 3	Dependent variable, experimental measurement	Modeled value $y_i = f(x_i, p)$	Deviation squared
1	$x_{1,1}$	$x_{2,1}$	$x_{3,1}$	y_1	$\tilde{y}_1$	$(y_1 - \tilde{y}_1)^2$
2	$x_{1,2}$	$x_{2,2}$	$x_{3,2}$	y_2	$\tilde{y}_2$	$(y_2 - \tilde{y}_2)^2$
3	$x_{1,3}$	$x_{2,3}$	$x_{3,3}$	y_3	$\tilde{y}_3$	$(y_3 - \tilde{y}_3)^2$
4	$x_{1,4}$	$x_{2,4}$	$x_{3,4}$	y_4	$\tilde{y}_4$	$(y_4 - \tilde{y}_4)^2$
N	$x_{1,N}$	$x_{2,N}$	$x_{3,N}$	y_N	$\tilde{y}_N$	$(y_N - \tilde{y}_N)^2$
						SSD

Table 2.2 Steady-state data and model with data rearranged

Run number, data set, trial	Independent variable 1	Independent variable 2	Independent variable 3	Dependent variable, experimental measurement	Modeled value $y_i = f(x_i, p)$	Deviation squared
N	$x_{1,N}$	$x_{2,N}$	$x_{3,N}$	y_N	$\tilde{y}_N$	$(y_N - \tilde{y}_N)^2$
3	$x_{1,3}$	$x_{2,3}$	$x_{3,3}$	y_3	$\tilde{y}_3$	$(y_3 - \tilde{y}_3)^2$
1	$x_{1,1}$	$x_{2,1}$	$x_{3,1}$	y_1	$\tilde{y}_1$	$(y_1 - \tilde{y}_1)^2$
4	$x_{1,4}$	$x_{2,4}$	$x_{3,4}$	y_4	$\tilde{y}_4$	$(y_4 - \tilde{y}_4)^2$
2	$x_{1,2}$	$x_{2,2}$	$x_{3,2}$	y_2	$\tilde{y}_2$	$(y_2 - \tilde{y}_2)^2$
						SSD

The last column in this table, "deviation squared" is a measure of badness, a penalty for the model not matching the data that increases with the square of the deviation and SSD, where the SSD is an overall measure of model badness for all of the experimental runs.

Since each data set is independent of the others, since the d^2 value in the rightmost column only depends on values of variables in its row, the SSD value is independent of the order that the data are listed, or sequence of which the experiments were run. The data rearrangement in Table 2.2 will return exactly the same SSD as Table 2.1.

Although the experimental designer might use a structured order to define the experimental conditions, the sequence of experimental conditions in runs in time (the chronological order of the experimental conditions) should be randomized to prevent autocorrelation of a controlled variable (one of the x values) with uncontrolled experimental conditions that would change in time. Uncontrolled conditions that change in time are atmospheric pressure and humidity, operator experience or distraction, warm-up of electronics, sensor calibration drift, process fouling, dust build-up, and so on.

However, after the experiment is run in a randomized sequence, the data can be sorted on any variable to re-present the information in a manner that helps human interpretation.

As a rule of thumb, a heuristic is that there should be at least three times the number of independent experimental data sets as there are model coefficients to be determined. Independent sets are not replicated, with repeated identical input values. Independent sets have unique input values. The input values in independent sets are not correlated to each other.

2.3.2 *Dynamic Models (Time-Dependent, Transient)*

Contrasting steady-state models, dynamic models describe how states evolve in time. Here is a classic elementary example. Water flows into a tank at a rate F_{in} and flows out at a rate F_{out}, which is dependent on the water level in the tank – the higher the level, the greater is the static pressure pushing the water out through the fixed restriction under ideal turbulent conditions. The model is a nonlinear first-order differential equation:

$$A\frac{dh}{dt} + k\sqrt{h} = F_{in} \tag{2.6}$$

If F_{in} makes a step increase, then the level in the tank, h, will begin to rise. As the level rises, the outflow will increase. When the level has risen to the point where the outflow rate matches the inflow rate, then the level will remain at the new steady value.

One can solve for $h(t)$ numerically, using Euler's method, an explicit finite difference approximation to the derivative:

$$h(t + \Delta t) = h(t) + \Delta t[F_{in} - k\sqrt{h(t)}]/A \tag{2.7}$$

Note that the new value of h, the value after a time interval of Δt, depends on the prior value of h. This is a key feature that differentiates steady-state and dynamic models. In steady-state models the prior state value is the same as the current state value. It does not change in time. This will be clear when one looks at a table of data from experimental runs.

Here F_{in} is the influence that causes the process state, h, to change from its initial to its final h value. For a single F_{in} value, there will be multiple response values during the transient. Contrasting a steady-state model, in a transient model there is not a one-to-one correspondence of influence to response.

A generic representation of a dynamic model would be

$$y_{i+1} = f(y_i, \underline{x}_i, \underline{p}, \Delta t) \tag{2.8}$$

where the subscript "i" is the time interval counter, $t = i\Delta t$.

In a typical experimental trial, an operator would implement particular x values, record y as it evolves in time, and after a while, change x values and continue recording the y response in time. The experimental data might appear as in Table 2.3.

Table 2.3 Dynamic data and model

Time interval counter	Independent variable 1	Independent variable 2	Independent variable 3	Dependent variable, experimental measurement	Modeled value $y_i = f(x_i, p)$	Deviation squared
1	$x_{1,1}$	$x_{2,1}$	$x_{3,1}$	y_1	$\tilde{y}_1$	$(y_1 - \tilde{y}_1)^2$
2	$x_{1,1}$	$x_{2,1}$	$x_{3,1}$	y_2	$\tilde{y}_2$	$(y_2 - \tilde{y}_2)^2$
3	$x_{1,1}$	$x_{2,1}$	$x_{3,1}$	y_3	$\tilde{y}_3$	$(y_3 - \tilde{y}_3)^2$
4	$x_{1,1}$	$x_{2,1}$	$x_{3,1}$	y_4	$\tilde{y}_4$	$(y_4 - \tilde{y}_4)^2$
5	$x_{1,2}$	$x_{2,1}$	$x_{3,1}$	y_5	$\tilde{y}_5$	$(y_5 - \tilde{y}_5)^2$
6	$x_{1,2}$	$x_{2,1}$	$x_{3,1}$	y_6	$\tilde{y}_6$	$(y_6 - \tilde{y}_6)^2$
7	$x_{1,2}$	$x_{2,1}$	$x_{3,1}$	y_7	$\tilde{y}_7$	...
8	$x_{1,2}$	$x_{2,1}$	$x_{3,2}$	y_8	$\tilde{y}_8$	...
9	$x_{1,2}$	$x_{2,1}$	$x_{3,2}$	y_{19}	$\tilde{y}_9$	...
10	$x_{1,2}$	$x_{2,2}$	$x_{3,2}$	y_{10}	$\tilde{y}_{10}$	...
11	$x_{1,2}$	$x_{2,2}$	$x_{3,2}$	y_{11}	$\tilde{y}_{11}$	...
12	$x_{1,3}$	$x_{2,2}$	$x_{3,2}$	y_{12}	$\tilde{y}_{12}$	...
13	$x_{1,3}$	$x_{2,2}$	$x_{3,2}$	y_{13}	$\tilde{y}_{13}$	...
N	$x_{1,N}$	$x_{2,N}$	$x_{3,N}$	y_N	$\tilde{y}_N$	$(y_N - \tilde{y}_N)^2$
						SSD

Note that the time interval counter has replaced the experimental run number. Note that the value of x_1 was held constant for the first four time intervals at the first value, $x_{1,1}$, and that x_1 and x_2 were both held at constant values until the eighth and tenth time samplings. Although the first four time samplings imposed the same independent variable values, the y values would change in time as y evolves toward its steady-state value corresponding to the x inputs. The first four y values are not expected to have the same values.

Note also that the third value of y depends on the second value of y. The modeled y value depends on the prior modeled y value. Accordingly, in a dynamic model, one cannot reorder the sequence of data by sorting on row values. The time sequence must be preserved.

The difference in relation between input and outputs requires separate treatment of steady-state and dynamic models in regression. Static or steady-state models have a unique output response value to the input value(s). Hundreds of independent input values will generate hundreds of output values with a one-to-one correspondence. A good model will match the measured values for each of the hundred data sets. By contrast, a transient (dynamic) model will define an evolution of the output from an initial to its final value. Even with fixed input values (variously labeled forcing function, influences) the model output will evolve in time to a steady-state value. Each sampling instant in time generates a new data point. Therefore, even one unique input held for a time will generate many different output values as time evolves, perhaps 30 samples for each input set of conditions. A good transient model will match the time-dependent trend on process output, for each of many influence combinations, leading to very many squared deviations to match.

With the heuristic that there should be at least three independent sets of influences for each model parameter being valued by regression, and perhaps 30 samples for each transient, dynamic models would have about 30 times more data than steady-state models.

2.4 Pseudo-First Principles – Appropriated First Principles

Often we attribute multiple mechanisms to a single pseudo-phenomena in a procedure variously termed *lumping* or *clubbing*. For example, the "dispersion" coefficient in reactor modeling combines the multiple influences affecting concentration mixing – molecular diffusion, transport by fluid turbulence, and the impact of fluid flow rate distributions. The multiple conceptual mechanisms are combined into one "pseudo" mechanism, and the ideal equation is appropriated (taken and redeployed without sanctioned permission – Webster's 7th *New Collegiate Dictionary*, 1963, uses *converting something to one's own use with questionable right*) to better match the reality.

Another example of lumping is the a and b terms in the van der Waals equation of state (Equation 2.2), which are simple representations of multiple and complex phenomena. The coefficient b represents the volume occupied by a mole of the molecules, leaving $V - nb$ as the volume permitting compressible gas behavior. However, molecular volume is a fuzzy concept. The electron cloud is soft, and part of the occupied volume is due to molecular vibration. Additionally, the coefficient a represents all of the attraction forces between molecules that would reduce the observed pressure. These include dipole moments, nuclear, magnetic, ionic, gravitational, and whatever else may become understood as part of the van der Waals forces. The ideal gas law has been appropriated; its basic form is used, but is heuristically or qualitatively adjusted to make it better match the data.

Engineers have many theoretical models. One of them is the Bernoulli equation, which can be rearranged to yield the orifice equation, which relates flow rate to the square root of the pressure drop due to the orifice restriction. However, the ideal Bernoulli analysis is for inviscid flow, which has a flat velocity profile and no friction losses. Within an orifice these conditions may be approximately true, but the "bullet" velocity profile, permanent friction losses, swirling, and nonideal location of the flange taps make the ideal square root relation not quite true. As a result, a better model relating measured pressure drop across the orifice and calculated flow rate would ease the half-power idealization, leading to a power law relation in which both the coefficient a and exponent b are to be evaluated from experimental data. The model in Equation 2.9 is grounded in theory, which provides a functional relation. However, the functional relation has been appropriated to better match the reality (Rhinehart *et al.*, 2011):

$$F = a(\Delta P)^b \tag{2.9}$$

We appropriate many equations resulting from an ideal analysis when lumping (or clubbing) too-difficult-to-describe phenomena into one coefficient. Although the value of that one coefficient does not scientifically describe a single mechanism, although it is not phenomenological or rigorous, it provides a value that reflects the combined impact of mechanisms and has a valid interpretation, which makes it useful in process understanding, design, analysis, control, and optimization.

The ideal Arrhenius form of a homogeneous, first-order reaction rate is

$$r = k_0 e^{-E/RT} c \tag{2.10}$$

However, a next level of comprehensiveness could include the transport limit of molecular diffusion to active sites when one in a series of elementary reactions is considered to be the rate-limiting step:

$$r = \frac{k_0 e^{-E/RT} c}{1 + k_1 c} \tag{2.11}$$

Other model adjustments, appropriations, include the use of an effective diffusion coefficient for diffusion in a multicomposition mixture, or effective tray efficiency or separation factor in distillation.

Linearization will be a final example here of the customary use of the liberty of appropriation. Equation 2.6 is nonlinear, an inconvenience for analysis or automatic control, and differential equations are often linearized by approximating the nonlinear term with a local linear approximation. Locally linearized about the value of h_0, Equation 2.6 would become

$$A\frac{dh}{dt} + kh/(2\sqrt{h_0}) = F_{in} - (k\sqrt{h_0})/2 \tag{2.12}$$

or if coefficient values were to be determined empirically and the influence and response are stated as deviations from an initial steady-state

$$\tau\frac{dh'}{dt} + h' = aF'_{in} \tag{2.13}$$

This represents an appropriated first principles model.

2.5 Pseudo-First Principles – Pseudo-Components

The use of pseudo-components is an essential tool in attempting to characterize many processes, as given in the following examples. (i) The make-up of petroleum liquids and coal is characterized by pseudo-components to develop yield models of the material response to cracking and devolatilization. (ii) Molecular weight of polymers in a chemical process and thermal energy of neutrons in a nuclear reactor have nearly continuous distributions, but to obtain tractable models, these continuous-valued properties are often grouped into bins or intervals representing pseudo-components of an average bin property (molecular weight, temperature) to reduce the number of states in a model. (iii) The use of a dead-zone as a modeling artifact in mixing represents an independent volume of fixed size, perhaps behind a baffle. Its contents are well mixed within the dead-zone, but this pseudo-volume has an independent composition from the well-mixed bulk of the mixer contents. Mass transfer between the dead-zone pseudo-volume and the active mixed zone is modeled as due to a concentration-driven mass transfer mechanism with a coefficient related to the rate of mass transfer between the zones. (iv) Viscoelastic films are often modeled as a parallel and series arrangement of pseudo-components – springs and dashpots in parallel and series, termed Maxwell and Kelvin–Voigt models.

Pseudo-component models, like phenomenological models, express the general functional relationship between variables, which leads to a fewer number of coefficients than do empirical models, which means fewer coefficient values need to be determined. Both the pseudo-components and coefficients have human interpretability and define a useful understanding of Nature.

2.6 Empirical Models with Theoretical Grounding

Often the natural process is too complicated to understand or to model. Therefore, we use empirical models, which are generally classified as steady-state or transient. Steady-state models are often called static or equilibrium models. Time-dependent models are often called transient or dynamic models. Phenomenological versions of either can be converted to empirical models with theoretically valid mathematical mechanisms.

2.6.1 *Empirical Steady State*

Steady-state models are commonly based on a truncated Taylor series that provides an infinite series expression for any function about a location x_0:

$$\begin{aligned} y = f(x) &= \sum_{i=0}^{\infty} \frac{1}{i!} \left.\frac{d^i f}{dx^i}\right|_{x=x_0} (x - x_0)^i \\ &= f(x_0) + \left.\frac{df}{dx}\right|_{x=x_0} (x - x_0) + \frac{1}{2!} \left.\frac{d^2 f}{dx^2}\right|_{x=x_0} (x - x_0)^2 + \ldots \end{aligned} \tag{2.14}$$

which can be rearranged in a power series form when recognizing that the derivative values are fixed at a particular x_0 value and when combining coefficients of powers of x. Equation 2.15

is for a single-input single-output process:

$$y = a + bx + cx^2 + dx^3 + \ldots \tag{2.15}$$

Equation 2.16 is for a multivariable response with two influence variables, indicating all powers of each input and all cross-product powers:

$$y = a + bx_1 + cx_1{}^2 + dx_1{}^3 + \ldots + ex_2 + fx_2{}^2 + \ldots + gx_1x_2 + hx_1{}^2x_2 + ix_1x_2{}^3 + \ldots \tag{2.16}$$

The generalization to a model with a greater number of input variables should be obvious. Here, output response is y, input is x, and coefficients are $a, b, c, d, \ldots$.

Consider that there is some set of equations that represents the process behavior. Even though the particular set may not be known, conceptually it could be derived or discovered. Mathematics proved that any function can be expressed as a Taylor series with an infinite number of terms and that a truncated series (with a finite number of terms) can express the phenomenon within a specified range, within a desired tolerance. A Taylor series can be rearranged to form a power series expansion as shown in Equations 2.15 and 2.16. So, even though those equations do not express the functional relation between y and x, a power series model has a legitimate mathematical basis and can approximate the function sufficiently over a limited range.

Alternately, Equations 2.15 and 2.16 can be stated as a sum of a sequence of polynomials such as Lagrange, Chebychev, wavelets, and so on. If Laguerre polynomials,

$$y = a + b[(-x + 1)] + c\left[\frac{(x^2 - 4x + 2)}{2}\right] + d[(-x^3 + 9x^2 - 18x + 6)/6] + \ldots \tag{2.17}$$

or it could be stated as a Newton's interpolating polynomial,

$$y = y_0 + a(x - x_0) + b(x - x_0)(x - x_1) + c(x - x_0)(x - x_1)(x - x_2) + \ldots \tag{2.18}$$

There are many mathematically equivalent rearrangements for the power series representation of a Taylor series expansion.

Steady-state empirical models are also based on dimensionless group correlations, such as the commonly used heat transfer relation in fluid systems that relates the Nusselt number to the Reynolds and Prandtl numbers:

$$Nu = aRe^b Pr^c \tag{2.19}$$

Fluidization models relate Reynolds and Archimedes numbers, and diffusion models through fixed media relate Fourier to Biot numbers and the aspect ratio. The dimensionless groups appear as a combination of variables or properties when the PDEs that describe a process, product, or procedure are converted to scaled or dimensionless variables to create a generic solution. The dimensionless groups become the variables or coefficients in the generic solution, representing the fundamental descriptors. However, other approaches, such as the Buckingham pi method can also reveal the dimensionless groups. The product of powers of groups as illustrated by Equation 2.19 is not the functional form of the solution to the PDE, but it is often found to be adequate as a model over a useful range of conditions for many engineering purposes. Values for coefficients a, b, and c, are determined by best fitting the equation to experimental data.

2.6.2 Empirical Time-Dependent

Derived from first principles these would be differential equations, which can be converted to finite difference algebraic equations, time series, autoregressive moving average (ARMA), z-transform, back-shift operator, and so on, representations.

ARMA models can be derived as a finite difference solution to a system of ODEs, and are often used in models of dynamic or transient time-series responses. If the derivative of a linear ODE such as Equation 2.13 is replaced with the finite difference representation

$$\tau \frac{\Delta h'}{\Delta t} + h' = aF'_{in} \tag{2.20}$$

and expanded at the ith time sampling,

$$\tau \frac{h'_{i+1} - h'_i}{\Delta t} + h'_i = aF'_{in,i} \tag{2.21}$$

Rearranged,

$$h'_{i+1} = \frac{\Delta t}{\tau} aF'_{in,i} + \left(1 - \frac{\Delta t}{\tau}\right) h'_i \tag{2.22}$$

and with combined coefficients,

$$h'_{i+1} = a_0 F'_{in,i} + b_0 h'_i \tag{2.23}$$

Equation 2.23 indicates that the new state value depends on the most recent past state and influence values. This is an ARMA(1,1) model. There is one autoregressive term (one past state value influences the next) and one moving average term (one influence term).

For a generic model with response variable y and influence variable u, and a higher-order differential equation and a delay, the differential equation can be linearized, discretized, and written as a ARMA(p,q) model:

$$y_{i+1} = a_0 u_{i-n} + a_1 u_{i-n-1} + a_2 u_{i-n-2} + \ldots + b_0 y_i + b_1 y_{i-1} + \ldots \tag{2.24}$$

In Equation 2.24, y_i is the response variable at the current sampling (time) and u_i is the current value of the influence. The "+1" and "−1," "−2," and so on, modifications to "i" represent argument values at the first future or recent past samplings. The n subscript in the u values indicates a delay of n samples, meaning that the ith y value is not influenced by the ith u value but by the u value that occurred n samples prior. Coefficients a and b represent combinations of coefficients in the differential equation, and if the differential equation is known, then values for the a and b coefficients can be determined. If the delay is known, then n can be determined as the number of samplings in the delay period. However, more often, the values for coefficients a, b, and n are determined by best fitting Equation 2.24 to experimental transient data.

Although these model forms are grounded in theory and although they represent what you could derive if you knew the fundamental differential equations, since the fundamental equations are not known, the user's choice for the structure is not grounded in theory. The selection of variables included in the equations, to linearize or not, deviation reference, the order of the equation (the highest power or number of discrete functionalities), and the number of past variables (or a delay) are all user choices to make the generic model adequately fit the data.

Notably, the coefficients a and b are usually continuously valued (real in computer programming, rational in number categories) variables. The value of n is an integer (discrete valued). An optimizer to determine both a and b and n values must handle both integer and real numbers – often termed mixed-integer programming. Further, the value of n must be non-negative (future influences cannot affect a present or past response) and combinations of the b values must reflect a stable response (if the process is stable), all creating constraints on the coefficient values.

Also notable, Equation 2.24 is linear in coefficient values a and b, but nonlinear in n. Linear optimization and linear regression cannot determine the value of n. A nonlinear, constraint-handling, mixed integer/real optimizer is required.

2.7 Empirical Models with No Theoretical Grounding

Useful modeling approaches do not have to have grounding in knowledge of the underlying phenomena. Artificial NNs (Hagan *et al.*, 2014) are often called a model-free way of modeling. This term means that there is no mathematical relation, such as the structure in Equations 2.1 through 2.24 that the user decides. Although the user must define the architecture (input variables, internal structure of the neurons, and the transfer function for each neuron), the NN has the flexibility to match nearly any functional relation without explicitly stating the functional relation. In contrast, Equation 2.19 specifies a power law product and would not perform well if the phenomena represented a summation function. NNs are not that complicated, but explaining them here would detract from the purpose of this textbook, as a tutorial on regression. If interested, see any text or tutorial on NNs. The bottom line is that the NN is a function that uses inputs to calculate an output, and regression (NN folks call it training or learning) is used to determine NN coefficient values (called weights) that best match the unknown phenomena represented in the data:

$$y = NN(\underline{u}, \underline{y}, \underline{p}) \tag{2.25}$$

In Equation 2.25 $\underline{u}$, $\underline{y}$, and $\underline{p}$ represent the vectors of inputs, responses, and model coefficients.

A particular feature of NNs and many other nonlinear models is that there may not be one unique "best" in the regression solution. There may be many local best values – valleys in the surface. In this case of multioptima, we need an optimizer that will find the global best, not get stuck in an almost good place.

2.8 Partitioned Models

Models are often partitioned, empirically, into regions where a particular relationship is relatively true. For example, if the flow is turbulent (if the Reynolds number is above 2100 for fluid flow within a pipe) use the Fanning or Moody–Darcy law to relate the fluid flow rate to the pressure drop in a horizontal pipe of uniform diameter. Alternately, if less than 2100, use the Hagen–Poiseuille relation. A general example is

$$\text{IF } [x \text{ } is \text{ } A] \text{ then } [y = f_A(x)] \tag{2.26}$$

In Equation 2.26 the antecedent condition [x *is* A] relates a variable to a category. This could be a simple less than, equal to, or greater than logic, for instance [$Re < 2100$], or it could be belongingness to a fuzzy category (child is hungry, temperature is hot). The consequent [$y = f_A(x)$] relates the state variables and there would be a different relation for each antecedent category. Coefficients in this model include those in both the antecedent as well as the consequence.

Assembled, a set of three rules might be

$$\begin{aligned} &\text{IF } [x < 17] \text{ then } [y = 4] \\ &\text{IF } [17 \le x < 26] \text{ then } [y = 4 + 2(x - 17)] \\ &\text{IF } [26 \le x] \text{ then } [y = 22] \end{aligned} \tag{2.27}$$

Such structures are used in fuzzy logic modeling, situation-weighted multimodel ensembles, TSK models, and so on. Even if the local models are linear (as above with model coefficient values of 4, 2, and 22), the overall model is nonlinear. Further, the partition coefficients (values 17 and 26 above) are nonlinear in regression. More than being nonlinear, partitioned or logical models will have a discontinuous surface, which adds a confounding effect on the optimizer that seeks to find coefficient values.

2.9 Empirical or Phenomenological?

Here is a perspective on the undesirable attributes of any of the empirical modeling methods. (i) Empirical models have little, or no, explicit representation of functional, mechanistic, cause-and-effect relations. (ii) They usually require additional coefficients to make the mathematical functionality of the terms describe the true mechanisms of the natural phenomena. (iii) The functional forms and coefficients have no interpretability to properties of the natural phenomena. (iv) The models cannot be expected to extrapolate (accurately or even rationally predict out of the data range used to develop them). (v) The models cannot be used to confirm, deny, or validate hypothesized cause-and-effect mechanisms.

Although all these features are undesired, empirical models remain popular and useful. They are simple to create. In contrast, the derivation of mechanistic models often requires extensive scientific or engineering analysis, and for many complex applications the effort is unwarranted. Additionally, justification to use empirical models comes from mathematics, which reveals that given enough terms, an empirical model can fit the true equation within a specified range and within a specified accuracy (but this does not address extrapolation, knowledge validation, etc.).

2.10 Ensemble Models

Models can be derived or selected to have different functional relations or degree of complexity. Often, we choose empirical models when the mechanism is too complex to attempt a phenomenological model, but we do not know what model structure is best. Therefore, we might choose several model structures and optimize their respective coefficients to best fit the data. In this case the models will have general, but not exact, agreement. Further, the models might have been regressed to different data sets, so even if the model structure is exact, the coefficient values will be different. As an example of different models, you may have heard the weather forecaster indicating that the European model and North American model are predicting slightly different tracks for a hurricane in the Atlantic. Ensemble models are commonly used in NN applications relating to human behavior and are used in process control as locally linearized versions of nonlinear processes.

I have seen these called multimodels.

Ensemble models are comprised of several distinct models and their prediction can be reported in any of many ways. For example, the average or the median could be reported as the expected value, along with an uncertainty based on the range of the separate predictions. Alternately, one model might be selected by fuzzy logic, related to the confidence that it provides a best answer. Here the model coefficients will include those related to how the separate models are to be combined to produce a single prediction value.

In any case, the model coefficients and the mixing rule coefficients must be adjusted to best fit the data.

2.11 Simulators

So far, we have been considering models that are explicitly visible as equations. However, often we use simulators, proprietary software that simulates a process (distillation, airplane flight). Here we cannot see the equations, but still would like to adjust coefficients of the simulator (tray efficiency, wing drag) so that the simulator best fits data. Often these are termed "black box" models, because the user cannot see the details of how they are computing values. They are just models and would have coefficients adjusted and results validated by the procedures of this book.

2.12 Stochastic and Probabilistic Models

Many material properties and process responses have a distribution of values, as shown by the following examples. (i) A rock crushing/grinding operation creates particles with a distribution of sizes. This example would result in size distribution of particles within a single sample. (ii) Human I.Q. has a distribution. This would be a distribution of values between individuals within a population. (iii) Sequential rolls of dice will result in a distribution of numbers. This would be a distribution of values over time.

Most often, engineering models are *deterministic*. They provide a single value for the process response, for example, temperature, T, as the result of compression of an ideal gas. However, the "temperature," the energy of each molecule in the ideal gas, has a Maxwell distribution. There is not one common temperature for all molecules. Although there is a *distribution* of molecular energies, on the macro-scale of engineering, we measure or calculate the average gas temperature, as though each molecule had the same deterministic value.

We use probability distributions to describe risk in financial investment, failure rates in products and processes, molecular weight variation within polymers, thermal energy of neutrons in a nuclear reactor, residence time in chemical reactors, traffic patterns at intersections, and many other applications.

There are common models of distributions, such as Gaussian (or the "normal" or "bell-shaped" curve), the binomial for dichotic events, Poisson, exponential, and many others.

Typically, such ideal distributions are characterized by a few (one to three) coefficient values related to mean, variability, skew, or range, for instance.

Regression is often used to find the values of the distribution coefficients that make the distribution best fit the data or to find coefficient values in underlying distributions that make a *probability model* best fit the outcomes.

Stochastic models return a single value, but in a replicate run they will return a different value. Like the role of a die, it may be a 3 on the first roll but a 5 on the second roll and a 2 on the third roll. With enough rolls (thousands) the histogram of values will approach the ideally expected distribution of outcomes. A stochastic model would need to be run thousands of times to generate a simulated realization for the distribution of outcomes. These are often used in gaming and reliability studies.

Deterministic models can provide an average or expected measurement. For example, use the substance Cp to determine the temperature rise upon adding heat. However, deterministic models can also indicate a probability distribution. The Maxwell temperature distribution indicates the range of energies within an ideal gas, but each time you ask it what fraction has a particular temperature range, it provides exactly the same, deterministic value.

By contrast, the output of stochastic models is influenced by random independent perturbations and each replicate provides a unique value. Each run of the model is kin to sampling from a population of values.

2.13 Linearity

Linear means that it is a straight line. *Nonlinear* means that it is not a straight line. The term nonlinear does not say what it is (quadratic, exponential, etc.). It just indicates that it is not a straight line. This distinction is a key selector for the appropriate optimizer for regression.

For steady-state (static) models, the concept is relatively clear. The value of the response (dependent variable, output, etc.) is linear with the value of the input (cause, dependent variable, etc.). In Equation 2.2 P is linearly dependent on T, but P is nonlinearly related to V.

However, for transient responses, the output evolves in time in response to a change in input (often called the forcing function in the mathematics nomenclature). Even when the input is held constant in time after its change that initiated a transient response, the output variable value evolves in time from the initial to the final value. During the transient response there will be many output values for the one input value. For distributions, there are also many response values, each with individual probability. Therefore, the straight line description for the linear concept is not comprehensive in its applicability.

Additionally, in regression, the role of variables and coefficients is switched. The "givens" represent the values of the variables from the experiments. These are fixed and regression seeks to change the coefficient values to best match the model output with the data. In regression, the "variable" values are fixed and the coefficient values are varied. In Equation 2.15, if the value of x is fixed, the value of y will be a linear response to coefficient c. Although Equation 2.15 provides a nonlinear response of y to x, in the search for coefficient values, y responds linearly to either a, b, c, or d.

In the context of this textbook, linear regression for steady-state models means that the second derivative of the response variable, y, with respect to a model coefficient, p, is zero:

$$\frac{\partial^2 y}{\partial p_i \partial p_j} = 0 \text{ for all } i \text{ and } j \tag{2.28}$$

A linear steady-state model means that the second derivative of the response variable, y, with respect to an influence variable, x, is zero:

$$\frac{\partial^2 y}{\partial x_i \partial x_j} = 0 \text{ for all } i \text{ and } j \tag{2.29}$$

In Equations 2.28 and 2.29 i and j may have the same values, which would represent the conventional second derivative.

Applying Equations 2.28 and 2.29 to the former model equations:

- y in Equation 2.15 is linear in coefficients a, b, and c, but nonlinear in independent variable x.
- P in Equation 2.2 is linear w.r.t. coefficient a, but nonlinear w.r.t. coefficient b.
- F in Equation 2.9 and Nu in Equation 2.19 are linear w.r.t. a and nonlinear w.r.t. b.
- y_i in Equation 2.2 is linear in each a and b coefficient and each u and y variable, but nonlinear w.r.t. n.

The test for nonlinearity of coefficients in dynamic models is more complicated and is confounded by the terminology in solving ODEs. In ODE terminology, linear means the state variable always appears to the first power, which permits superposition of solutions in ODEs. It means that if $y_1(t)$ is a solution and $y_2(t)$ is a solution then a linear combination, $ay_1(t) + by_2(t)$ is also a solution. However, in regression, we are not seeking a solution to the ODE, we are seeking coefficient values that make the solution best match the data.

The ODE could be linear in a coefficient but the solution is not. Consider a "linear" first-order ODE,

$$\tau \frac{dy}{dt} + y = y_{SS}, \quad y(0) = y_0 \tag{2.30}$$

Analytically, the solution is

$$y(t) = y_{SS} - (y_{SS} - y_0)e^{-t/\tau} \tag{2.31}$$

Since τ is in the exponent, attempting to adjust τ to fit $y(t)$ to data is nonlinear regression.

Numerically the solution is approximated as

$$y_{i+1} = \frac{\Delta t}{\tau} y_{SS} + \left(1 - \frac{\Delta t}{\tau}\right) y_i \tag{2.32}$$

where τ appears as an inverse, a nonlinear coefficient.

In regression it is how the model coefficient affects the model calculation, either the analytical or numerical embodiment. In either case τ in the exponent or as an inverse is a nonlinear coefficient.

Further, even though the SS limit may be linear in coefficients, the dynamic response will not be.

Linear also depends on how the equation is presented. For example, the time constant appears nonlinearly as the coefficient in Equation (2.32), but a new coefficient could be defined, $\beta = {}^{\Delta t}/_{\tau}$, which linearizes the Equation (2.32) model to $y_{i+1} = \beta y_{SS} + (1 - \beta)y_i$. As another example, $y = (a + x)/b$ is nonlinear in one of the coefficients, b, but choosing $c = a/b$ and $d = 1/b$ makes the equation $y = c + dx$ linear in both coefficients.

2.14 Discrete or Continuous

The delay variable, n, in Equation 2.24 is discrete. It cannot make infinitesimal changes in value. Accordingly, when it does change, it makes a jump in the y_i response to past u and y values. This makes the least squares regression surface have discontinuities.

The categories of the antecedent in Equation 2.26 or 2.27 may not mean that there is a surface discontinuity as x changes from one category value to another, but probably mean at least a slope discontinuity in the regression surface.

Many optimizers have difficulty in coping with either slope or value discontinuities.

2.15 Constraints

When conditions impose restrictions on the model coefficient values, they are called *constraints.*

The a and b coefficients in Equation 2.2 must have nonnegative values. They represent occupied space and attraction forces, and negative values do not make sense with the physical mechanism concepts. This would be a simple constraint condition on the coefficient.

The delay of Equation 2.24, the value of n, can only have positive values. The air temperature today cannot be influenced by the cloud that will blow by tomorrow. Further, for sampled data systems or for numerical simulation the value of n is an integer; it cannot have rational values. Therefore, the model coefficient, n, must be constrained to have non-negative and integer values. This would be a slightly more complex condition on the coefficient value, but still a condition in the coefficient value.

If the dynamic process of Equation 2.24 is stable then the b values must be constrained by conditions that make the model stable. This relation on the b values would impose a collective or joint condition on the coefficients. However, it still remains a constraint on the model coefficient values.

In more complex models, there could be constraints on other model variables (not necessarily the model coefficients). For instance, in using data to create a model of a distillation process, adjustable model coefficients may represent tray efficiency and ambient heat losses, but within the tray-to-tray calculations, mole fractions need to sum to unity, and both liquid and vapor need to be present.

The optimizer for regression, the "engine" that finds the best model coefficient values, must be able to accommodate a diverse variety of constraints. The constraints can be simple: non-negative values, or the constraint can be any number of relations between variables, and include discontinuities and integer conditions.

2.16 Model Design (Architecture, Functionality, Structure)

Should an empirical power series model go to third order or fourth order or higher? This is a choice regarding model design, which is variously termed architecture, functionality, or structure. For a linguistic (fuzzy logic) model, the design choices would include the number of linguistic categories and the membership function type (trapezoidal, radial basis). For a neural network model, design includes the number of neurons in a layer and the transfer function in a neuron. Some of the choices are numerical (number of items) while others are functional in structure. These are empirical models.

However, some models are intuitively created. In mixing liquids, temperature of the mixture is the weighted average of the temperature of the two samples. This experience about one aspect of mixing might direct one to think that all mixed properties are similarly computed. For instance, a mixing rule for viscosity would be more accurately modeled as related to the log of the composition, a mixing rule for plugging value (the amount of material to plug a filter) might be better modeled as a reciprocal relation, or freeness (the time to dewater fibers) might be better modeled as a quadratic relation. However, in an initial modeling approach, the developer might intuitively choose a linear relation.

Once a model structure has been defined, the regression algorithm is easy to apply. The issue is how the user decides which design is right. Chapters 16 through 20 address that.

2.17 Takeaway

Which variables are chosen to represent cause and effect or independent and dependent depend on the situation. Mathematical models provide a method to calculate values of the response of a process, outcome of a procedure, or property of a product. There are many types of responses (static and dynamic, deterministic and distribution), and many valid modeling approaches for each. Your choice of variables and equation form can lead to a static or dynamic model, a linear or nonlinear situation, with either discrete or continuous-valued variables, predicting either a value, a category, or a distribution.

The model properties w.r.t. coefficients (linear or nonlinear, constrained coefficient values or not, single or multivariable, discrete or continuous, single or multioptima, discontinuous or smooth surfaces) direct the appropriate optimization method for regression.

Exercises

2.1 Confirm these statements

(a) y in Equation 2.15 is linear in coefficients, a, b, and c, but nonlinear in independent variable, x.

(b) P in Equation 2.2 is linear w.r.t. coefficient a, but nonlinear w.r.t. coefficient b.

(c) F in Equation 2.9 and Nu in Equation 2.19 are linear w.r.t. a and nonlinear w.r.t. b.

(d) y_i in Equation 2.2 is linear in each a and b coefficient and each u and y variable, but nonlinear w.r.t. n.

2.2 Wikipedia reports that the phenomenological model for temperature dependence of resistivity in noncrystalline semiconductors is $\rho = Aexp(T^{-1/n})$, where $n = 2, 3$, or 4 depending on the dimensionality of the system. If you knew this functional relation, you would have one coefficient, A, to fit the model to data. Is this linear or nonlinear regression? Explain.

2.3 Provide examples of models that are:

(a) Steady-state and transient
(b) Phenomenological and empirical
(c) First principles and rigorous
(d) First principles and appropriated
(e) Deterministic and stochastic
(f) Deterministic and distribution parameters
(g) Value and probability of a value range
(h) Linear and nonlinear in the independent variable, but linear in the coefficients
(i) Nonlinear in the coefficients
(j) Continuous valued and discrete
(k) Continuous valued and rank
(l) Continuous valued and category
(m) Explicit and implicit
(n) Direct and inverse
(o) Single valued and multivalued
(p) Single variable and multivariable
(q) Single equation and partitioned
(r) Single equation or ensemble of equations
(s) Unconstrained and constrained
(t) Primitive variables and dimensionless variables
(u) Tractable and intractable

2.4 Differentiate between the concepts of human conceptual understanding, mathematical model, and the truth about Nature. How do you classify the knowledge that we teach in school (such as the ideal gas law).

2.5 Show that Equation 2.15 is equivalent to a Taylor Series expansion, and therefore fundamentally acceptable as a generic model.

2.6 Show that Equation 2.16 is equivalent to a Taylor series expansion and therefore fundamentally acceptable as a generic model.

2.7 Show that Equation 2.17 is equivalent to a Taylor series expansion, and therefore fundamentally acceptable as a generic model.

2.8 How does one justify the number of terms in a Taylor Series expansion or in a representation such as Equations 2.15 to 2.18?

2.9 Show that an ordinary differential equation, with derivatives expressed as finite difference is equivalent to an ARMA model.

2.10 Apply the criteria for determining if a relation is linear to the following:

(a) $PV = nRT$. Is P a linear response to n, R, T, or V?
(b) $r = ke^{-E/RT}c$. Is r a linear response to k, E, T, or c?

(c) $F = a(\Delta P)^b$. Is F a linear response to a, b, or ΔP?
(d) $y = a + bx_1 + cx_2 + dx_1x_2$. Is y linear in a, b, c, d, x_1 or x_2?
(e) $y = a + bx_1 + cx_1^2 + dx_1^3 + \ldots + ex_2 + fx_2^2 + \ldots + gx_1x_2 + hx_1^2x_2 + ix_1x_2^3 + \ldots$ Is y linear in a, b, c, $\ldots$, h, i, x_1, or x_2?
(f) $y_{i+1} = a_0u_{i-n} + b_0y_i$. Is the response linear in a, b, n, u, or y?
(g) In parallel resistors in an electrical circuit $1/R_T = 1/R_1 + 1/R_2$. Is R_T linear in R_1?

Part II

Preparation for Underlying Skills

3

Propagation of Uncertainty

3.1 Introduction

Propagation of uncertainty is a fundamental concept in several aspects related to modeling.

- Uncertainty in data values (often termed experimental error) leads to uncertainty in model coefficient values and uncertainty in model coefficient values leads to uncertainty of the in-use model-calculated result.
- Prior to regression, in design of experiments propagation of uncertainty is applied to experimental measurements to ensure adequate precision in the experimental design, to define the number of data sets required for adequate precision, and to determine experimental conditions that need additional focus.
- In regression, propagation of uncertainty can be used to select regression convergence criteria.
- In model development, uncertainty analysis is useful in understanding the impact of linearization, in assessing model utility, and in relating data uncertainty to model coefficient uncertainty.
- In model validation, uncertainty analysis is used to compare modeled residuals to expected trends and magnitudes.
- In model use, propagation and reporting uncertainty on model-calculated results is important to the user, who must accommodate uncertainty of calculated values to make decisions.

Consider how a decision maker needs to understand the uncertainty in numerical values. We take action based on numerical values and since that action has an effect on human aspirations, any decision should include the possible "short fall." Whoever balances all of the management issues and makes the decision needs to know the uncertainty in each numerical value. This includes calculated values from equations and models, whether it relates to the flow rate needs to support anticipated product sales, the size of a heat exchanger when considering the uncertainty in fouling factors, the required number of distillation trays, or the rate of return on investment.

Nonlinear Regression Modeling for Engineering Applications: Modeling, Model Validation, and Enabling Design of Experiments, First Edition. R. Russell Rhinehart.

If you perform experiments to determine values for reaction rates, diffusion coefficients, pump efficiency, and so on, or use models to predict temperature, flow rate, lift, or composition, someone will use those calculated or measured values in design or policy decisions. That someone will want to know the uncertainty associated with the numerical values that you report.

For example, you may run a lab test that indicates a new drug has 3% fewer side effects than those drugs presently allowed. Reporting only the calculated result, decision makers may accept the new drug as safer and allow its use. However, if the "error" on your test is ±20% side effects, then the truth is that the new drug might be 3 + 20 = 23% better, or it might be 3 − 20 = 17% worse, and a right decision would be to, "Do more testing. Do not release it on the public yet. The uncertainty in the results is too high to make any confident comparison."

The right decision needs to be made understanding the uncertainty as well as the nontechnology context of the situation. For example, a heat exchanger design might have a ±25% size uncertainty, and any of several business choices could be right. One could be, "Money is tight right now, so we will buy the small size. If it is too small, we can limp along until we can afford a bigger one." By contrast, another decision could be, "The process operation is critical, and start-up timing is urgent. Resources are available. Buy the large one, so that we are sure that we can operate as needed." Yet another could be, "This is for a new process making a new product. We probably will run at reduced capacity for the first years so buy the smaller exchanger and allocate floor space to permit an additional one. The single small one will be big enough for the reduced initial production rate, and if we need the larger one later, we have plenty of time (and space) to install additional capacity later. However, if this product line never makes it, then we will not waste money by buying the larger one." The decision and the action taken considers the uncertainty of the calculation within the situation context.

Whether in industry or government, whether dealing with economic or human resources, an engineer's job is to generate numerical values that managers use to make decisions. These decisions have an impact on the success of human endeavors. To make rational decisions, the managers need to understand the uncertainty of your numerical values. An engineer's ethical responsibility to society is to provide an honest representation of the work, so report numerical values in a manner that appropriately acknowledges the uncertainty of the value.

To propagate uncertainty, you have to know why it arises and how to propagate it in your calculations. Upon understanding error sources and propagation, you can design experiments to minimize uncertainty and make decisions that account for uncertainty.

Material in this chapter parallels the developments in Bethea and Rhinehart (1991) and Rhinehart (2016).

3.2 Sources of Error and Uncertainty

We have many names for the uncertainty of numerical values. These include error, fuzz, noise, bias, fluctuation, and variance. Since the term "error" often connotes a mistake and "bias" denotes a systematic deviation, the term uncertainty is a better representation. Sources of uncertainty are not necessarily human mistakes. They include the naturally occurring variability on measurements due to either systematic bias or random fluctuation. They also include process-to-model mismatch that results from idealizations, truncation, or the use of tabulated data. This book will use the term uncertainty; however, both the terms "error" and "propagation of error" remain as commonly used labels. Following are a few sources of uncertainty (errors) encountered in engineering analysis.

3.2.1 Estimation

Often the basis of a calculation is an estimate. "Oh, I expect we'll be able to sell 25 metric tons per year." However, a plant that large may be a large capital risk for the company resources. Accordingly, the investment managers will want to know the likelihood of only being able to sell 15 or 20, or of the potential to sell 30 or 35 metric tons per year. Whether termed "givens" or "basis," such estimates of the situation are uncertain.

3.2.2 Discrimination

No measurement device is capable of infinitely small measurements. For instance, a pressure gauge may be marked in increments of 10 psi and one might be able to estimate a gage reading to the nearest 2 psi. Since the width of the pointer and markings is 2 psi, obtaining a more precise reading would be impossible. Discrimination limits the reading to ± 1 psi. Similarly, if a 0–12 800 gpm flow rate reading is processed by an 8-bit computer, the discrimination ability is 2^{-8}, about 0.4% of full scale, about 50 gpm. Consequently, as the flow steadily increased from 900 to 1000 gpm, the computer would continue to report 900 gpm during the period in which the flow was changing from 900 to 910, to 920, and so on. At 950 gpm it would report (jump-to and hold) 950 gpm until the flow reached 1000 gpm. Similarly, discrimination error in reading numerical data from charts and diagrams depends on the thickness of the pencil line or the scale of the axis. From tabulated data discrimination error is related to the number of digits displayed in the data. As a common example, my cell phone indicates time in hours and minutes. It displays 7:23 a.m. until it becomes 7:24. Even though time progressively increases, the reading remains at 7:23 until it jumps to 7:24.

3.2.3 Calibration Drift

A metal ruler lengthens and contracts due to temperature change. A wooden ruler also changes length due to humidity. Temperature affects spring stiffness in pressure gauges or weigh scales. Temperature changes the resistance and capacitance in the electric circuitry of sensors and transmitters, and may be due to either ambient conditions or unit warm-up in uses. In general, calibration drift is affected by instrument age, temperature, humidity, pressure, and catastrophic events such as dropping the device.

3.2.4 Accuracy

Sensors and transmitters are not exactly linear. Calibrations are usually performed by adjusting a transmitter zero and span so that the instrument reports a value "close enough" to the "true" value of two standards. Then one assumes that the instrument response between the standards is linear even though the response is not exactly linear. (In fact, the local calibration standards are not perfect, either.) The *accuracy* of a device, its reading deviation from the true value, is often called *systematic error* or *bias*.

3.2.5 Technique

The measurement procedure may measure the wrong thing. For example, the outside circumference of a tank might be easier to measure than the inside diameter. Diameter can be calculated from circumference. However, if the objective is to calculate tank volume, a measure of outside diameter is inappropriate. As another example, if knowledge of the steam pressure is required to size a reboiler, one must have the pressure downstream of the steam flow control valve and not the steam header pressure. As a third example, a thermowell might be measuring the temperature in a hot spot in a furnace, but this would not represent the average temperature over all of the space that drives the heat transfer.

3.2.6 Constants and Data

Most calculations involve fundamental constants, but their values come from previous measurements that are not exactly known. Examples of constants include the gas law constant, the gravitational constant, the speed of light, and the molecular weight of sodium. Typically, these values are known to errors of only 0.001 or 0.01%. Examples of data include a tabulated viscosity or thermal conductivity. Typically these data are known only to 0.1–1% error. Often, data are obtained from correlations or graphs, such as pipe flow friction factors, thermodynamic properties of mixtures, and convective heat transfer coefficients. Such values may have a 10–30% uncertainty.

3.2.7 Noise

Process flow turbulence, changing electromagnetic fields around equipment, equipment mechanical vibrations, and so on, may cause the measured value to randomly vibrate or fluctuate about its average value. Due to the fluctuation, any one instantaneous reading may not reflect the time-local average. One has to average many values to temper the uncertainty of random noise. Noise reduces your ability to obtain a precise measurement.

3.2.8 Model and Equations

Nonidealities are usually not expressed in models of phenomena and, accordingly, inaccuracy is reflected in the equations that are used to calculate the results. For instance, the ideal gas law ($PV = nRT$) is a model of particle dynamics that neglects the effects of intermolecular forces and molecular size. Although the law is often used for gases at ambient conditions, it can introduce a 5% error at those conditions and up to an 80% error at conditions near the critical point of real gases. The volume of a tank may be calculated by $V = \pi r^2 h$, an equation that is based on a right-circular cylinder model. Such a model neglects real surface irregularities such as the effects of dents or ribs, sensor intrusions into the side of the tank, and the curved bottom and the length of drain pipe above the bottom discharge valve. The square-root relation between differential pressure and flow rate, derived from the Bernoulli equation, is commonly accepted in orifice calibration equations and may lead to a 5% error. The Beer–Lambert relation for the concentration-dependent light absorption is commonly accepted in spectrophotometric devices. Models are often incorrect because they are incomplete; therefore, the calculated values (whether part of the measurement device or off-line) are also imperfect.

3.2.9 Humans

Humans are often part of the data valuation process. We might transliterate, reversing the position of adjacent digits. We might decide when a process has achieved steady-state for sampling, but a desire to finish the project might override waiting long enough for the transient to settle. A noisy signal might be biased by using a convention that reports the mental average of the upper (or lower) extreme that was observed. A human might judge that a data point is faulty because it is inconsistent with expected trends in the data, and may discard that point, when in fact the data was good and the understanding of the process was wrong.

3.3 Significant Digits

Digits in a number that are *significant* are the digits that have values of which we are fairly confident. For instance, in a circumference calculation, the diameter might be reported as 100 $\pm$ 1 inches. This means that the diameter might be any number between 99 and 101 inches. Nominally, the calculated circumference has a value of $C = 314.159265\ldots$ inches, with a range of $\pm 3.14159265\ldots$ inches. We are sure that the numerical value of the first digit for the circumference is 3 and that the second is 1. However, the third digit, 4, could have been a 7 or a 1 due to measurement uncertainty. The third digit is "fuzzy." Since we cannot know exactly what the value of the third digit is, we cannot pretend to know the numerical value of the fourth digit, which represents an order of magnitude smaller value. The error on the fourth digit, 1, is ± 31. In this example, the third digit from the left, the first digit with an uncertain value, is the last reasonably reportable digit. All of the digits following the third represent at least one order of magnitude smaller contribution than the uncertainty, and are thus insignificant for this example. The circumference could be reported as 314 inches.

Whether explicitly stated or implied, you should reveal the uncertainty in reporting numerical values. The following is a reporting convention for integers and real numbers. An integer has no decimal point. It is used to represent the number of whole events or whole things. For example, there are 4327 people on the payroll. A real number has a decimal point and is used to represent the value of a variable that can have fractional values. An example is the factor for conversion of mass units, 0.4536 kg/lbm. Some uncertainty is associated with both integers and real numbers. For instance, if you tried to count 2000 items, you might lose track, and might have a count that is off by 10 or so. The number of items should then be reported as 2000 ± 10 to explicitly acknowledge the counting precision. Similarly, the speed of light in a vacuum is reported to be $2.997925 \pm 0.000003 \times 10^8$ m/s.

By custom, the uncertainty in numerical values is usually not explicitly reported, but is implied by the number of digits reported. The last reported digit is the digit with uncertainty. When the precision is implied, we do not explicitly know whether the last digit is accurate to ± 1 or ± 2 or ± 3 or ± 4. If the accuracy were ± 5 with a range of 10, the next digit to the left would be uncertain and would have been the last reported digit. If ±0.2, the last reported digit, would be known with certainty, and the next digit to the right, the first fuzzy digit, should be reported. Therefore, the uncertainty on the last digit reported could range from ±1 to ±4. Without specific guidance, we will assume a mid-value for the uncertainty of ±2.5. For continuously valued numbers, the last reported digit from the left, even if zero, is the fuzzy number. For example, a length may be reported as 100.0 inches, which implies an error of about ±0.25 inch. Similarly, the atomic weight of hydrogen = 1.0079 g/g mole implies that the value is known with an error of about ±0.00025 g/g mole.

For integers, the last nonzero digit from the left represents the fuzzy number. For example, a past census reported the population of the United States as 227 000 000. The last of the three nonzero digits, 7, is fuzzy and reflects a counting error of about 2 500 000 people. This method is the convention for reporting integers, but it has the unfortunate aspect that significant zeros are not identified. For instance, if there are 2000 ± 10 items, a reader might interpret a reported 2000 as 2000 ± 250.

Both reporters and readers must use care in reporting and interpreting implied integer precision.

If the last reported digit was the last digit with a certain value, then the uncertainty would be ±0.25.

The ±2.5 is the average error if the last digit is the first (from the left) fuzzy (uncertain) digit. The maximum error would be twice that. Roughly, the sigma is the maximum uncertainty divided by 2.5. For example, if a table indicates a value of 0.624 and the "6" and "2" are certain, but the "4" is the first uncertain digit, then the nominal or average uncertainty on the data is expected to be ±0.0025, and the maximum uncertainty on the data value might be ±0.005. If the uncertainty value was larger, then the 2 would be uncertain. Therefore, the data value might be between 0.629 and 6.19. However, the deviation is not always the maximum value. Sometimes the deviation might be very small, so on average the uncertainty of the value is ±0.0025, or the data value is between the limits of 0.6265 and 0.6215. The average error would be ±0.0025. The maximum error would be ±0.05. The standard deviation would be about 0.002.

However, if the table reports the last certain value, the analysis is different. If the "4" is certain in the 0.624 value, the average uncertainty on the data is −±0.00025.

3.4 Rounding Off

When reporting significant figures with implied precision, the convention, according to ASTM Standard E29-88, is to round off the number to the nearest significant digit. For base 10 digits, the procedure is as follows: select the last (rightmost) reportable digit. If the digit immediately to the right of the last reportable digit is less than 5, truncate to the last reportable digit. If the digit immediately to the right of the last reportable digit is either greater than 5 or is a 5 followed by a nonzero digit, increase the last reportable digit by 1. If the digit immediately to the right of the last reportable digit is either exactly 5 or is a 5 followed by only zeros, then either truncate if the last reportable digit is even or increase the last reportable digit by 1 if it is odd.

As examples, the following numbers are to be rounded to report only significant digits. The underscored digit in the original number is the last reportable digit based on the explicit error statement (Table 3.1).

Table 3.1 Rounding examples

Original number	Rounded value
$41\ \underline{2}76 \pm 350$	41 300
$6\underline{1}50.02 \pm 73.61$	6.2×10^3
$11\underline{7}.5 \pm 1.2$	118
$11\underline{6}.5 \pm 1.2$	116
$6\underline{1}49.99 \pm 207.13$	6.1×10^3

Rounding should only be performed on the final reported value. Do not round intermediate results. Each rounding operation changes a numerical value; hence, it introduces discrimination error. If several intermediate roundings are performed, this error accumulates through the calculation procedure and could have a significant effect on the results. Of course, the command, "Do not round intermediate results," is impossible when a calculator display only shows eight digits. In practice, it is fully sufficient to maintain two digits past the fuzzy digit for intermediate calculations.

In these examples the numbers are truncated to the first (from the left) uncertain digit. Accordingly, the implied uncertainty is ±2.5 on the last digit in the rounded value.

3.5 Estimating Uncertainty on Values

Before we can propagate uncertainty through calculations and report the uncertainty on calculated (or dependent) variables, we need to know the uncertainty on the numerical values that we use within the calculation. There are many legitimate ways to determine uncertainty on independent variables.

We often get data from tables (viscosity, thermal conductivity, density, etc.). If the table does not explicitly report an uncertainty and you believe that the last reported digit is the uncertain digit, use ±2.5 on the rightmost reported value as an average uncertainty. This means that ±5.0 on the rightmost reported value is the maximum uncertainty. Alternately, if you believe that the last reported digit is a certain digit, use ±0.25 on the rightmost reported value as an average uncertainty.

If it is a measurement, replicate it enough times to be able to calculate the standard deviation. Usually, 10 replications will be ample to provide a definitive estimate, but 5 replications are often minimally adequate and better than 3, which provides an uncertain estimate. If the distribution is Gaussian (most are close enough) then about 99% of the values fall within $\pm 2.5\sigma$ of the mean. Use $\pm 2.5\sigma$ as the uncertainty. See Figure 3.1 and note several scale features. (i) The standard deviation, σ, is neither the error range, nor the maximum error, but it is proportional to those. (ii) Although not defined in this manner, the one-sigma deviation from the average marks the inflection point in the curve – the point at which it changes from curving downward (concave) to curving upward (convex). (iii) The length indicated as "error" is not the absolute highest a normal measurement could be from the average. There is a tail on the curve that extends further to the right. If 99% of the data fall within the data range described by the average $\pm 2.5\sigma$, then 1% fall in the extremes, leaving 0.5% in the higher extreme values and 0.5% in the lower extreme values.

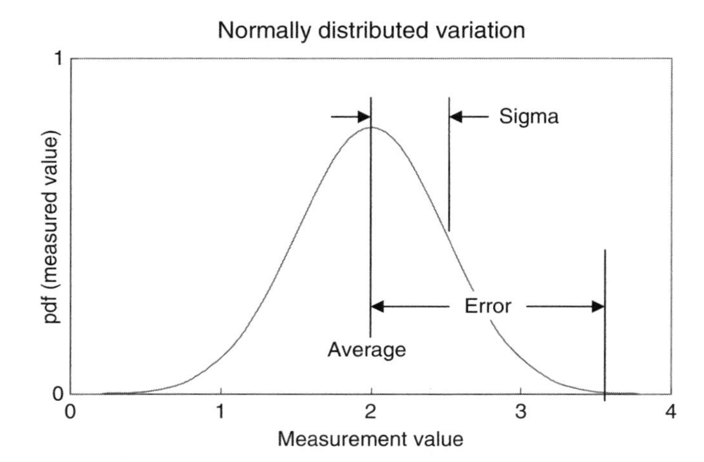

Figure 3.1 Illustration of variability in measurements and relation of standard deviation and error

If the instrument has a calibration record or manufacturer specifications that indicate precision, use that for repeatability. (If it reports accuracy, use that for systematic error.)

If the numerical value is calculated from a linear model with coefficients determined in a least-squares regression or correlation, use the "standard error of the estimate" as the standard deviation on the model coefficient value.

If you bounded the optimum use convergence criteria $\epsilon_{x_i} = {}^1\!/_2\, \Delta x_i$ threshold.

Use your judgment and experience to estimate the possible error. For instance, in using a stopwatch to time the flow collection in a bucket, you might estimate the start and stop times so each have a 0.5 second error. For another instance, in reading data from a curve, you can estimate the error that might happen due to pencil line thickness, your lines not being exactly parallel to the axis lines, discrimination ability on the two axes, or curvature of the graph due to photocopying.

3.5.1 Caution

There is a common mistake in interpreting the standard deviation needed for the sigma in the propagation of uncertainty. Do not use the standard deviation of all of your data. Use only those data that should have the same value.

Here is an example: I measure the height of each grandchild when they come to visit. I stand them next to the wall, eye-ball level the pencil from the top of their head to the wall, and mark the wall. If the pencil is not perfectly horizontal, or their socks are thick, or I don't start from top-dead-center on their head, then the mark on the wall has some error to it. Perhaps each mark is $\pm 1/8$ inch from true. Each mark for each grandchild for each visit is off by a maximum of about 1/8 inch. However, not every mark is off that much, some are closer to the exact value, but 0.125 inch maximum error on any one means about $0.125/2.5 = 0.05$ inch standard deviation. If I were to replicate one kid's measurement 100 times and look at the distribution of marks, I expect the sigma of the variation for any one measurement would be about 0.05 inch.

At one point in time, Kennedy was 63, Jamaeka 62, Parker 61, Conor 45, Ashton 40, and Kain 33 inches tall. Landon was not old enough to stand up yet. The average of 63, 61, 62, 45, 40, and 33 is 50.67 inches, but none of them are 50.67 inches tall. The standard deviation of the same data is 13.003 inches, but my measurements are not in error by 13 in.

What you need in propagation of uncertainty is the uncertainty in a particular measurement (the 0.05 inch), not the standard deviation of all the data in your experimental conditions (the 13.003 inches). To experimentally estimate uncertainty, use only replicates, independent trials that should provide the exact same result.

3.6 Propagation of Uncertainty – Overview – Two Types, Two Ways Each

There are two common types of propagated uncertainty. One is the "maximum uncertainty" on the calculated value, and the other is the "probable uncertainty" on the calculated value. Both measures are important and the concepts of propagation are similar for both. Within both, there is a numerical method and, if nonlinear effects are ignored, an analytical simplification. As the propagation of maximum uncertainty is more easily developed, it will be presented first.

Here is an introduction to the concept of propagation of error in a calculation. Consider a calculation of the surface area of a right parallelepiped. The formula is

$$A = 2S_1S_2 + 2S_1S_3 + 2S_2S_3 \tag{3.1}$$

The measured lengths of the sides are S_1, S_2, and S_3. The surface area A is calculated from a phenomenological model (the equation) for which the coefficients 2, 2, and 2 are exactly known. In general, one may have an equation in generic terms

$$y = f(a, b, c, \dots, x_1, x_2, x_3, \dots) \tag{3.2}$$

in which $f(\dots)$ represents some model or equation that we pretend is exactly correct; a, b, and c represent coefficients whose values we are pretending are exactly known, and x represents a value of a variable having a true, but unknown, value of μ. The unknown measurement error on x is δ:

$$\delta = \mu - x \tag{3.3}$$

The calculated value is y, and the true, but unknowable, value of y is Ψ. This notation will be used in the descriptions that follow.

3.6.1 Maximum Uncertainty

Maximum uncertainty refers to the deviation on the calculated value of y, when the uncertainty on each variable in the calculation is at its maximum level and all errors are coordinated to push the value of y in the same direction.

3.6.1.1 Propagation of Maximum Uncertainty – Numerical Approximation

To use the numerical method to calculate the maximum uncertainty on y due to errors on x_i, calculate values for y for each extreme combination of the x's. Then search through the 2^N values of y for the maximum and minimum. This is an exhaustive search through all combinations of extreme x values.

The error on x_i must be estimated from a technique in Section 3.5.

Sometimes the maximum and minimum errors on any one variable are not symmetric. For instance, the value of a variable may be 10, and the upper and lower limits on the uncertainty in the value of 10 might be 10.3 and 9.5 (+0.3 but −0.5). Sometimes the uncertainty range in one variable will be influenced by the value of another variable. For instance, when $x_2 = 3.7$ the uncertainty on $x_5 = \pm 0.05$, but when $x_2 = 25.1$ the uncertainty on $x_5 = \pm 0.01$. Either of these cases can be handled in the numerical method.

3.6.1.1.1 Caution

If the uncertain range on an x value is large or there is a functional relation near a minimum or maximum, it could be that the extreme y value is associated with an intermediate x value, not some combination of the extreme x values. Some people consider an exhaustive search through all combinations associated with three values for each of the N x variables (low, nominal, and high) in a search through the 3^N values.

3.6.1.2 Propagation of Maximum Error – Analytical Approximation

Alternately, you can use analytical expressions for the sensitivity of the calculated value to errors on the independent variables.

Consider the following calculation:

$$\tilde{y} = \alpha x_{nominal} \tag{3.4}$$

which might represent the specific equation, for example, circumference of a circle, $C = \pi D$. Since

$$x_{nominal} = x_{true} \pm \varepsilon_x \tag{3.5}$$

where ε represents the magnitude of the uncertainty, then

$$\tilde{y} = \alpha x_{true} \pm \alpha \varepsilon_x \tag{3.6}$$

or

$$\tilde{y} = y_{true} \pm \alpha \varepsilon_x \tag{3.7}$$

The true value of y lies within a range of the calculated value plus or minus the coefficient times the error on x. In general, the error, ε_x, is not known. If ε_x were known, then x_{true} would be known and then y_{true} would be known. The error must be estimated from experience with the measurement technique. If ε is the maximum expected error on x and α has a positive value, then

$$\tilde{y} - \alpha \varepsilon_x \leq y_{true} \leq \tilde{y} + \alpha \varepsilon_x \tag{3.8}$$

The maximum error on x has been propagated (its effect on y reflected) as the maximum error on y.

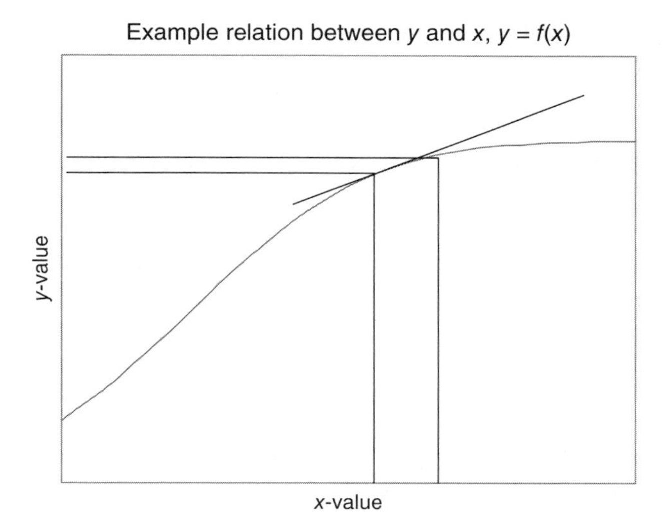

Figure 3.2 Illustration of the relation between uncertainty on $y = f(x)$ w.r.t. uncertainty in the value of x

The more complicated the arithmetic operation, the more complicated the arithmetic error analysis becomes. However, using differential notation, one can systematically generalize the error propagation rules. (Diverse alternate derivations come to the same result.) Figure 3.2 illustrates the concept of sensitivity used in the following analysis. The illustration shows some variable, y, as a function of another variable, x. If the value of "x" is certain, then the value of "y" is known. However, if the value of "x" is changed, then the value of "y" will change. The magnitude of the change in "y" is approximately the product of the change in the value of "x" times the slope of the y–x curve at that point.

Given a model of some general functional form that calculates y from the x's, represented as

$$\tilde{y} = f(a_1, a_2, a_3, \ldots\ x_1, x_2, x_3,\ \ldots) \tag{3.9}$$

or

$$\tilde{y} = f(\underline{\alpha}, \underline{x}) \tag{3.10}$$

you can take the total derivative (from calculus) or, equivalently, use a truncated Taylor series approximation of the y model and obtain the impact of small changes in the x values on the calculated y value:

$$d\tilde{y} = \sum \left(\frac{\partial f}{\partial x_i} \right) dx_i \tag{3.11}$$

The dx_i differentials represent small changes in x_i and dy represents the resulting change in y. Note that Equation 3.11 represents only the linear, first-order x effects considered. Note also that the dx's are assumed independent of each other. Assuming that the uncertainty on each x is symmetric (the "+" maximum error is equivalent to the "−" maximum error) and allowing

ε to represent the differential uncertainty on y and x_i, you obtain the maximum error as

$$|\varepsilon_y| = \sum \left| \frac{\partial f}{\partial x_i} \right| |\varepsilon_i| \tag{3.12}$$

For instance, suppose $y = \ln(x)$ for the calculations. Then, using Equation 3.12,

$$|\varepsilon_y| = |\varepsilon_x / x| \tag{3.13}$$

Note that for linear models, with symmetric and uncorrelated errors, the "analytical" method of Equation 3.13 and the "numerical" exhaustive search method will yield exactly the same maximum error on y. However, for nonlinear models the "numerical" and linearized "analytical" methods will not exactly agree. Strictly speaking, the "analytical" approach to propagation of maximum error, Equation 3.13, is only valid when linearization is permissible (when the errors are small relative to their variable impact) and where the variability on the independent variables can be assumed independent and symmetric. However, if the errors on the independent variable values are relatively small (and good practice should make them so), then the two methods yield equivalent results. Further, considering that values for the uncertainties on the independent variables are often just experienced "guestimates" there is often no engineering justification to prefer the numerical method (fewer implied assumptions, but greater numerical work) over the analytical (linear assumptions, likely mathematically tedious or intractable).

Note that this linearized "analytical" method has the convenience of not doing any computer programming. It also provides the engineer with useful insight. Because the partial derivatives indicate the sensitivity of the uncertainty on y to the precision of the independent variables, you can design experiments to make the "right" variables precise enough to limit the uncertainty on y to within desired limits.

Although this is termed the "analytical" method, one does not have to use calculus to obtain symbolic expressions for the partial derivatives. You can estimate the sensitivity numerically by a forward finite difference approximation

$$\frac{\partial f}{\partial x_i} \cong \frac{\Delta f}{\Delta x_i} = \frac{f(x_1, x_1, \ldots, x_i + \Delta x_i, \ldots, x_N) - f(x_1, x_1, \ldots, x_i, \ldots, x_N)}{\Delta x_i} \tag{3.14}$$

Or, better still, by a central difference approximation

$$\frac{\partial f}{\partial x_i} \cong \frac{\Delta f}{\Delta x_i} = \frac{f(x_1, x_1, \ldots, x_i + \Delta x_i, \ldots, x_N) - f(x_1, x_1, \ldots, x_i - \Delta x_i, \ldots, x_N)}{2\Delta x_i} \tag{3.15}$$

However, also note that the analytical method does not accommodate nonsymmetric uncertainty, or correlation of uncertainty in one variable with the value of another. It also requires that the linear assumption is valid; therefore, to be applicable for nonlinear functions uncertainty must be small in comparison to the values. For a heat exchanger design with a normal error on U of about $\pm 25\%$ of U, the analytical method will distort the error results and the numerical method might be preferred.

Knowing the issues connected with each method, you can choose which is appropriate for your treatment.

Example 3.1 The surface area of a rectangular box is A = 2hl + 2hw + 2lw, where the 2, 2, and 2 represent model coefficients, and the height, h, length, l, and width, w, represent the

Table 3.2 Numerical propagation of maximum uncertainty

Case	Height (m)	Length (m)	Width (m)	Area (m^2)
Nominal	0.50	1.00	0.50	2.50000
1	0.51	1.01	0.51	2.58060
2	0.51	1.01	0.49	2.51980
3	0.51	0.99	0.51	2.53980
4	0.51	0.99	0.49	2.47980
5	0.49	1.01	0.51	2.51980
6	0.49	1.01	0.49	2.45980
7	0.49	0.99	0.51	2.47980
8	0.49	0.99	0.49	2.42060

independent variables. If the h, l, and w values are 0.5, 1, and 0.5 m the surface area is 2.5000 m^2. If the uncertainty on each measurement is 1 cm, then the height, for instance, could be between 0.49 and 0.51 m (Table 3.2).

Table 3.2 represents the numerical search for the area associated with the nominal and each of the $2^3 = 8$ combinations of the extreme length values for the three dimensions.

The extreme calculated values for the area are 2.5806 and 2.4206. Propagation of maximum uncertainty by this primitive search of extreme combinations would report the nominal value of 2.5 m^2 with the possible range of 2.58–2.42 m^2. Here the calculated nominal area was rounded to reflect significant digits (the value of 5 in the nominal area of 2.5 m^2 is uncertain) and the range rounded to one additional digit.

Propagate maximum uncertainty using the analytical method for the prior surface area example.

For the area example above:

$$\epsilon_A = |2l + 2w||\epsilon_h| + |2h + 2w||\epsilon_l| + |2h + 2l||\epsilon_w| = 0.08 \text{ m}^2$$

Note that, for linear models, with symmetric and uncorrelated errors, the analytical method of Equation 3.12 and the numerical method will yield exactly the same maximum error on y. However, for nonlinear models the numerical and linearized analytical methods will not exactly agree. The area model uses the product of independent variables and is not linear. The analytical idealization provided an area uncertainty of ± 0.08 m^2, while the numerical approach revealed +0.0806 m^2 and–0.0794 m^2. The value 0.08 is not identical to 0.0806, but it is effectively similar.

Example 3.2 What is the propagation of error if the ideal relation for a homogeneous reaction rate is used

$$r = ke^{-E/RT}[c]$$

in place of the more rigorous model that includes the square-root temperature dependence?

$$r = k_0\sqrt{T}e^{-E/RT}[c]$$

Here $k = k_0\sqrt{T_{nominal}}$ and the error is in using the fixed value of k, evaluated at some nominal value.

Applying Equation 3.12

$$\varepsilon_r = r\frac{\varepsilon_T}{2T}$$

If the temperature range is 50 °C then the model choice has an 4% impact on the calculated reaction rate (at a nominal 300 K).

Example 3.3 The following nonideal relation is presumed for an orifice flow meter calibration relation:

$$\dot{Q} = a(i - i_0)^b$$

The calibration procedure sets a flow rate and measures $\dot{Q}$ and i. After perhaps 10 measurements, the values of coefficients a and b are determined by regression. Use propagation of maximum error to determine the convergence criterion on the changes in the coefficient values. Choose to stop the optimization iterations when the impact of changes in coefficient values on the reported flow rate is much less (two orders of magnitude smaller) than the uncertainty in the flow rate w.r.t. the model.

Propagating uncertainty of coefficient values on the modeled value

$$\varepsilon_{\dot{Q}} = (i - i_0)^b \varepsilon_a + a(i - i_0)^b ln(i - i_0)\varepsilon_b$$

Equating the uncertainty of the prediction to the root mean squared (rms) error of the model-to-data

$$rms = \sqrt{\frac{1}{N}\sum(\dot{Q}_{data} - \dot{Q}_{model})^2}$$

The stopping rule for convergence would be to stop when

$$(i - i_0)^b \Delta_a + a(i - i_0)^b \ln(i - i_0)\Delta_b < 0.01 \text{ rms } \forall\, i$$

where Δ represents the incremental change in coefficient value.

3.6.2 Probable Uncertainty

Note that for propagation of the maximum error, we have taken the worst case and added each worst case deviation, although we acknowledge that:

1. The independent variable errors are independent of each other (some are positive valued, some are negative), so it is improbable that all will push the error on y in the same direction to some extreme confluence.
2. The independent variable errors are not always at their largest possible value, represented by ε_x, so the maximum error on any particular independent variable is improbable.

The probable (or likely) cumulative error on the answer will be less than the maximum error. The probable error on y can also be calculated. Again, it can be done in two ways, analytically or numerically. The analytical method is easier. In either way, one propagates variance and then uses the variance to calculate probable error.

3.6.2.1 Propagation of Variance – Analytical Approximation

First, one must have an estimate of the standard deviation of the variables with uncertainty. Preferentially, you should obtain σ from replicate measurements. Otherwise, you may estimate σ based on guidance in Section 3.5, your experience, implied precision, or perhaps from the maximum range, R_x, associated with a variable using

$$\sigma \sim R_x/5 = 2\varepsilon_x/5 \tag{3.16}$$

Equation 3.16 is not truth. It expresses that, for a normal (Gaussian) population, a 5σ range centered on the average encompasses about 99% of the data.

Classic propagation of variance (when correlation and nonlinear effects can be ignored and the errors on independent variables are independent, random, symmetric, and of equivalent magnitude) produces

$$\sigma_y{}^2 = \sum \left(\frac{\partial f}{\partial x_i}\right) \sigma_{xi}{}^2 \tag{3.17}$$

This can be derived in a number of ways and has strong similarities to propagation of maximum uncertainty, Equation 3.12, but it sums squared contributions rather than absolute values. As a parallel, it is like the Pythagorean theorem of diagonal distances in space; when the x_i contributions are orthogonal (linearly independent) the resultant is a standard norm.

If the sources of uncertainty on y are many, independent, and of equivalent magnitude, then they should create a Gaussian (normally) distributed impact on y. Then 95% of the errors will be within 1.96 ... standard deviations from the mean error (=0). If we term the probable uncertainty as the range that encompasses 95% of the errors, the 95% probable error on y is then

$$\varepsilon_{95\%\ probable\ on\ y} = \pm 1.96\sigma_y \tag{3.18}$$

It is often more convenient to use the maximum error on x than its standard deviation.

Ideally 95% of the x values fall within $\pm$ 1.96σ of the mean, and approximately 99.7% are within $\pm 3\sigma$ and 99% within $\pm 2.5\sigma$. Then, roughly, considering the 99% limits,

$$2.5\sigma_x = \epsilon_x = \frac{1}{2} Range \tag{3.19}$$

Therefore, if you estimate either the range on x or ϵ_x you can estimate

$$\sigma_{x_i} \cong \frac{1}{2.5}\epsilon_{x_i} = \frac{1}{5} Range_{x_i} \tag{3.20}$$

Similarly,

$$\sigma_y \cong \frac{1}{2.5}\epsilon_y \tag{3.21}$$

Combining Equations 3.17 and 3.21

$$\sigma_y \cong \frac{1}{2.5}\sqrt{\sum_{i=1}^{N} \left(\frac{\partial y}{\partial x_i}\right)^2 \epsilon_{x_i}^2} \tag{3.22}$$

Then the 95% probable error is

$$\epsilon_{y.95} \cong \frac{1.96}{2.5}\sqrt{\sum_{i=1}^{N}\left(\frac{\partial y}{\partial x_i}\right)^2 \epsilon_{x_i}^2} \tag{3.23}$$

There are many assumptions in Equation 3.23: (i) values for ϵ_{x_i} are probably estimates; (ii) the opinion of probable for probable uncertainty may range from 90 to 99%; (iii) the ideal situation of many, independent, similar-sized perturbations is not true; (iv) linearization of $y(\underline{x})$ is not true; (v) the functional form of the model itself is a human construct; and so on. Further, (vi) the 1.96/2.5 value is not very different from unity. Finally, (vii) only the first digit in the estimate of uncertainty is useful for identifying the uncertainty on the calculated y value. Consequently, it is often good enough to use

$$\epsilon_{y_{probable}} = \sqrt{\sum_{i=1}^{N}\left(\frac{\partial y}{\partial x_i}\right)^2 \epsilon_{x_i}^2} \tag{3.24}$$

Here $y_{calculated} \pm \epsilon_{y_{probable}}$ should define the uncertainty on y about 95% of the time, where $\epsilon_{y_{probable}} \cong 95\%$ probable uncertainty in the half-range.

3.6.2.2 Propagation of Variance – Numerical Approximation

This uses nested loops, but in a Monte Carlo approach. It is more generic than the analytical approach. It can cope with nonsymmetric errors, with conditional relations, and with nonlinear influences. However, the analytical approach provides a reasonable estimate, and since this approach requires computer programming, is usually one step more complicated than a user desires. Chapter 5 provides some details. Here is an overview.

First know the functional form of the cumulative probability distribution of errors on each independent variable, cdf(x) (cumulative distribution function). It is likely to be Gaussian, and you can get the mean and standard deviation from replicate measurements. Then, numerically, for each x_i, randomly sample an x_i value from its cdf and use those values to calculate a value for y. Place that value of y in a histogram. Repeat the experiment enough times so that you can determine the shape of the histogram of realizable y values. (Usually 1000–100 000 trials are needed, but this is no problem for the computer, once programmed.) From the histogram create the cdf of y values and read the 95% probable maximum (at 97.5%) and minimum (at 2.5%) values for y. Again, the numerical method is not limited by the analytical method idealizations. Nonlinearities are included, the actual cdf(x) can be included, and correlations in errors can be included in sampling conditional cdf($x_i|\underline{x}_j$) distributions.

3.6.3 Generality

The model might be an equation that converts raw data to the y measurement or it might be the model used in design. Either way in the above representations, x represents the variables.

However, in regression, there are is also uncertainty on the model coefficient values. It arises from several sources. First, errors (perturbations) in the data lead to corresponding errors in the model coefficient values required to best fit the data. Second, convergence criteria in optimization lead to imperfection in the model coefficient values. Third, decisions about the objective function in regression (minimize vertical least squares, maximize likelihood, etc.) lead to different model values.

In the propagation of uncertainty equations above, the symbol x does not have to be limited to an experimental or design variable. If a model coefficient value changes, the model-calculated value will change. If the model coefficient value is uncertain, the model-calculated value will be uncertain. The x symbol throughout Section 3.6 could represent either a model coefficient or a variable.

However, even if the errors in the x values are not correlated, the coefficient values are correlated to the x errors and to other coefficient values. Therefore, in nonlinear regression, when seeking to propagate uncertainty of data-to-coefficient and coefficient-to-model prediction, use the technique called bootstrapping, given in Section 3.8.

3.7 Which to Report? Maximum or Probable Uncertainty

Obviously, in the use of a propagation of variance to obtain a probable error on a calculated variable, Equation 3.24 is different from the use of a propagation of maximum error, Equation 3.12. Which should you use? The basis for either of the analytical methods is that the error on a measurement must be much less (an order of magnitude or more) than the measurement itself. Be sure this is true.

Further, the propagation of variance to obtain a probable error is based on the use of many independent measurements. It can be justified for one particular calculation if many such calculations are performed over a lifetime or if the calculated value is subject to many independent and equivalent errors. Therefore, when either incidents of use or the number of independent variables is few, or when risk is high, use the propagation of maximum error to obtain a maximum error on y. If the number of uncertain variables is large enough so that "+" influences will probably be balanced by "−" influences, use the propagation of variance to estimate a probable range on the calculated value of the maximum error. If the risk is high, you can use the multiplier 3.29 instead of 1.96 in Equation 3.18 and report a 99.9% probable error, or make similar changes to report alternate confidence intervals.

Acknowledge which error estimation method you use. It is likely that you will prefer to report the smaller 95% probable error than the maximum error, because it reflects better on the precision of your work. However, your reader should be offered the opportunity to understand your work, including the way the errors were estimated. It would be unethical to misrepresent the precision of your calculated values.

3.8 Bootstrapping

Bootstrapping is a technique to estimate the uncertainty in model prediction values, using randomized sampling from the experimental data. Normally, you collect data and then use regression to determine model parameter values that make the model best match the data. This results in the base-case model. However, another set of replicate data would represent the realization of new noise and new uncontrolled events on the measurements, and the new values would lead to different model coefficient values. The new coefficient values would result in different model predictions. If the experimental procedure is valid, the deviations between replicate sets would be small and the model-to-model prediction variability would be small, but it would exist.

If the model matches the underlying phenomena, then ideal natural experimental vagaries (the confluence of many, small, independent, equivalent sources of variation) should result in residuals that have a normal (Gaussian, bell-shaped) distribution. If this is the situation where (i) the model matches phenomena, (ii) normally distributed residuals, (iii) model coefficients are linearly expressed in the model, and (iv) experimental variance is uniform over all of the range (homoscedastic), then analytical statistical techniques have been developed to propagate experimental uncertainty to provide estimates of uncertainty on model coefficient values and on the model. It gives the 95% probable range, or so, for the model. However, if the variation is not normally distributed, if the model is nonlinear in coefficients, if variance is not homoscedastic, or the model does not exactly match the underlying phenomena, then the analytical techniques are not applicable. In this case numerical techniques are needed to estimate model uncertainty. Bootstrapping is the one I prefer. It seems to be understandable, legitimate, simple, and is widely accepted.

Bootstrapping is a method to determine this impact without creating replicate experimental data sets. In bootstrapping, consider that the experimental data you have represents the population, the whole of all data possible. Here it will be called a surrogate population, because it is not the true entire population. Then randomized sampling from the surrogate population represents an experimental realization. Perform regression on sample data from the surrogate population to provide one realization of model coefficients and one model realization. Repeat many times (perhaps $N = 100$–1000) to generate N sets of model coefficient values.

The N models represent the diversity that could be expected due to the vagaries in the experimental data and, also, the mismatch between model and truth about Nature. This set of N models is termed an ensemble. To estimate the uncertainty in the model prediction determine the model prediction from all N models at a desired set of conditions and use the standard deviation of the model-predicted values (or alternately 25/75 quartiles, 95% confidence interval, or range) as a measure of uncertainty.

The bootstrapping technique samples with replacement. This means, for example, if it selects the 83rd data set from your original data, that data set remains in the original data, and random sampling might select the same 83rd data set several times.

The bootstrapping technique uses the same number of data sets in the sample as there are in the surrogate population. For example, if you had 106 experimental runs, 106 sets of data, 106 points for the regression to match, then sample (with replacement) 106 times. Since the central limit theorem indicates that variability in an average reduces with the square root of the number averaged, keeping the number of data in the sample the same as in the population normalizes the sample-model standard error of prediction to that of the base case model (from all the data in the population).

The advantage of bootstrapping over conventional propagation of uncertainty is that you do not have to estimate the uncertainty (error) or make assumptions about error distributions on individual elements in the model. Bootstrapping uses the uncertainty in the data, as Nature decided to present it, and provides model-prediction uncertainty corresponding to the data uncertainty.

As a caution: if the model does not match the data (if the data rejects the model) then bootstrapping provides the measure of variability of your bad model; it does not indicate the 95% range about your bad model that encompasses the data.

Although the concept is that data uncertainty leads to model coefficient uncertainty, which leads to model prediction uncertainty. The model coefficient values are correlated to each other, so a fundamental condition of the Section 3.6 methods of propagation of uncertainty is not true for model coefficient variation. Bootstrapping accounts for model coefficient uncertainty.

However, bootstrapping does not account for the component of uncertainty that would be contributed by your estimates of coefficient values (such as the gas law constant, the speed of light, the value of pi, a tabulated viscosity, etc.) or givens (heat exchanger duty, production rate, etc.). Use bootstrapping to determine the impact of experimental uncertainty on the model prediction, ε_y. Then use propagation of uncertainty to combine that with a propagation of uncertainty from model parameters and givens to generate an estimate of total model error.

Bootstrapping generates a set of model coefficient values, one for each data sampling realization. The variability or range in individual coefficient values can be an indication of the sensitivity of the coefficient to the data. A model coefficient that has a large range, perhaps relative to its base case value, could indicate any of several features. (i) The model parameter has little impact on the model, so the specific phenomena that it represents should be reconsidered, and either modeled differently or the inconsequential phenomenal removed. (ii) The model parameter is sensitive to the data variability and experimental design should be reconsidered to generate data with sufficient precision.

In bootstrapping, the model coefficient values will be correlated. In a simple case consider a linear y–x model, $y = a + bx$. If a best model for a sampling has a high intercept, it will have a low slope to compensate and keep the model in the proximity of the other data. Since the parameter values are correlated, one cannot use the range (or alternate measures of variability) of the parameter values from bootstrapping to individually estimate the uncertainty on the model due to the parameter value. Estimate model uncertainty from the ensemble – each of the N model predictions from each of the N sets of coefficient values.

3.9 Bias and Precision

There are two general categories of measurement uncertainty – bias and precision. Sometimes, their error propagation is treated individually. Consider measurement with an instrument that is not calibrated correctly. If 1000 people use that instrument, they will report a spectrum of values that express random events in the measurement process. The average of those measurements will tend to eliminate the random error and give the average measurement of the improperly calibrated instrument. However, that average will be different from the true value.

Bias is the systematic error due to instrument calibration or measurement technique. According to ASTM Standard E177-86, bias is a generic term that represents a measure of the consistent systematic difference between results and an accepted reference.

The ASTM standard states that precision is a generic term, which is a measure of the closeness of repeated measurements. Whereas σ, an indicator of precision, can be calculated from multiple measurements, an evaluation of bias must be obtained either by calibration to an absolute standard or by estimation and judgment. Once done, ASTM E177-86 directs that estimated bias error is individually propagated as a variance, as was the random error in Equation sets 3.17 and 3.18.

There have been several attempts to develop methods to combine bias and precision, as propagated through calculations, into a single probable error measure. However, the method must be explicitly stated, and any single measure reduces the information communicated. In concurrence with the ASTM Standard E177-86 view that no formula combining precision and bias is likely to be useful, use separate statements of probable errors due to precision and to bias.

Example 3.4 Here is an optimization application of propagation of uncertainty. The objective is to define a cubic curve (a path) from point "A" to point "B" on a portion of the Earth's surface and seek the curve coefficient values that minimize the path length. The path (x, y) (longitude, latitude) relation is

$$y = a + bx + cx^2 + dx^3$$

and elevation is

$$z = elevation = f(x, y)$$

However, since the path must go through points A and B,

$$y_A = a + bx_A + cx_A^2 + dx_A^3$$
$$y_B = a + bx_B + cx_B^2 + dx_B^3$$

The specific optimization statement is

$$\min_{\{c,d\}} J = S = \int_A^B ds = \int \sqrt{dx^2 + dy^2 + dz^2} \cong \sum \sqrt{dx^2 + dy^2 + dz^2}$$

subject to

$$y_A = a + bx_A + cx_A^2 + dx_A^3$$
$$y_B = a + bx_B + cx_B^2 + dx_B^3$$

By searching for values of c and d, the two equality constraints define a and b values

$$\begin{bmatrix} 1 & x_A \\ 1 & x_B \end{bmatrix} \begin{bmatrix} a \\ b \end{bmatrix} = \begin{bmatrix} y_A - cx_A^2 - dx_A^3 \\ y_B - cx_B^2 - dx_B^3 \end{bmatrix}$$

which yields

$$a = \frac{x_B(y_A - cx_A^2 - dx_A^3) - x_A(y_B - cx_B^2 - dx_B^3)}{x_B - x_A}$$

$$b = \frac{(y_B - cx_B^2 - dx_B^3) - (y_A - cx_A^2 - dx_A^3)}{x_B - x_A}$$

In numerical optimization the algorithm stops in the vicinity of the optimal values, with uncertainty on the decision variables related to the convergence criterion. Since values of a and b depend on c and d, uncertainty on c and d propagate to uncertainty on a and b. Using propagation of maximum uncertainty (which is probably more reasonable since having only two variables violates the "many" assumptions that underlie propagation of probable error),

$$\epsilon_a = |x_A\, x_B|\epsilon_c + |x_A\, x_B(x_A + x_B)|\epsilon_d$$
$$\epsilon_b = |x_A + x_B|\epsilon_c + |x_B^2 + x_A\, x_B + x_A^2|\epsilon_d$$

Here is a table showing the propagated maximum error on ε_a and ε_b due to locations of points A and B and uncertainty on coefficients c and d.

x_A	x_B	ϵ_c	ϵ_d	ϵ_a	ϵ_b
1	*8*	*0.001*	*0.001*	*0.080*	*0.082*
1	*8*	*0.0001*	*0.001*	*0.073*	*0.074*
1	*8*	*0.001*	*0.0001*	*0.015*	*0.016*
4	*5*	*0.001*	*0.001*	*0.2*	*0.07*
4	*5*	*0.0001*	*0.001*	*0.18*	*0.062*
4	*5*	*0.001*	*0.0001*	*0.04*	*0.015*
9	*10*	*0.001*	*0.001*	*1.80*	*0.29*
9	*10*	*0.0001*	*0.001*	*1.72*	*0.27*
9	*10*	*0.001*	*0.0001*	*0.26*	*0.046*

At the (x_A, x_B) pair of (4, 5), there is substantial uncertainty on coefficients a and b even if the uncertainty on c and d are small. Propagation of uncertainty can be used to define stopping criteria on ΔDV to attain a desired certainty on other coefficients defined by constraints, on the OF value, on others used in the model equations. For example, if x_A and x_B are close together (4 and 5 or 9 and 10) then accuracy on variable d is more important to ε_a and ε_b than accuracy on c. Stopping criteria on d should be tighter (smaller Δd threshold) than that on c. Therefore, one way to make ε_a and ε_b have tolerable values is to reduce the convergence tolerance on ε_c and ε_d.

Example 3.5 The computer measures the mA electric current transmitted by the orifice sensor and calculates a flow rate from it. It displays this calculated flow rate, which is called the *measurement*, but it is really a calculation from a phenomenological model. If we accept that this model is an accurate representation of the orifice phenomena, then

$$F_m = a(i - i_0)^b$$

We will determine a and b values that make the model best match the experimental flow rates:

$$\min_{\{a,\ b\}} J = \sum (F_{e,i} - F_{m,i})^2$$

If we accept that there is no error on the a and b coefficient values, then propagation of variance provides an estimate of the uncertainty on the "measured" flow rate due to uncertainty on the noisy i values:

$$\sigma_{Fm}{}^2 = 2(ab(i - i_0)^{b-1}\sigma_i)^2$$

Here, it is assumed that the uncertainty on i and i_0 are the same.

This value would let operators know the uncertainty on the flow rate "measurement"

$$\epsilon_{Fm,\ 0.95} = 2\sqrt{\sigma_{Fm}^{\ 2}}$$

Example 3.6 Consider calibrating an orifice flow meter and measuring the flow rate by the bucket-and-stopwatch method,

$$F_e = \frac{M_f - M_e}{(t_2 - t_1)\rho}$$

Here, the dependent variable, the calculated value of the flow rate, is a function of five variables. The value of each variable has an uncertainty that is dependent on the method of getting its value.

Propagation of uncertainty reveals how the collection time period, $(t_2 - t_1)$, affects the uncertainty in the calculated, experimental flow rate:

$$\sigma_{Fe}^{\ 2} = 2\left(\frac{1}{(t_2 - t_1)\rho}\sigma_M\right)^2 + 2\left(\frac{M_f - M_e}{(t_2 - t_1)^2\rho}\sigma_t\right)^2 + \left(\frac{M_f - M_e}{(t_2 - t_1)\rho^2}\sigma_\rho\right)^2$$

Here, I assumed that the error in M and t is the same for each value, so there are only three terms, but two are multiplied by two. F_e will then be substituted for the $\Delta M/\Delta t$ terms:

$$\sigma_{Fe}^{\ 2} = 2\left(\frac{1}{(t_2 - t_1)\rho}\sigma_M\right)^2 + 2\left(\frac{F_e}{(t_2 - t_1)\rho}\sigma_t\right)^2 + \left(\frac{F_e}{\rho^2}\sigma_\rho\right)^2$$

Note that as the time interval gets larger, the uncertainty (estimated variance) in the calculated experimental flow rate gets smaller for the two experimental data collection terms. At very large collection times the uncertainty depends on the uncertainty in density. Therefore, you could choose a long enough collection time and a density calculation method precise enough to get a desired value for the propagated uncertainty on the calculated flow rate. Experimental design is choosing the experimental conditions.

Example 3.7 You could decide on a desired F_e variance from the application of propagation of uncertainty on the in-use equation. If the F_e values have substantial error, then the a and b calibration coefficients will be corrupted by that error and the propagation of uncertainty in the model cannot assume that there is no uncertainty on the a and b values. Therefore, to get precise a and b values, you want the uncertainty of the calibration experiments to be much less than the uncertainty of the in-use equation. You want

$$\sigma_{F_e}^{\ 2} \ll \sigma_{F_m}^{\ 2}$$

If much less is accepted as less than a tenth, then you would choose the collection time interval long enough so that

$$2\left(\frac{1}{(t_2 - t_1)\rho}\sigma_M\right)^2 + 2\left(\frac{F_e}{(t_2 - t_1)\rho}\sigma_t\right)^2 + \left(\frac{F_e}{\rho^2}\sigma_\rho\right)^2 <\sim 2(ab(i - i_0)^{b-1}\sigma_i)^2$$

Rearranging:

$$\Delta t > \sim \sqrt{\frac{2\left(\frac{1}{\rho}\sigma_M\right)^2 + 2\left(\frac{F_e}{\rho}\sigma_t\right)^2}{2(ab(i-i_0)^{b-1}\sigma_i)^2 - \left(\frac{F_e}{\rho^2}\sigma_\rho\right)^2}}$$

You cannot use the above equation until you know the a and b values. Therefore, start with any reasonable Δt value and get two points to start estimating a and b. Then with your a and b values and the F and i values at the new conditions, determine the required Δt. After each experiment, your a and b estimates will improve (change). Use the most recent a and b values along with new F and i values; each calibration point will have a unique minimum Δt.

Example 3.8 Determine the variance reduction when using a first-order filter to temper noise on a measurement. The equation for a first-order filter is

$$x_{f,i} = \lambda x_i + (1-\lambda)x_{f,i-1}$$

where

x_i = measurement, a noisy process variable
i = sampling interval
$x_{f,i}$ = filtered value
λ = filter factor (a user's choice) = $(1 - e^{-\Delta t/\tau})$ if the use specifies a filter time-constant

Propagating variance on the filtered value,

$$\sigma_{x_{f,i}}{}^2 = \lambda^2 \sigma_{x_i}{}^2 + (1-\lambda)^2 \sigma_{x_{f,i-1}}{}^2$$

If the process is stationary (the variance on x is unchanging) then the variance on the filtered value is the same as the prior variance on the filtered value, and the subscript, i, can be dropped and the equation rearranged to express the reduction in standard deviation (noise level):

$$\sigma_{x_f} = \sqrt{\frac{\lambda}{2-\lambda}}\sigma_x$$

3.10 Takeaway

Understand the sources of uncertainty in your data and their impact on the model. Report uncertainty so that a user can make decisions that are compatible with the uncertainty. Recognize that the linear approximation of the propagation of uncertainty relations and the estimates required for the underlying uncertainties make the estimate of the probable or maximum error somewhat uncertain. Prepare to use propagation of uncertainty in the design of experiments chapter.

Exercises

3.1 A model has the functional relation, $y = ae^{bx}$. Uncertainty on the calculated value of y depends on the uncertainty of the values of model coefficients a and b, and on the uncertainty of the independent variable value of x. Use propagation of maximum error to derive an equation indicating how ϵ_y depends on $a, b, x, \epsilon_a, \epsilon_b, \epsilon_x$.

3.2 Determine the uncertainty on the Moody–Darcy friction factor, f, for turbulent flow in full circular pipes using the Swamee–Jain relation.

$$f = \frac{0.25}{\left[\log_{10}\left(\frac{\epsilon}{3.7d} + \frac{5.74}{Re^{0.9}}\right)\right]^2}$$

where ε is the surface roughness and d is the pipe inside diameter. The dimensionless ε/d ratio is typically between 0.0001 and 0.01, and the dimensionless Reynolds number is typically between 104 and 107. Choose a value for ε/d and Re, and determine the value of f. The coefficient values of 0.25, 3.7, 5.74, and 0.9 were empirically determined and reflect the number of significant digits that can be reported. How does uncertainty in the coefficient values relate to uncertainty on the calculated f? Use propagation of maximum error and use propagation of variance to determine the 95% probable error. As a helpful hint, use a numerical approximation for the partial derivative, not the analytical value.

3.3 For the following equation, which represents an optimum duration, to keep a furnace fan running after the fire is stopped:

$$t^* = -c \ln\left(\frac{ac}{b}\right)$$

Analytically, (i) determine the maximum uncertainty in t^* due to the uncertain range on coefficients a, b, and c and (ii) the 95% probable uncertainty in t^* due to the uncertain range on coefficients a, b, and c.

3.4 Determine the maximum uncertainty on the moles of gas calculated from the ideal gas law (compressibility version). You choose reasonable values for P, V, and T, and reasonable measurement precision. The compressibility factor, z, read from a chart, has a value of 0.97:

$$n = \frac{PV}{RTz}$$

4

Essential Probability and Statistics

4.1 Variation and Its Role in Topics

There is a desire for reproducibility. If an experiment is exactly duplicated it will return exactly identical results. Like a calculation of 3×5, it does not matter which day, when, or who does the same calculation, the result is reproducibly 15. However, experiments are not exactly reproducible. Sensors age (warm up, corrode, fade) and there are uncontrolled influences on both the experiment (humidity, barometric pressure, human personality, raw material composition) and on its measurement (electromagnetic currents, vibrations, reading parallax, lab analysis, sample aging). Replicate experiments (attempts at exact duplication) will generate a diversity of values, not a single repeated value.

Since regression is an attempt to model the truth about nature, not the vagaries of perturbations; techniques for regression need to recognize the inherent variability in the data. This chapter presents some common distributions related to experimental data or its analysis. This chapter also summarizes key statistical and simulation techniques that provide tools for analyzing variability and for generating data that can be used for testing regression procedures.

4.2 Histogram and Its PDF and CDF Views

Replicated results have expectedly identical values; but, because individual data are subject to independent fluctuation, replicated results will not have identical values. Commonly, we represent the distribution of replicated values by creating a histogram, a chart of the number of samples within small intervals. The intervals are termed bins (that collect data of a certain size range, like sorting small, medium, and large nails into bins). The bins could be numbered as first, second, and so on, but are centered on the bin interval and labeled by bin center value.

The bin intervals are usually of uniform width and there might be 10 or 20 bins between the high and low values. With too few a number of bins (perhaps 5), the histogram is too coarsely graded to visualize the shape details. With too many bins, the number of bins approaching the

Nonlinear Regression Modeling for Engineering Applications: Modeling, Model Validation, and Enabling Design of Experiments, First Edition. R. Russell Rhinehart.

number of samples, there will just be one, or two, or no samples in a bin, and again the shape of the distribution cannot be visualized.

With too few number of data replicates, the number of samples within a bin will be too infrequent to be confident in the number or probability that would fall in a bin on average. There needs to be enough population to have a histogram become a confident representation of the data distribution. When creating a histogram from experimental data, the standard deviation of the number of samples in a bin is the square root of the number in the bin. Accordingly, with 10 bins and a sufficient number per bin so that the standard deviation is about 10% of the value on average, there needs to be on the order of 1000 or more samples to adequately populate the bins.

Figure 4.1 illustrates a histogram in which the number of data samples in each bin have been normalized by the total number of samples. This scales the ordinate (the vertical axis value) and provides the property that the sum of all bin values is unity. The values on the abscissa (horizontal axis) represent the variable value. As opposed to using the bar height to represent the bin population, the graph connects the center of the bin tops with line segments.

If a value exceeds the bin range, it is added to the count in the extreme bin. In Figure 4.1 the count in the left-most bin is nearly zero. That the counts in the 10th bin are not nearly zero indicates that the bin width or number of bins should have been a bit larger to see the details in the tails better.

Alternately, bin intervals might be of an unequal width, planned to anticipate collecting an equal number of events in each bin. In such a case the height of each bin will be about the same, which distorts the histogram. To correct for this visual misrepresentation, normalize the in-bin count by the bin width.

For any bin center value, x_i, some samples have smaller x values and others larger. The fraction of values that is smaller is the sum of the bin values for all bins to the left of the ith bin. The cumulative distribution, Figure 4.2, is a plot of the fraction of samples with values in a particular bin or a lower bin.

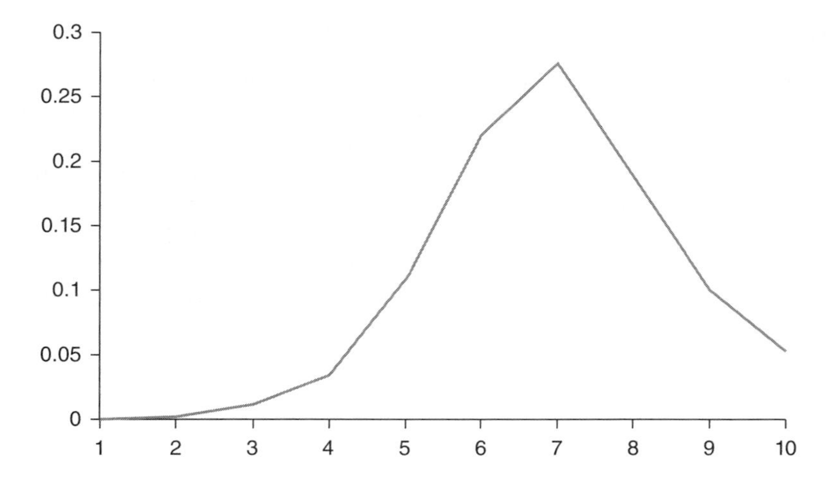

Figure 4.1 Histogram represented by line segments that connect the bin tops at bin center

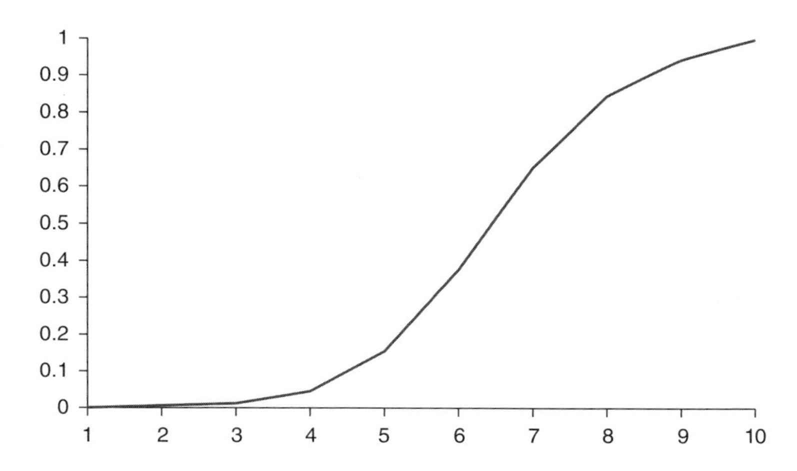

Figure 4.2 Cumulative distribution of the data in Figure 4.1

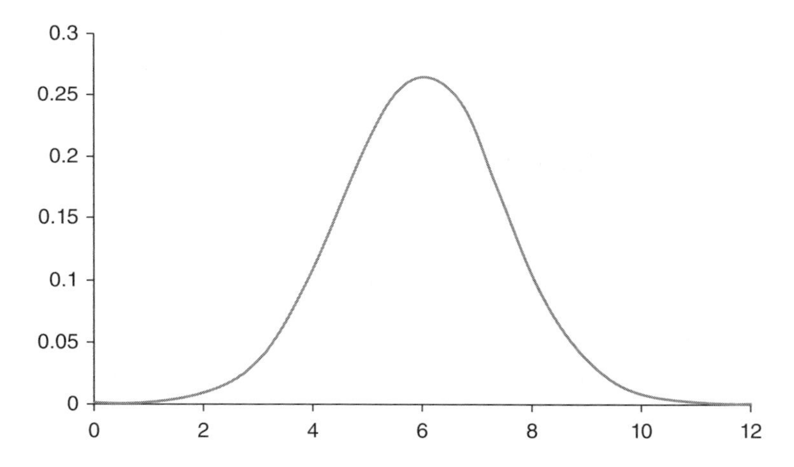

Figure 4.3 Probability distribution function for normally distributed data

If the data are continuously valued, then the bin intervals could be made smaller to have a finer distinction in the distribution, but the number of samples would need to be very large to provide representative fractions in each bin. In the limiting case of infinitesimal bin intervals and infinite samples, the histogram becomes a smooth curve, as represented in Figure 4.3. This is termed the probability distribution function (PDF) and is not necessarily symmetric. Figure 4.3 illustrates the Gaussian distribution with a mean of 6 and a standard deviation of 1.5.

The area under the entire PDF curve is unity and the fraction of the population with a lower value than x_i is the integral from the far left to the value x_i. The fraction of the population with a lower value can be graphed w.r.t. x and is termed the cumulative distribution function (CDF).

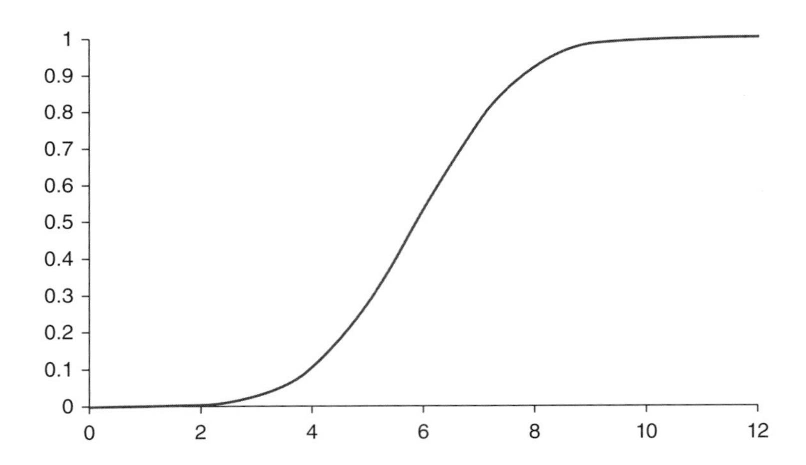

Figure 4.4 Cumulative distribution function for normally distributed data

Figure 4.4 is a representative CDF for the Gaussian distribution of Figure 4.3. The CDF goes from zero to unity as the x value spans the entire range of population values.

4.3 Constructing a Data-Based View of PDF and CDF

To use data to construct the PDF or CDF graphs:

1. Sort the N number of data by its numerical value.
2. Assign a counter to the sorted data of $i = 1$ to the smallest and $i = N$ to the largest.
3. Scale count by N to get the CDF value, $CDF = i/N$.
4. Plot data CDF w.r.t. the data x value. This is the CDF graph of Figure 4.5, similar to that in Figure 4.2, but in greater detail.
5. The PDF (to be shown) is the derivative of the CDF w.r.t. the x value. You can calculate the derivative with a forward finite difference:

$$PDF_i = (CDF_{i+1} - CDF_i)/(x_{i+1} - x_i) = 1/(N \ \Delta x_i) \tag{4.1}$$

This uses successive measurements to determine $N - 1$ number of PDF values. However, this has the undesirable result of creating a noisy representative CDF. Instead, use every nth interval to generate $N - n$ number of PDF values:

$$PDF_i = n/[N(x_{i+n} - x_i)] \tag{4.2}$$

Choose n to be large enough to temper the noise, but small enough to see the PDF shape. This is a user-preference. Several choices are illustrated in Figure 4.6.

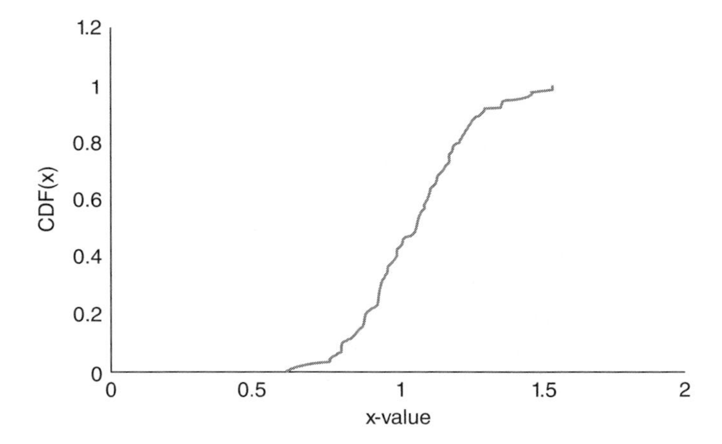

Figure 4.5 Experimental CDF plot

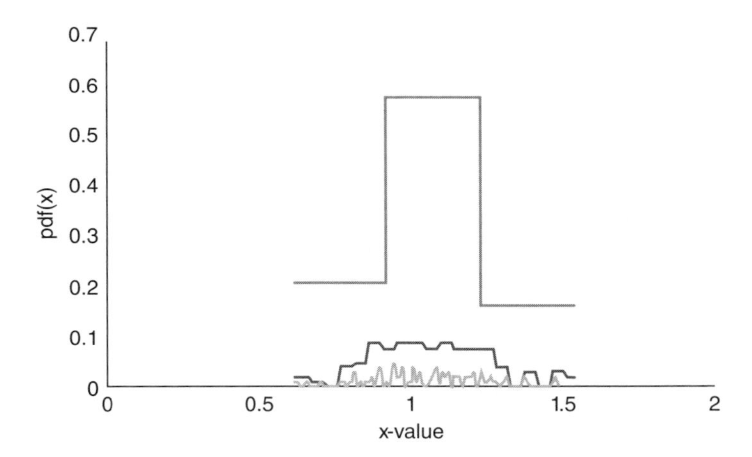

Figure 4.6 Experimental PDF plot illustrating the impact on the user's choice of n

4.4 Parameters that Characterize the Distribution

The common characterizations for distributions are measures of location (which include average, center, mean, median, mode) and measures of dispersion (which include range, width, variance, standard deviation, confidence interval). However, there are also measures of symmetry (such as skewness) and many other higher-level moments (such as kurtosis, a measure of flatness or peakedness of the PDF distribution). These parameters are called statistics. Which statistic is the appropriate measure? This depends on the assumptions made about the distribution of the data and of the feature that is being tested. Commonly, for continuous-valued real numbers with a normal distribution, average and standard deviation are appropriate measures of location and dispersion. For nonsymmetric distributions it is common to report the average and the 25th and 75th quartile values. However, for data that represent ranks, median is an appropriate measure for location.

4.5 Some Representative Distributions

4.5.1 Gaussian Distribution

The *Gaussian* distribution is also known as the normal distribution or bell-shaped curve. This is a mechanistic representation of a situation in which many, independent, random fluctuations with equivalent magnitude of impact have an influence on the outcome. It matches well the variation in most experimental data, has become the standard basis for statistical tests, dominates the introductory lessons, and is the default right distribution to use.

The PDF is

$$f(x) = \frac{1}{\sqrt{2\pi\sigma}} e^{-1/2\left(\frac{x-\mu)}{\sigma}\right)^2} \tag{4.3}$$

where σ^2 represents the variance in the data and μ represents the mean. These parameter values are estimated by standard deviation and average of the data (either x or y values)

$$\mu \cong \bar{x} = \frac{1}{N}\sum x_i \tag{4.4}$$

$$\sigma \cong s = \sqrt{\frac{1}{N-1}\sum(\bar{x} - x_i)^2} \tag{4.5}$$

To simulate data with Gaussian distributed fluctuations, I use the Box–Muller method for its balance of simplicity and fidelity to the ideal. In it r_1 and r_2 are independent random numbers, uniformly distributed with $0 < r \leq 1$:

$$d = \sigma\sqrt{-2\ \ln(r_1)}\sin(2\pi r_2)\pi \tag{4.6}$$

Here, d is the deviation to be added to a true value, ψ, to simulate a measurement, x:

$$x = \psi + d \tag{4.7}$$

Recognize that, in a simulation, the true value is obtained from a model.

4.5.2 Log-Normal Distribution

Many natural processes generate data that are approximated by a log-normal distribution. The Gaussian distribution has features that might not be consistent with some data. It permits negative values and has a symmetric shape. However, particle sizes and daily rainfall amounts cannot be negative, and the distribution tail on the high side reaches much farther from the average than the tail on the low side. Figure 4.7 illustrates the PDF of a log-normal distribution.

When the x axis of the log-normal PDF or CDF distribution is log-transformed, it appears normal, like the Gaussian distribution. This is illustrated in Figure 4.8. First, scale the x value by the x average, $x' = x \div \bar{x}$, and then log-transform. The upper right quadrant of Figure 4.8 reproduces the PDF of the log-normal distribution. Overlaid on the right half is the curve

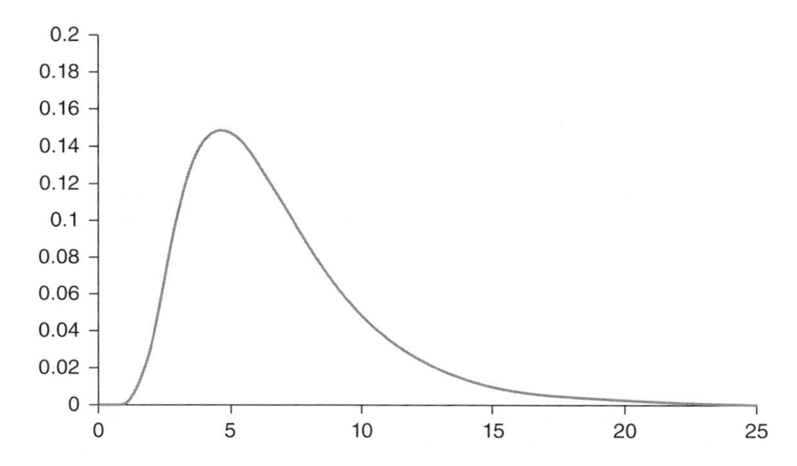

Figure 4.7 Probability distribution function for a log-normal distribution

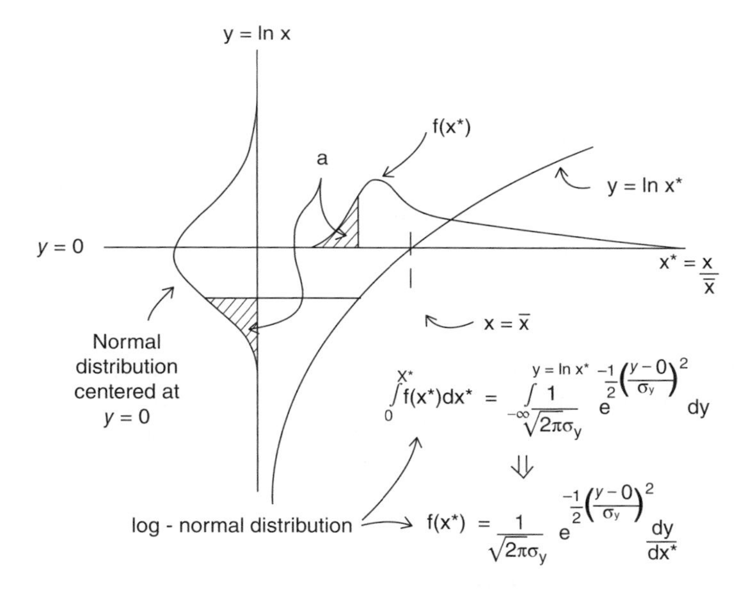

Figure 4.8 Analysis of a log-normal distribution

representing the log of x'. The 90° rotated normal PDF in the left half represents the PDF($y = \ln(x')$). Note that when $x = \bar{x}$, $x' = 1$ and $y = \ln(x' = 1) = 0$; as illustrated, the log-transformed PDF is centered on zero. Also note that as x' approaches zero, then $\ln(x')$ approaches $-\infty$; therefore, although the limit of the x values is zero, the tails of the log-transformed PDF go to infinity. Thus the PDF(y) is centered on zero with a standard deviation of $\sigma_y = \sigma_{\ln(x')}$.

The shaded area in Figure 4.8, a, represents a cumulative area from the lowest value to either x' or y. The CDF value must be the same for any corresponding y and x', $y = \ln(x')$. Defining the PDF(y) as Gaussian with a mean of zero and equating CDF(y) to CDF(x') leads to the PDF for a log-normal distribution:

$$f(x) = \frac{1}{\sqrt{2}\sigma_{\ln(x)}} e^{-1/2\left(\frac{\ln(x)-\ln(\bar{x})}{\sigma_{\ln(x)}}\right)^2} \tag{4.8}$$

To determine a value for $\sigma_y = \sigma_{\ln(x)}$ choose a y value, recognize that the cumulative area must be equal on both the y and x' graphs and equate it to the corresponding x' value. Using a conventional mean plus two-sigma value for y, representing the 97.5% area, the corresponding x^* value is the one that represents the 97.5% highest. Then

$$\sigma_{\ln(x)} = \frac{1}{1.96} \ln\left(\frac{x_{0.975\ upper\ limit}}{\bar{x}}\right) \tag{4.9}$$

The log-normal distribution is not fundamentally derived from a mechanistic concept. It is an appropriation to the normal distribution to provide a PDF that appears to match many natural process distributions.

To simulate independent and random data that have a log-normal distribution with the Box–Muller method, use

$$x = \bar{x}e^{\sigma_{\ln(x)}\sqrt{-2\ln(r_1)}\sin(2\pi r_2)} \tag{4.10}$$

4.5.3 Logistic Distribution

The logistic model is often used as a sufficient approximation to a cumulative distribution. It is simple to use, flexible, and when the true distribution is unknown, the simplicity and effectiveness of the logistic model is often justified:

$$p = CDF(x) = \frac{1}{1 + e^{s(x-c)}} \tag{4.11}$$

The center, c, represents the x value for which there is a 50% chance of a higher or lower value. The sign of the scale factor, s, determines whether the function goes from zero to unity or unity to zero, and the magnitude of s relates to how sharply the CDF changes with x.

Figure 4.9 reveals that the logistic model (center $c = 10$ and scale factor $s = 0.75$) is very similar to the normal CDF.

4.5.4 Exponential Distribution

The *exponential* distribution describes the asymptotic residence time in an ideal mixer with continuous flow through it. It also describes any continuous feed random selection process. The concept is that N units (molecules, grams, items) enter a control volume (tank, warehouse, waiting room) and N units simultaneously leave. There is no change in inventory quantity or number within the control volume. The units internal to the control volume are continuously relocated (mixed, randomized), so that they do not pass through in an orderly manner. There is a chance that an item that enters will be immediately selected to leave and a chance that it will reside for a long time without being selected to leave. Since mixing is perfect, the chance that any one item is selected to leave is purely random. There is a distribution of residence times of the items selected to leave. If the residence time is related to some desired feature such as

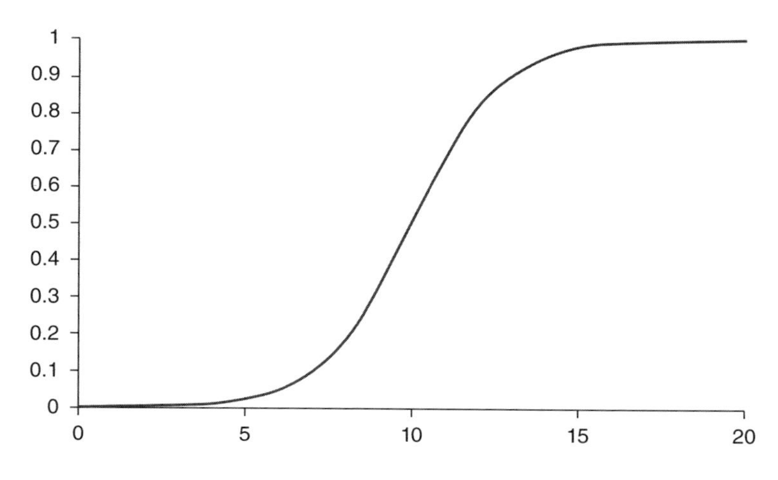

Figure 4.9 Logistic model

a chemical reaction, cooling, microbial growth, sterilization, humidification, and so on, then there will be a distribution of those features, as a consequence of the distribution of residence times.

If a control volume is initially filled with identical units, then the residence time distribution is uniform. Once the inflow and outflow starts the residence time distribution will begin to asymptotically approach the exponential at steady-state.

The PDF and CDF for an exponential distribution are

$$PDF(t) = \frac{1}{\bar{t}} e^{-t/\bar{t}} \tag{4.12}$$

$$CDF(t) = 1 - e^{-t/\bar{t}} \tag{4.13}$$

where $\bar{t}$ is the average residence time, which would be the control volume divided by the flow rate or the inventory divided by the rate that items enter and leave. The standard deviation of the residence time, t, is

$$\sigma = \frac{1}{\bar{t}} \tag{4.14}$$

Figure 4.10 illustrates the PDF of the exponential distribution for an average residence time of 8.

4.5.5 Binomial Distribution

The *binomial* distribution characterizes an event or outcome that is dichotomous (two-valued), such as head–tail, on–off, yes–no, included–other, and so on. One condition is termed a success, the other a fail. The probability of any trial leading to a success is p and the complementary probability of a fail is $q = 1 - p$. If there are N trials, the number of successes will be an integer, n. The probability density function for a discrete distribution is termed a point

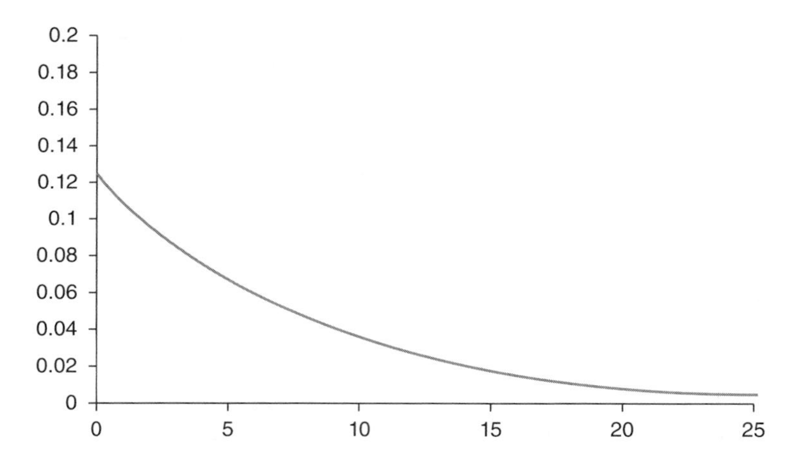

Figure 4.10 Probability distribution function for an exponential distribution

density function. It is still represented by the acronym PDF, but only permitting integer n and N values. For the binomial distribution it is

$$PDF(n|N) = \frac{N!}{n!(N-n)!}p^n q^{N-n} \tag{4.15}$$

and the CDF is

$$CDF(n|N) = \sum_{n=0}^{N} PDF(n|N) = \sum_{n=0}^{N} \frac{N!}{n!(N-n)!}p^n q^{N-n} \tag{4.16}$$

The mean for the expected number of successes in N trials for the binomial distribution is

$$\mu = pN \tag{4.17}$$

and the standard deviation of the expected number of successes is

$$\sigma = \sqrt{pqN} \tag{4.18}$$

From this, if you wish to have a large enough number of trials so that the standard deviation in number of successes is less than a tenth of the expected number of successes, then Equations 4.17 and 4.18 lead to

$$N \geq 100(1-p)/p \tag{4.19}$$

which is a basis for the estimate of needing a minimum of about 1000 trials for a 10-bin histogram in Section 4.2 (if the nominal value is $p = 0.1$).

4.6 Confidence Interval

We often wish to know what values represent the 95% extremes of replicated data, defining a range for which there is only a 5% chance of a more extreme value. The extreme could be above the upper or below the lower 95% limits, and customarily the probability of a more

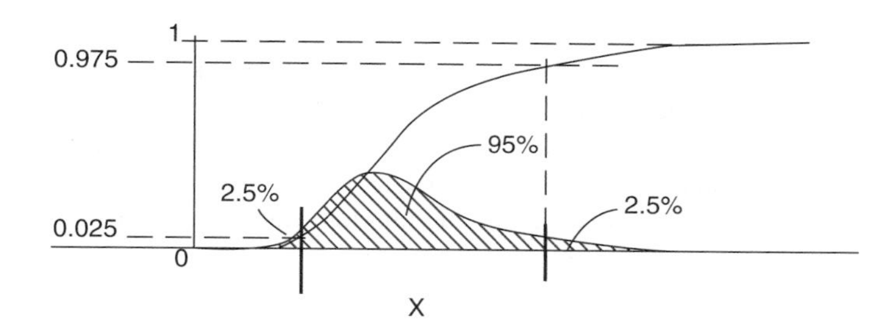

Figure 4.11 Illustration of a confidence interval

extreme value is equally allocated to the upper and lower possibility. If seeking the range of values that encompasses 95% of the data, then there will be 2.5% in the upper extreme and 2.5% in the lower extreme. Relating this to either a PDF or CDF distribution, as illustrated in Figure 4.11, this means, "Find the x values that have CDF values of 0.025 and 0.975."

The conventional symbol for the extreme area is α, termed the level of significance. The fraction of values that includes the central section is $(1 - \alpha)$, the confidence interval. The CDF value of the lower x value is $\alpha/2$ and the CDF value of the upper x value is $1 - \alpha/2$. Conventionally, for economic decisions, the 95% confidence interval, $\alpha = 0.05$, is accepted. However, for safety or other high-stakes situations, one might want to know the 99.9% limits, or $\alpha = 0.001$. By contrast, the upper and lower quartiles might be a sufficient indicator of precision, $\alpha = 0.25$.

Be careful to discriminate between a two-sided or a one-sided test in either reading or reporting. In a two-sided test, the extreme region is allocated to both the upper and lower extremes. In a one-sided test, only one extreme is of concern. For instance, in designing a flagpole, only the high extreme wind needs to be considered, the low extreme is inconsequential. In this case the central section is still $(1 - \alpha)$, but the high CDF is at $(1 - \alpha)$, not $(1 - \alpha/2)$. Alternatively, in designing a reservoir, both the high and low extremes of rainfall (flood and drought) need to be accommodated. Then CDF($\alpha/2$) and CDF($1 - \alpha/2$) are both important.

4.7 Central Limit Theorem

Increasing the number of samples decreases the uncertainty about the location of the average. If the population represented in Figure 4.3 were to be sampled individually, a histogram of the individual values would look the same with an average of about 6 and standard deviation of 1.5. Figure 4.3 reveals a reasonable probability, about 10%, of getting a sample with a value between 8 and 9, a high value. However, if the population were to be sampled five times and the average of the five samples taken as one measurement, there is hardly any chance of getting an average in the 8–9 range. For this to happen, all five samples must be in the upper region where the probability is only about 10%. Equation 4.15 indicates that the chance of five out of five successes with $p = 0.1$ is 0.001%. An average of 8.5 would be extremely rare. Averaging multiple samples tends to make the data have less variability; the average will still fluctuate about the mean, but with less variability.

The central limit theorem predicts the reduction in variability that results from averaging data that are independently sampled from the same population. "Independent" means that the value of the sample is independent of the value of prior samples – that there is no autocorrelation in the data sequence – that if a prior value was high, what made it high has no influence in the subsequent sample. The "same population" means that the mean, variance, and other descriptions of the population distribution do not change from sample to sample – that the population distribution is stationary. The standard deviation of the average of N samples will be the standard deviation of the individuals scaled by the square root of N:

$$\sigma_{\bar{x}} = \sigma_x / \sqrt{N} \tag{4.20}$$

This can be extrapolated to model uncertainty. Regression determines model coefficient values that make the model best fit the data. However, a new sampling realization will generate data with different values (representing the inherent data variability), which would lead to new model parameter values. There is uncertainty on the model location, the predicted y value, due to the uncertainty that data variation propagates to model coefficient values. Paralleling Equation 4.20, the uncertainty on a model prediction will decrease with the square root of the number of data that were involved in the regression. If you want to halve the model uncertainty, quadruple the number of data used in regression.

4.8 Hypothesis and Testing

There is a procedure in statistical testing. First, consider which variable or attribute would reveal the difference between treatments. If there are several responses that could reveal the difference between treatments, select the one most important to the application or the one expected to reveal the largest difference relative to the inherent variability. Second, determine the normal variability in that metric, the variation resulting from replicate trials. Third, assume that the two treatments have no difference. This is the null hypothesis; the two treatments will have identical outcomes. Fourth, implement the two treatments and measure the responses. Fifth, compare the responses. If the difference is greater than expected from statistical variability then reject the null hypothesis – there was a difference. Alternately, if the difference in outcomes is within the range that is normally encountered, do not reject the null hypothesis, accept the null hypothesis – no definitive difference was discovered.

Although this seems independent of human choices there are many. The human must choose the response metric, decide what distribution is the right one to characterize the metric, and choose a level of significance appropriate to the decision. It is important to be self-critical in these choices.

The general procedure for a statistical test is:

- Start with the null hypothesis (the two treatments are identical, the model matches the data, there is no difference in the mean).
- Define a quantifiable metric, the statistic, that would distinguish a difference if it existed. This could be the average difference, it could be the number of data in a run, or it could be a variance ratio. Depending on the metric, it might be that too small a value and/or too large a value would indicate the difference. There would be an ideal value for the statistic if the null hypothesis is true. For instance, the difference in averages should be zero, the ratio of variances should be unity, the number of "+" residuals should be equal to the number of "–" residuals. (Chapter 16 reveals applications of several useful statistics.)

- Because of the inherent variability in the data, the experimental value of the statistic will not have the ideal value, but it will have normal variation about the ideal value.
- Define the normal variation for the statistic and limit(s) or critical values that define a confidence interval. (The null hypothesis distributions of the statistics described here are well known and published for easy access.) Typically, this is the 95% limit. (You will make this choice.)
- Acquire experimental data. (You define the experimental procedure and number of trials.)
- Compute the value of the statistic from your experimental data. (You do this.)
- If the value of the statistic exceeds the critical value, reject the null hypothesis. Alternately, accept the null hypothesis.

Note that rejecting the null hypothesis does not necessarily mean that the model was wrong. It is possible to flip a fair coin 10 times and get 10 heads. However, if I lost 10 times in a row, I would reject the fair-coin hypothesis. It is possible for experimental data to generate an extreme confluence of results, a possible but very rare pattern that makes a true model appear to be untrue. Therefore, rejecting the null hypothesis is not a definitive action. Rejection would be at the confidence limit that was used to define the critical value.

A complete statement related to rejecting a model would be "If the model was a true representation of the phenomena then ___ statistic would have an ideal value of ___. The actual value of ___ is beyond the 95% critical value. Accordingly, if the model is true, there is only a 100% − 95% = 5% chance that this could happen. On the other hand, if the model is a wrong representation then there is a very high chance that the extreme value of the statistic would be encountered. We'll bet on the situation with the higher probability and reject the model."

The short form is "Using data, reject the model at the 95% level of confidence."

Conversely, accepting the null hypothesis means the model was not rejected by the data. It does not mean that the model is true. A "not guilty" verdict does not mean innocence. It means that there was not enough evidence to confidently make the guilty verdict. Flip a trick coin once and it shows an H. Well, a fair coin could have done that. Therefore, one flip does not provide enough experimental evidence to claim it was rigged. A weighted die might have a 0.2 probability of showing a 1, which could not be detected by counting the number of times it shows a 1 in 10 rolls. Similarly, a model that does not represent the data might appear acceptable. It might take a ton of data to see that it is not right. Thus, not rejecting a model does not mean that it is true.

A complete statement related to accepting a model would be "If the model was a true representation of the phenomena then ___ statistic would have an ideal value of ___. The actual value of ___ is not beyond the 95% critical value. Accordingly, if the model is true, there is a 95% chance that this could happen. There is inadequate evidence to confidently reject the model. (However, I know my model is wrong.)"

The short form is "The model appears to match the data, within a 95% level of confidence."

In shades of gray, a test that rejects something as white does not infer that it is black. It could be gray. In the jury system the verdicts are not guilty or *innocent*, they are guilty or *not guilty*. Unfortunately, the statistical tradition uses the terms reject or *accept*. Accept does not mean that it is true. Not-white does not mean it is black. *Accept* simply means that we tentatively accept the null hypothesis because there was not sufficient indication in the data to reject it.

4.9 Type I and Type II Errors, Alpha and Beta

A Type I statistical error is the conclusion to reject the null hypothesis when in fact it is true. As with the coin flipping example, it is possible to flip a fair coin and get the same value of a flip seven times sequentially. The first flip defines the outcome (H or T) and the probability of any subsequent flip being the same is 0.5. The probability of all sequential flips being the same is the probability of the second, and the third, and the fourth, ... The AND conjunction means multiply the individual probabilities, so the P(7 like flips in a row) is $(0.5)^6$, which is 0.016. In 1000 trials of 7 flips of a fair coin, only 16 trials are expected to have identical outcomes. It is possible, but a rare possibility. If you start a game and encounter 7 out of seven identical outcomes, you would have a strong suspicion that the coin was not fair. The probability of not seeing such a pattern is $100(1 - 0.016) = 98.4\%$, which is greater than 95%. Therefore, you would reject the null hypothesis and claim the coin was foul. However, the data do not provide definitive evidence of the situation (such as checking the coin); the experiment provides data corrupted by probability and vagaries of natural processes.

In any statistical test, when the null hypothesis is true, there is a probability that the statistic will have a value more extreme than the critical value. The probability of this happening is the level of significance of the statistic, α, that defines the critical value, which is the complement to the level of confidence $\alpha = (100 - c)/100$. The probability of a Type I statistical error is α, the level of significance.

Figure 4.12 illustrates the probability distribution of a statistic when the null hypothesis is true. If $\alpha = 0.05$, then the critical value demarks the point where the extreme area (the probability of a more extreme value of the statistic) is 0.05 fraction, or 5% of the area under the curve.

In this illustration, the extreme value is to the right – too large a value. However, depending on the meaning of the statistic, too small a value might also be cause for suspicion. Then the critical value would be at the left extreme. These are called one-tailed tests, because it is only concerned with too large or too small a value. However, it might be that either too small or too large a value is of concern. Then there are two critical values, one in either tail.

You do not want to make Type I errors. You do not want appearances in the data that result from random chance to reject the null hypothesis when it is true, to find a person guilty when in fact they were innocent. In the extreme, if you decided to reduce the Type I error to $\alpha = 0.000\,000\,001$, you would be waiting until data let you be 99.999 999 9% confident that the null hypothesis was not true. In the coin flipping example this would require a series of about 30 flips with identical outcomes. Normally, for economic decisions, we use the 95% confidence, $\alpha = 0.05$. However, the user should choose a value of α that is appropriate to the situation.

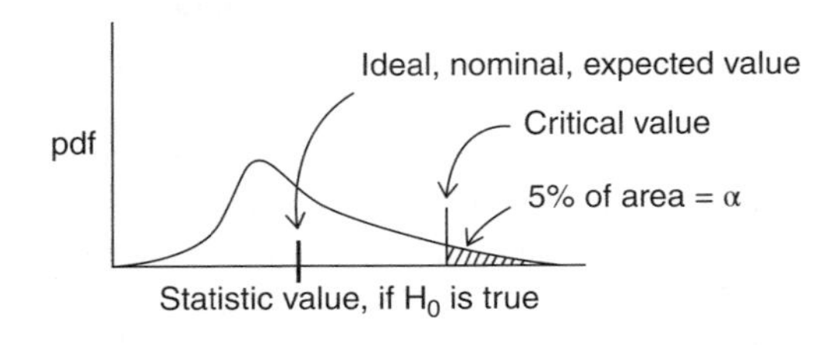

Figure 4.12 PDF of a statistic when H_0 is true

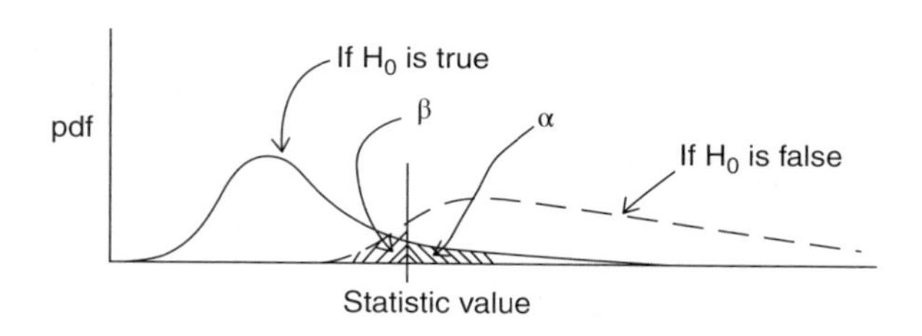

Figure 4.13 PDF of a statistic when H_0 is true or not true

Contrasting the Type I error is the Type II statistical error. This is accepting the null hypothesis, when in fact it is not true. The equivalent is claiming a guilty person is not guilty. Figure 4.13 reveals the distribution of a statistic when the null hypothesis is not true, superimposed on the statistic if H_0 is true. Mostly, when H_0 is false, the value of the statistic exceeds the critical value and H_0 would be rejected. However, there is a chance that a statistic value from the H_0 false distribution will be within the not extreme values if H_0 is true. The area under the false curve labeled β is the probability that the data from an H_0-false situation will generate a value of the statistic that does not exceed the reject H_0 critical value. The probability of a Type II statistical error is β.

The value of β depends on the critical value. If α were smaller, shifting the critical value to the right, then β would be larger. However, the value of β also depends on the location of the H_0-false distribution. In comparing models and data the model might be close but not quite right or it might definitely misrepresent the data. In one situation the distribution of the data might be marginally different from the H_0-true distribution and in the other it might be markedly different. Figure 4.14 compares the statistic from H_0-true to the two H_0-false situations, revealing the difference in the probability of the Type II error with larger *N*.

You do not want to make Type II errors. You do not want to accept H_0 when it is false. By increasing α you decrease β. However, this means increasing the probability of a Type I statistical error.

The method to reduce the probability of both Type I and Type II errors is to increase the number of samples, the number of data comparisons to the model. The central limit theorem reveals that the width of a distribution, the standard deviation, reduces with the square root of *N* (Equation 4.20). Figure 4.14 shows what happened to the distributions in Figure 4.13 when *N* is increased. With larger *N*, there is less chance of an extreme statistic when H_0 is

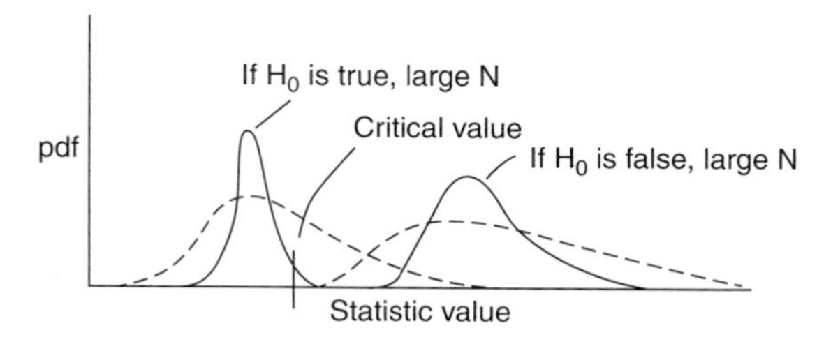

Figure 4.14 Increasing *N* in Figure 4.13

true, so the critical value, when the same $\alpha = 0.05$, has shifted toward the ideal value. The H_0-false distribution has also narrowed and, as illustrated, the probability of a Type-II error, β, is imperceptible.

However, increasing *N* increases the experimental cost and time. Even with more *N*, a model that is just barely in error might still lead to a Type II error. The user needs to balance the cost of the experimental data with the consequences of making Type I and Type II errors. This decision is grounded in the human interpretation of the situation, not statistical analysis.

4.10 Essential Statistics for This Text

Essential statistics for model validation and experimental design include:

- t-test for bias in residuals – parametric test of mean
- Wilcoxon signed rank for bias in residuals – nonparametric test of median
- Chi-square – contingency test of tabulated data versus expectations for model validation of class variables and for impact of different treatments
- Kolmogrov–Smirnov – nonparametric test of CDF from modeled distribution (test for normalcy in residuals), test data w.r.t. hypothesized distribution, modified KS to test between two distributions (data and stochastic simulation)
- F-test for changes in variance – parametric
- Runs – nonparametric test of autocorrelation
- r-lag-1 – parametric test of autocorrelation
- Binomial – estimating *N* required for precision in design and testing proportions
- *R*-statistic for SS or TS.

There are several statistics that are useful in model validation. If the model is true (matches the natural phenomena) and the experimental data are subject to independent and normal variation then there are several measures that you expect to see in the data:

- The average of the residuals (deviation between data and model) should be zero. Use a t-test (on the average) or a rank test (on the median) to see if it is significantly different from zero.
- The sequence of signs in the residuals should reveal random independent fluctuations, not sections of data with a run of like signs. Use either an r-lag-1 or a runs test to see if there are unexpected sections of like sign.
- The classification in histogram bins should match the distribution. Use either a chi-square contingency test or a Kolmogrov–Smirnov test to compare distributions.
- Steady-state data should not show a trend in time or iteration. Use a ratio of variance test to detect a trend.

Other statistics are useful in experimental design.

- There should be enough data to see trends with confidence. Use the binomial distribution to determine *N*.
- Steady-state models need to be regressed to steady-state data. Use the *R*-statistic to select steady-state periods.

There are many other tests useful in linear modeling such as correlation coefficient, regression r-square, standard error of the estimate, and slope tests. The ones described here are useful for validation of nonlinear models.

The tests can be classified as parametric or nonparametric. Parametric tests are based on theoretical analysis of a presumed underlying distribution. Typically this is a normal (Gaussian) distribution, in which the residuals would be expected to have a zero mean and normal (nonskewed, independent) perturbations about the mean of zero. Typically parametric tests do not need as many data sets to reach a conclusion as nonparametric tests. Where you can accept the assumption about the underlying distribution, the parametric test will be stronger than a nonparametric test. However, there may be evidence in the data that negates the parametric assumption and requires the non-parametric test.

4.10.1 t-Test for Bias

This is a parametric test, meaning that the statistic is based on the residuals having a known distribution – in this case Gaussian (normal or bell-shaped).

The null hypothesis is that the average of the residuals is zero. We will reject the null hypothesis if the average value is either excessively positive or excessively negative. However, what defines an excessive value is relative to the inherent variability in the data, and the more data you have the more certain you can be about the deviation. The t-statistic is therefore defined as

$$t = \frac{\bar{r} - 0}{s/\sqrt{N}} \tag{4.21}$$

where $\bar{r}$ is the average of the residuals, s is the standard deviation in the residuals, and N is the number of data.

If H_0 is true, the calculated t-statistic will be within a certain expected range about zero (between about -2 and $+2$), called a *critical value*. The critical value depends on the number of residuals minus one, called the *degree of freedom*, $\upsilon = N - 1$. Choose a *level of confidence*, typically 95% in economic decisions. For a two-sided test (either too positive or too negative an extreme), this leaves 2.5% of the area ($\alpha = 0.025$) in the distribution tails. Look up the critical value, t-critical, for your values of α and υ. For $N = 15$ ($\upsilon = 14$) and 95% confidence ($\alpha = 0.025$) the critical t value is 2.145. Any elementary statistics reference book should have a t-table.

If the absolute value of the calculated t-value from Equation 4.21 is greater than t-critical, then reject the null hypotheses at the 95% confidence level – you are 95% confident that the bias is not zero.

Alternately, if the absolute value of the calculated t value is not greater than t-critical, you cannot reject the null hypothesis with a 95% confidence. This does not mean that the bias is zero. Do not say that the bias is zero.

4.10.2 Wilcoxon Signed Rank Test for Bias

This is a nonparametric test, meaning that there are no assumptions about the underlying data distribution. An advantage is that nonparametric tests are more broadly applicable, but a disadvantage is that a greater number of data is needed to be able to make conclusive statements.

Table 4.1 Example of calculating the Wilcoxon signed rank statistic

Residual	ABS (residual)	Rank	Rank of "+"
−0.8567	0.85673	1	
0.55631	0.55631	2	2
0.19543	0.19543	3	3
−0.168289	0.168289	4	
−0.10554	0.105536	5	
0.096781	0.096781	6	6
0.08605	0.086052	7	7
−0.06321	0.063211	8	
−0.05117	0.051169	9	
			Sum of "+" data rank = $2 + 3 + 6 + 7 = 18$

The Wilcoxon signed rank test tests for the median, not the mean (average). Regardless of whether the median or the mean is not zero, there is bias. The null hypothesis is that the median is zero. This is a two-sided test.

First, sort the residuals by their absolute value. Then rank the residuals from 1 to N, either from the lowest magnitude to the highest magnitude, or the reverse. If there is a tie in residual value, use the average of the ranks for the tied individuals. Sum the values of ranks of "+" residuals to generate the statistic "T." If the sum is either greater than the upper critical value or less than the lower critical value (for N and α) then reject the null hypothesis that the median residual is zero – there is a significant bias. Otherwise accept that the bias is not significant enough to reject. Critical value tables for the signed rank test are common, but not universal, in statistics texts.

Example 4.1 Table 4.1 reveals a brief example of the signed rank method for nine data points. The sum of the "+" ranks is 18, which is between the critical values (6 and 39) for $N = 9$ at the 95% confidence level. In this example, the Wilcoxon signed rank test cannot reject the hypothesis that the median is zero, with a 95% confidence. Consequently, accept the no-bias hypothesis.

4.10.3 r-lag-1 Autocorrelation Test

Autocorrelation means that if one residual is positive then what made it so persists, and the next residual will also tend to be positive. If something makes the model underpredict in a range, then all data in that range will tend to be on one side of the model. This is often termed serial autocorrelation because it refers to one variable and considers sequential terms in the series.

The r-lag-1 autocorrelation test is a parametric test. This makes it more efficient than non-parametric tests, but requires uniform variance and a Gaussian distribution of residuals. Since the autocorrelation depends on the sequence of residuals, this test needs to be performed when residuals are organized with respect to y *and* in turn for each x-variable *and* run number (chronological order or time). The r-lag-1 statistic is defined as

$$r_1 = \frac{\sum r_i r_{i-1}}{\sum r_i r_i} \tag{4.22}$$

First, calculate the residual, r_i, for each data point. List residuals in ascending order of y (and, subsequently, for each x and for run number). Sum $r_i \times r_i$ (deviation-squared) (N items). Sum $r_i \times r_{i-1}$ ($N - 1$ items). Calculate r_1 from Equation 4.22. If $r_1 >$ the critical value (for N and the 95% confidence) then reject – there is a significant skew or curvature. Otherwise, accept – the skew or curvature is not significant.

This uses a one-sided test to reject the model if the r_1 value is too large and positive. Therefore, at a 95% confidence, $\alpha = 0.05$.

The subscript, 1, indicates that one residual is compared with the immediately preceding residual. However, one residual could be compared with second, tenth, or such preceding value to investigate either persistence of an event or some mechanism that might be affecting every third sample value (such as the day-shift). For model validation, we will assume that there are no skip-over mechanisms so that the r-lag-1 statistic is adequate.

This presumes that the average residual is zero. First use either a t-test of the mean or the Wilcoxon signed rank test for the median. If the average residual is not zero, then the model does not fit the data. The model can be rejected and there is no need for the r-lag-1 test for autocorrelation.

The lag-of-one could be one time step, one stage, or one place in a sequence or ordered list. It means that each residual is compared to the previous residual in the sequence. The numerator product of $r_i \times r_{i-1}$ is the basis for the sequential comparison. If there is no bias in the model, the r-lag-1 statistic can be interpreted as a ratio of variances. Since the residual is the y deviation from the average, the denominator is $(N - 1)$ times the conventional variance, which is based on n squared terms. The numerator is $(N - 1)$ times the sum of $(N - 1)$ products of sequential deviations, instead of like deviations.

There are $N - 1$ terms in the numerator and N in the denominator. If all residuals had equivalent values then r_1 is bounded between $\pm(N - 1)/N$.

Unfortunately, nomenclature has developed multiple uses for the same symbol, r. Here, r_i *and* r_{i-1} indicate the residual or process-to-model mismatch, and r_1 is a statistic to quantify autocorrelation in the residuals. Neither is the r-square statistic used to measure the reduction in SSD due to the model or the ratio statistic for the steady-state convergence criterion.

The denominator represents the variance in the residuals, but not scaled by $N - 1$. Regardless of the sign of the residual, squared, the residual makes a positive contribution to the denominator sum and the sum will increase with N. If the residuals are not autocorrelated, if each sequential residual is independent of the previous residual, then the product in the numerator will have positive values as often and as large as it has negative values, and the sum will tend to remain near to zero. The numerator and denominator have the same units, and the ratio is normalized to be independent of the magnitude of the residuals. Consequently, the value of the ratio will be within the extreme limits of $-1 < r_1 < +1$. Multiple sources report that the one-sided 95% confidence limits for the autocorrelation coefficient of lag 1 is given by

$$r_{1,\,0.95} = \frac{-1 + 1.645\sqrt{N - 2}}{N - 1} \tag{4.23}$$

and the two-sided 95% CL is

$$r_{1,\,0.95} = \frac{-1 + 1.96\sqrt{N-2}}{N-1} \tag{4.24}$$

Note that both of the coefficient values in Equations 4.23 and 4.24 are the standard normal z-statistic values for $\alpha = 0.05$.

Certainly, those values are illogical for small N, so I will surmise that the critical values are only valid if $N \geq \sim 10$.

In the limit of very large N, the two-sided confidence limit reduces to

$$r_{1,\ 1-\alpha/2} = \frac{z_{1-\alpha/2}}{\sqrt{N}} \tag{4.25}$$

Minitab, apparently, calculates the limits as

$$r_{1,\ 1-\alpha/2} = \frac{t_{\nu=N-1,1-\alpha/2}}{\sqrt{N}} \tag{4.26}$$

If only interested in positive correlation (the expected indication of a bad model, successive residuals have the same sign), then use Equation 4.23 for the critical value. If anticipating either positive or negative correlations (successive residuals have unexpectedly alternating signs), then use Equation 4.24. The coefficient values represent the standard normal statistic, z, for confidence = $(1 - \alpha)$ or confidence = $(1 - \alpha/2)$. If the absolute value of r_1 from Equation 4.22 is greater than $r_{critical}$ from Equation 4.23 you can claim, at a 95% confidence level, that the data reject that portion of the model.

It is notable that I cannot confirm the validity of Equations 4.23 and 4.24; z is the standard normal statistic when the mean and variance are known and represents the probable deviation from the mean scaled by the standard deviation. However, the r-lag-1 statistic represents a squared value divided by $(N - 1)$ times the experimental variance (assuming zero mean and NID values for sequential residuals). Therefore, the meaning is not congruous; one seems to be the square of the other. The r-lag-1 statistic should be centered on zero and have symmetric deviations to either the + or to the – side. However, the -1 in the numerator of Equations 4.23 and 4.24 make the critical values not symmetric. Therefore, the shape is not as expected. The z statistic has values between + and – infinity, but the r-lag-1 is constrained to be $-1 < r\text{-lag-}1 < +1$. Therefore, the range on the critical value is not consistent. However, it seems that in the limit of large enough N, perhaps $N > 10$, the equations for the critical values are reasonably valid. When I perform Monte Carlo simulations the value from Equation 4.23 usually rejects more than about 6.5% of the trials with no autocorrelation. That is close to 5% and perhaps good enough for any application.

The autocorrelation test presumes both zero bias and uniform variance throughout the data sequence, which may not be true. If the data fails the test for bias, then the model is rejected, and there is no reason to consider autocorrelation. If the model is not rejected for bias, then accept that the bias is zero and inspect the data (as they are independently arranged on each graph) for uniform variance.

If the variance is one region is high, then the data in that region dominate the numerator and denominator terms in Equation 4.22, making the r_1 value focus on that region. Therefore, only use the autocorrelation test if variance is uniform throughout the range of the y and x variables. Visual inspection may be an adequate test for uniform variance.

4.10.4 Runs Test

The runs test also seeks to see if there is an unusual pattern in the data on one side of the model. A run is a sequence of residuals with the same sign. If there are too few number of runs, this indicates that residuals are clustered on one side of the model.

The runs test is a nonparametric test. This makes it more generally applicable than a parametric test, but requires more data to be able to make confident conclusions. Since the pattern of runs in the data depends on the sequence of residuals, this test needs to be performed when residuals are organized with respect to y *and* in turn for each x variable *and* time.

A run is a contiguous section of data between zero-crossings (normally data with the same sign). For example, in the series $+ + - + - - - - -$, there are four runs. The first has a length of 2, the second and third each have lengths of 1, and the fourth has a length of 5. If the value of a residual happens to be exactly zero, include it in the previous run. A run stops with a zero-crossing, not hitting zero. For example, the series $+ + 0 + - 0 + + - - - -$ has four runs, containing 4, 2, 2, and 4 elements. If ordered in reverse direction, it also has 4 runs, but of lengths 4, 3, 1, 4. Since the number of runs, not run length distribution, is assessed, the sequence does not matter.

Sort the residuals by y, then by each x variable, and then by chronological sequence. For each structure, count the number of runs. If the number is smaller than the critical value then reject the model (it confidently has skew or curvature). Otherwise, accept the model.

Use a one-sided test to reject the model if the number of runs value is too small. Thus, at a 95% confidence, $\alpha = 0.05$.

Too many runs, like alternating data points giving a large negative r_1 value, should create suspicion about data legitimacy.

This presumes that the average residual is zero. First use either a t-test of the mean or the Wilcoxon signed rank test for the median. If the average residual is not zero, then the model does not fit the data. The model can be rejected, and there is no need for the runs test for autocorrelation.

4.10.5 Test for Steady State in a Noisy Signal

If a signal is at steady state, but corrupted with independent and normally distributed perturbations, then the variance can be calculated from the differences in data from the mean (the conventional approach) and from sequential differences in the data. In the asymptotic limit of many samples, the ratio of the two variances will be unity. However, with a finite number of samples the ratio statistic will have a value that tends around 1.0 with a normal range from about 0.8 to about 1.5. It has an F-like distribution, limited to non-negative values and a tail toward the right.

The time series of data could represent a process signal that relaxes in time to a noisy steady state, or it could represent the result of an optimizer that progressively approaches the optimal value with iterations.

There are three steps in the operation that require data. First calculate the average in the observation window, the past N samples. Then compute the variance (i) as a deviation from the average and alternately (ii) from sequential data deviations. In either case, the normal approach to determining variance is computationally burdensome. It requires the storage, updating, and processing of N data values, representing an observation window of N samples. A computationally simpler approach is to use a simple incremental update procedure. Depending on the community it would be termed a first-order filter or an exponentially weighted moving average, or an exponentially weighted moving variance.

A filtered value (not an average) provides an estimate of the data mean:

$$X_{f,i} = \lambda_1 X_i + (1 - \lambda_1) X_{f,i-1} \tag{4.27}$$

where X is the process variable, X_f is the filtered value of X, λ_1 is a filter factor (effectively lambda is the reciprocal of the number of data in the analysis window), and i is the sampling index.

The first method to obtain a measure of the variance uses an exponentially weighted moving "variance" (another first-order filter) based on the difference between the data and the filtered value:

$$\upsilon^2_{f,i} = \lambda_2 (\mathrm{X}_i - \mathrm{X}_{f,i-1})^2 + (1 - \lambda_2)\upsilon^2_{\mathrm{f,i-1}} \tag{4.28}$$

where $\upsilon^2_{f,\,i}$ is the filtered value of a measure of variance based on differences between data and filtered values and $\upsilon^2_{f,\,i-1}$ is the previous filtered value.

Equation 4.28 is a measure of the variance to be used in the numerator or the ratio statistic. The previous value of the filtered measurement is used instead of the most recently updated value to prevent autocorrelation from biasing the variance estimate, $v^2_{f,i}$, keeping the equation for the ratio simple.

The second method to obtain a measure of variance is an exponentially weighted moving variance (another filter) based on sequential data differences:

$$\delta^2_{f,i} = \lambda_3 (\mathrm{X}_i - \mathrm{X}_{i-1})^2 + (1 - \lambda_3)\delta^2_{f,i-1} \tag{4.29}$$

where $\delta^2_{f,i}$ is a filtered value of a measure of variance and $\delta^2_{f,i-1}$ is the previous filtered value.

This will be the denominator measure of the variance.

The ratio of variances, the R-statistic, may now be computed by the following simple equation:

$$\mathrm{R} = \frac{(2 - \lambda_1)\mathrm{v}^2_{f,i}}{\delta^2_{f,i}} \tag{4.30}$$

The calculated value is to be compared to its critical values to determine SS or not SS. Neither Equation 4.28 nor Equation 4.29 compute the variance. They compute a measure of the variance. Accordingly, the $(2 - \lambda_1)$ coefficient in Equation 4.30 is required to scale the ratio, to represent the variance ratio with ideal values about unity.

The initial values for the three filtered variables could be set to characteristic values representing a lead-in data analysis to determine the initial average and variances. However, a simpler approach is to set those initial values to zero and let them incrementally approach values that represent the data. If the signal is at an initial steady state then this approach will require about $2.5/\lambda$ number of samples to relax the initial filtered values to their steady-state values and accept the steady-state hypothesis.

Cao and Rhinehart (1997) provide some critical values for the R-statistic. With recommended λ values of 0.1, reject the steady-state hypothesis if $R > 1.8$.

4.10.6 Chi-Square Contingency Test

This is useful for testing distributions, to determine whether a histogram of experimental results matches the expected distribution. Chose bin intervals and assign data to bins. Count the number of samples in each bin. The statistic is based on the observed count, O, and the expected count, E:

$$\chi^2 = \sum \frac{(O_i - E_i)^2}{E_i} \tag{4.31}$$

where i indicates the bin index.

The chi-square test is a parametric test based on a variance distribution from normally distributed data. For a wide range of underlying distributions, the statistic from the chi-square contingency test is approximately ideally chi-square distributed when there are at least five counts in each bin. Therefore, choose bin intervals and total number of samples so that there are at least five counts in any bin. Preferentially, use 10 or more bins.

The degrees of freedom for the chi-square statistic is the number of bins minus 1, minus the number of population parameter values determined from the experimental data, $\nu = N - 1 - p$. If the distribution was derived from a theoretical analysis then $p = 0$.

The null hypothesis is that the expected distribution matches the observed. In this case the value of the chi-square statistic should be zero. Reject the null hypothesis if the calculated chi-square value is larger than the critical value. This is a one-sided test.

4.10.7 Kolmogorov–Smirnov Distribution Test

The chi-square test requires a user to define the bin intervals, which could raise questions about user bias in setting up the test. It also desires a minimum of 50 samples (10 bins with at least 5 in each bin), but likely many more than 50 if the bin intervals are chosen so that the histogram visually represents the distribution. The Kolmogorov–Smirnov goodness-of-fit test is a nonparametric test applied to the cumulative distribution, not the histogram. It seeks the maximum deviation in the expected and actual CDF. Ideally, at all points the experimental cumulative distribution matches the modeled distribution and the deviation is zero. If the null hypothesis is true, the hypothesized distribution matches the actual distribution. The K–S statistic is the largest deviation over the CDF. If unexpectedly large, reject the null hypothesis.

To generate an experimental CDF, sample N times and sort the observed response values in ascending order, $\{x_1, x_2, \ldots, x_k, \ldots, x_N\}$. Here the index on x does not mean chronological order, but order in the ascending sequence. The empirical $CDF(x)$ is k/N, the fraction of samples with a value equal or lower than x_k. The empirical $CDF(x)$ will be a step function that holds its value of k/N for all x values greater than x_k until the $(k + 1)$th sample is reached. Then it steps to $(k + 1)/N$:

$$CDF(x) = \frac{k}{N}, \quad x_k \le x < x_{k+1} \tag{4.32}$$

The K–S statistic is D, the absolute value of the maximum deviation from the experimental to the hypothesized CDF over the entire range of possible x values, not just within the experimental range and not just at the experimental values. The maximum deviation might occur just before a sample value.

If D is greater than D-critical for the desired level of significance and N, then reject the null hypothesis.

4.10.8 Test for Proportion

The null hypothesis is that a theoretically derived proportion matches the measured proportion. In this case the difference in the measured and expected should be zero and the statistic will be the difference scaled by the expected standard deviation. Grounding the statistic in the theoretically expected "known," there would be no uncertainty on the standard deviation and the z-statistic would be

$$z = \frac{p_{meas} - p}{\sqrt{p(1-p)/N}} \tag{4.33}$$

where p is the theoretically expected proportion and N is the number of samples used to determine the measured proportion.

Alternately, if the theoretical distribution is being tested against actual data, with the actual data providing the basis for the standard deviation estimate, the t-statistic would be

$$t = \frac{p_{meas} - p}{\sqrt{p_{meas}(1-p_{meas})/N}} \tag{4.34}$$

This is a two-sided test since the null hypothesis would be rejected if the measured proportion were either too high or too low relative to the expected. In either case, if the absolute value of either z or t exceeds the critical value, reject the null hypothesis.

4.10.9 F-Test for Equal Variance

An underlying assumption in the parametric tests (t-test, r-lag-1, steady state, chi-square) is that the variance in the data residuals is uniform over all of the experimental range. This is termed homoscedastic.

However, variance may change over time as the experimental technique changes. It could either improve with experience or become sloppy with the routine. Variance might also be related to the operating conditions. For example, increasing the flow rate increases turbulence and would improve mixing uniformity. Alternately, a region of high gain in a process (for example, pH near neutrality) may amplify the data response underlying sources of variability (such as mixing or temperature). This is termed heteroscedastic.

An F-test will provide a statistical test of equal variances. However, often a graph of residual w.r.t. data chronological order or one with respect to the response variable, or each graph of residual w.r.t. each input variable provides a fully adequate visual determination whether the residuals are homoscedastic or not. However, a visual opinion does not carry the unbiased credibility of an F-test. The F-test outcome greatly depends on the user's choices of how to segregate the data and of what level of significance to use. Therefore, even if it pretends to be unbiased, it still substantially incorporates the experimenter's opinion.

The F-statistic is a ratio of variances scaled by degrees of freedom:

$$F = \frac{{\sigma_1}^2/\nu_1}{{\sigma_2}^2/\nu_2} \tag{4.35}$$

where σ is the standard deviation of residuals in group 1 and group 2, and υ is the degrees of freedom, the number of data minus 1, in groups 1 and 2.

The F-value will be a positive value. If the F-value is unexpectedly large, this indicates that group 1 has a larger variance than group 2. However, if too small, it means the opposite. Therefore, reject the homoscedastic hypothesis if the F-value is either too large or too small. This is a two-tailed test.

Calculate the F-statistic value of the data using Equation 4.35. Choose a level of confidence and look up the critical F-value for υ_1 and υ_2. If F-data is either larger than the larger F-critical or smaller than the small F-critical, then reject homoscedasticity. Otherwise, accept.

4.11 Takeaway

Plot your residuals to visualize trends with chronological order, operating conditions, and response. Create CDF of residuals to characterize the distribution. Often the visual judgment is fully adequate (the variance is uniform, there is no trend in the residuals, the model goes through the middle of the data, etc.). However, the statistical tests will add legitimacy and credibility.

Exercises

4.1 The following table presents test data for a steady-state or batch process, and has been sorted by the input (influence, control, independent variable) value, x-data. The runs were executed in a randomized order, but runs 1, 5, 9, 16, and 18 replicated a medium input value of 5.0. The modeler has explored linear, quadratic, power-law, and log-function models, with the following as best fit using vertical least squares:

Linear: $y = 2.3070755 + 0.305373514x$

Quadratic: $y = 2.080771502 + 0.441832935x - 0.011920524223x^2$

Power: $y = 2.55098862x^{0.304932248}$

Log: $y = 0.96592025267 + 2.727601393\ \ln(x)$

(a) Estimate the standard deviation of the measurement.
(b) Visually inspect the data to determine if the variance (residual deviations) seem constant over the experimental conditions.
(c) Use a t-test on residuals to see if the average residual for each model is zero-ish.
(d) Use an F-test to determine if the variance on the residuals is consistent with the expected variance in the data.
(e) With data sorted by the x value, use an r-lag-1 test to see if there is autocorrelation in the residuals for each model.
(f) With data sorted by the y value, use an r-lag-1 test to see if there is autocorrelation in the residuals for each model.
(g) With data sorted by run number, use an r-lag-1 test to see if there is autocorrelation in the residuals for each model.
(h) With data sorted by the x value, use the Wilcoxon signed rank test to see if there is bias in the residuals for each model.

Run	x-data	y-data
10	0.3	1.70
17	0.4	1.84
2	0.5	2.09
14	0.6	2.91
15	1.1	3.09
3	2.1	3.20
19	2.9	3.04
7	3.1	3.63
12	4.7	4.42
1	5.0	3.57
5	5.0	3.85
9	5.0	4.05
16	5.0	3.63
18	5.0	3.96
8	6.7	4.63
13	8.4	5.09
21	9.3	5.04
4	9.9	4.91
11	10.4	5.50
6	11.0	5.56
20	11.8	5.78

4.2 The following data are sampled from a distribution and modeled with a logistic function to best fit the CFD: $CDF(x) = [1 + e^{-5.04633(x-5.134533)}]^{-1}$. Graph the experimental and the modeled $CDF(x)$. Use a K–S test to determine if the data can reject the model.

5.04
5.76
5.05
6.02
5.35
5.05
5.14
4.75
4.93
4.85
5.55
5.42
4.99
5.36
5.05

5

Simulation

5.1 Introduction

Simulation is an essential tool for visualizing the results of data processing techniques, and readers are encouraged to use simulation to generate their own data, apply techniques of this book to the data, and experience the procedure and outcome. Hopefully, simulation will validate procedures.

If a simulation generates data from a known model, then regression should generate the same model values. However, considering experimental variability, the model coefficient values will not exactly match those of the simulation generator. There will be an uncertainty on the regression coefficient values, which is related to the variability in the simulated data. Therefore, if the 95% confidence interval on regressed model parameter values does not include the simulator value that generated the data, then the regression technique can confidently be rejected as being faulty. If two alternate regression approaches both generate model coefficient values with uncertainty that encompasses the true value, then the technique that leads to better precision (lower uncertainty) would be the preferred technique.

This chapter reveals simulation techniques that were found to be useful for methods of exploration and validation. There are two basic concepts in the following. The first is that the simulator represents the truth about Nature. In general, the exact rules and coefficient values that Nature uses to generate data are unknowable, but in a simulator you will have written the equations and know exactly the coefficient values. This absolute knowledge of the truth about your simulator provides a basis for validating techniques that hope to see the truth within the noisy data. The second concept is that the simulation needs to include mechanisms that generate the vagaries of the experimentation, in a manner consistent with Nature.

5.2 Three Sources of Deviation: Measurement, Inputs, Coefficients

There are three types of variables that have variation. The common view is that an experiment is exactly controlled and that the measurement contains the fluctuation. In this view, add the perturbation to the deterministic simulator output to obtain the simulated measurement. However, inputs to an experiment are not exactly controlled. Inputs are the result of the final

Nonlinear Regression Modeling for Engineering Applications: Modeling, Model Validation, and Enabling Design of Experiments, First Edition. R. Russell Rhinehart.

control element (valve, knob, dial), which is physically set to some position or manipulated to make an input reading (flow rate, temperature) have a desired value. Either way, the true value of the input is not the desired value; it is the measured value adjusted for the unknowable measurement error. Third, there are also uncontrolled external influences on a process, such as disturbances from ambient temperature, composition of an input reagent, tire inflation pressure, air density, and so on.

To create a simulator using this model, first define the equation or procedure to obtain y_{true}, the truth about Nature, from x_{true}, the unknowable value of the influence:

$$y_{true} = f(x_{true}) \tag{5.1}$$

Then perturb the values by the disturbances and noise.

Figure 5.1 illustrates how to add variation to the basic deterministic simulator to simulate experimental results. The input labeled $x_{nominal}$ would represent the experimenter's chosen input value. This would be a setting on a device or a set point for a controller. However, what is actually influencing the process will be an alternate value. For instance your oven temperature display may read 350 °F or your speedometer may read 60 mph, but are the processes at those values exactly? In this case, the illustration shows that d_1 is added to $x_{nominal}$ to generate x_{true}, the value of the input to the simulator. Because it is a simulator, you could ask the computer to display the value of x_{true}. However, in reality, the value of x_{true} is unknown and unknowable. Although only one input is illustrated, the method can be used for multiple inputs. The other inputs could represent either additional controlled variables or disturbance influences such as those represented by ambient temperature or pressure, variations in raw material composition, and so on.

Although the disturbance, d_1, is added here, it might be more appropriate to make it a multiplicative factor when you anticipate that the noise level or perturbation magnitude scales with the input value.

The figure also illustrates how to add variation to the basic deterministic simulator to simulate uncontrolled drifts in the process behaviors. In it $\mathrm{p}_{\mathrm{nominal}}$ is the base case value of a model coefficient. Model coefficients might represent chemical reactivity, fouling, or friction, and recognizing that these attributes of a real experiment will change in time or with sampling, d_2 is added to $p_{nominal}$ to generate the p_{true} that actually influences the simulation. As before, p_{true} is unknown and unknowable. Although only one uncontrolled process characteristic is indicated, many may be chosen and, although an additive disturbance is indicated, a multiplicative relation might provide a mechanistically more appropriate response for your particular application.

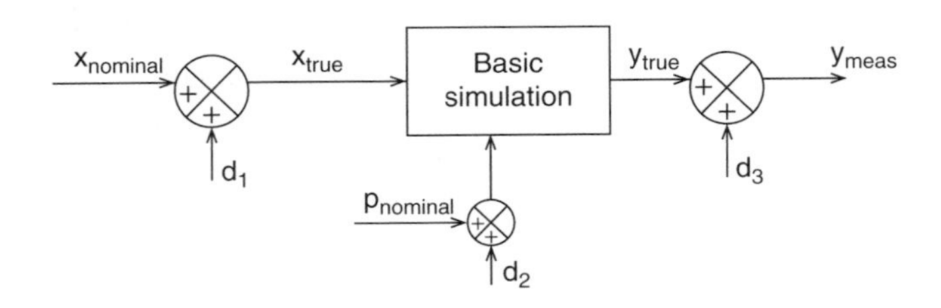

Figure 5.1 Simulation of experimental data

The figure also illustrates how to add variation to the basic deterministic simulator to simulate variation in the measurement. In it y_{true} is the calculated value from the basic simulator model, d_3 is the error related to the measurement, and $y_{measured}$ represents the experimental outcome. As before, y_{true} is unknown and unknowable. Although only one output is indicated, all modeled outputs that represent experimental measurements should be perturbed and, although an additive disturbance is indicated, a multiplicative relation might provide a mechanistically more appropriate response.

5.3 Two Types of Perturbations: Noise (Independent) and Drifts (Persistence)

Looking at measurement sequence or time, there are two basic concepts of variation. One is independent variation, in which sample-to-sample variation is uncorrelated. This is often termed noise. Here, what causes one perturbation from ideal does not persist and the next perturbation is the result of independent causes. The other mechanism for variation reflects a cause that has persistence, in which sequential perturbations are correlated. Although disturbances are labeled as d_i in Figure 5.1, we will now use the nomenclature d_i to represent autocorrelated, time-dependent trends, or drifts, in an influence or model coefficient. The symbol n_i will represent uncorrelated, independent, noise.

Representing an independent influence, normally distributed, with a mean of zero, the value of n_i can be calculated by Equation 4.6. Equation 4.7 reveals how to add it to the nominal or true value to generate the actual or measured values. If mechanistically appropriate, you could scale the σ value in Equation 4.6 in a relation to the base value or to the change in time.

Independent perturbations have no persistence; the value of one n_i is independent of the proceeding values. However, many influences on experiments have persistence. For example, if a cloud covers the sun, the shadow persists for a while. If the blocked sunlight would influence a temperature disturbance, then a negative d value at one sampling (in the shadow of the cloud) should generate a negative d value at the subsequent sampling (the shadow persists). In this case, if the clouds were randomly scattered in the sky, with randomly assigned density, the random drivers would influence the persisting effect.

Simulating the additive autocorrelated d and uncorrelated n influences, Equation 5.1 becomes

$$y_{meas,i} = f(x_{nominal,i} + d_{1,i} + n_{1,i}, d_{2,i}) + d_{3,\,i} + n_{3,i} \tag{5.2}$$

where the subscript i represents a time counter index, a sampling number.

Alternately, one could consider that some value of x_{true} is given to the process, such as a flow rate, but its value is known by measuring it. Then the measured value is corrupted by drifts and noise. This model is represented as

$$x_{meas,i} = x_{true,i} + d_{1,i} + n_{1,i} \tag{5.3}$$

$$y_{meas,\,i} = f(x_{true,\,i}, d_{2,i}) + d_{3,\,i} + n_{3,i} \tag{5.4}$$

For either regression analysis or simulation, the concepts in Equation 5.2 and the sets (5.3) and (5.4) are equivalent. Solving Equation 5.2 for x_{true}, equating the concepts of $x_{nominal}$

and x_{meas} which both represent what you think is the input, and substituting Equation 5.3 in Equation 5.4 results in

$$y_{meas,i} = f(x_{meas,i} - d_{1,i} - n_{1,i},\ d_{2,i}) + d_{3,\ i} + n_{3,i} \tag{5.5}$$

Because disturbances and noise could have either positive or negative values, deviating from the nominal zero case, the signs on $d_{1,i}$ and $n_{1,i}$ are inconsequential. Equation 5.5 is equivalent to Equation 5.2. This book will use the Equation 5.2 convention.

The values of $d_{1,i}$, $d_{2,i}$, and $d_{3,i}$ would nominally be constant, or would change slowly with i, the experimental set number. Alternately, $n_{j,i}$ would have an independent value for each sampling. Here the index j represents the variable or influence. The Gaussian (normal, bell-shaped) distribution is commonly accepted as a representative model of noise and represents the confluence of many random influences of equivalent impact. I have been very pleased with the 1958 Box–Muller model of generating Gaussian distributed independent perturbations with mean of zero and standard deviation of σ, NID(0, σ) (normally and independently distributed).

$$n_i = \sigma \sqrt{-2\ \ln(r_{1,i})}\ \sin(2\pi r_{2,i}) \tag{5.6}$$

Here, the variables $r_{1,i}$ and $r_{2,i}$ represent uniformly distributed independent random numbers on the 0–1 range.

This is a basic approach to creating simulated experimental results for steady-state simulators. The disturbances are often called bias or systematic error, and the noise is often called random error. The simulation creator would choose values of σ_x and σ_y and d_x and d_y that represent the modeled phenomena of measurement uncertainty and bias.

For dynamic, time-dependent simulators, a common method to simulate the manner in which disturbances (bias or systematic error in measurements, as well as uncontrolled influences on the process) change in time is to consider that they are driven by random events that have persistence. Again, for example, a cloud passes over the sun blocking radiant heat and an object starts to cool, but it does not cool instantly because the past accumulation of heat needs to dissipate in time. Then random patterns in the clouds randomize the on–off sun source, but also randomly temper the intensity due to cloud thickness.

A drifting jth influence with a first-order persistence driven by NID(0, σ) noise is modeled as

$$\tau \frac{dd_j}{dt} + d_j = n(t) = \sigma \sqrt{-2\ \ln(r_{1,i})}\ \sin(2\pi r_{2,i}) \tag{5.7}$$

Expanding this differential with a forward finite difference and rearranging leads to the autoregressive moving average (ARMA) model for d_j:

$$d_{j,i} = (1 - \lambda) d_{j,i-1} + \lambda\sigma \sqrt{-2\ \ln(r_{1,i})}\ \sin(2\pi r_{2,i}) \tag{5.8}$$

where

$$\lambda = 1 - e^{-\Delta t/\tau} \tag{5.9}$$

and Δt is the simulation time step.

If $\Delta t \ll \tau$, which is usually the case in a simulation, then the simplified approximation $\lambda = \Delta t/\tau$ can be used. In creating a simulation, the user would choose a time constant for the persistence that is reasonable for the effects considered and a σ value that would make the disturbance have a reasonable variability. Since the first-order persistence tempers the response to the random influences, the σ-driver needs to be greater than the desired response σ_{d_j}.

How should one choose values for λ and σ in the simulator? First consider the time-constant, τ, in Equation 5.7. It represents the time constant of the persistence of a particular influence. In an ideal mixer it would be volume divided by flow rate. Roughly $\tau \approx 1/3$ of the lifetime of a persisting event. Therefore, if you considered that the shadow of a cloud persists for 6 minutes, then the time constant value is about 2 minutes. Once you choose a τ value that matches your experience with Nature and decided for a time interval for the numerical simulation, Δt, calculate λ from Equation 5.9.

To determine the value for σ in Equation 5.8, propagate variance on the equation. This indicates that the σ value depends on the user choices of σ_{d_j} and λ (which is dependent on Δt and τ):

$$\sigma = \sigma_{d_j}\sqrt{\frac{2-\lambda}{\lambda}} \tag{5.10}$$

Choose a value for σ_{d_j}, the resulting variability on the disturbance. To do this, choose a range of fluctuations of the disturbance. You should have a feel for what is reasonable to expect for the situation that you are simulating. For instance, if it is barometric pressure the normal local range of low to high might be 29–31 inches of mercury, if the outside temperature in the summer it might be from 70 to 95 °F, or if the catalyst activity coefficient it might be from 0.50 to 0.85. The d_j value is expected to wander within those extremes. Using the range, R, as

$$R = HIGH - LOW \tag{5.11}$$

and the standard deviation, σ_{d_j}, as approximately one-fifth of the range, then

$$\sigma = \frac{R}{5}\sqrt{\frac{2-\lambda}{\lambda}} \tag{5.12}$$

The reader is welcome to use other methods for generating NID(0, σ) values, alternate models for disturbance persistence, or other distributions that are deemed appropriate. For example, in particle size results the distribution could be log-normal, or in queuing or radiation intensity it could be Poisson. Section 5.5 indicates how to generate noise and disturbances with alternate distributions.

The standard deviation of the noise could be proportional to the measurement, as in the orifice flow rate measurement, in which fluid turbulence causes pressure drop perturbations and turbulence is proportional to the flow rate. The disturbances could have a higher order ARMA behavior mimicking environmental conditions or blended raw material properties. Additional disturbances could be impacting model coefficients related to fouling factors, efficiency factors, and so on.

Equation 5.8 is variously termed an exponentially weighted moving average, an autoregressive moving average model of orders 1 and 1, ARMA(1, 1), a first-order filter, or a first-order lag. Since computer assignment statements update the new variable value from the old, in implementation the subscripts are not needed:

$$d := \lambda\sigma u + (1-\lambda)d \tag{5.13}$$

Using the symbol ":=" to represent a computer assignment, the structured text would be

```
u := SQRT(-2*LN(rand()))*SIN(2*Pi*rand())
d := lambda*sigma*u + (1-lambda)*d
```

An illustration of one 100-sample realization with tau = 25 and range = 5 is shown in Figure 5.2. Note that the drift into positive values lasted for about 60 samplings and the drift into low values lasted for about 20 samplings. The lifetime of any one particular event is approximately three time constants, and should be about 75, but more recent events mask the action of past events. Note that the range of the data in these 100 samples is about 4, when the model was designed for a range of 5. However, in this limited realization of 100 samples neither the full range of the disturbance value nor the full range of its persistence are expected to be revealed.

This approach modeled a drifting influence as a first-order response to a Gaussian distributed driver. One certainly could model drifts as higher-order or driven by alternate forcing function distributions. However, my view is that this adequately mimics how I consider randomly driven drifts in natural disturbances to respond, and provides a relatively simple method to both execute and parameterize.

Certainly, noise and drifts can be combined to mimic the perturbation of Nature and, instead of perturbing just the variable *d* or *n*, use both.

```
u := SQRT(-2*LN(rand()))*SIN(2*Pi*rand())
d := lambda*sigma_drift*u + (1-lambda)*d
n := sigma_noise*SQRT(-2*LN(rand()))*SIN(2*Pi*rand())
y_measured := y_true + d + n
```

5.4 Two Types of Influence: Additive and Scaled with Level

Some sources of variation are expected to be related to the variable value. For instance, turbulence in fluid flow increases with flow rate and the magnitude of energy in turbulent eddies in

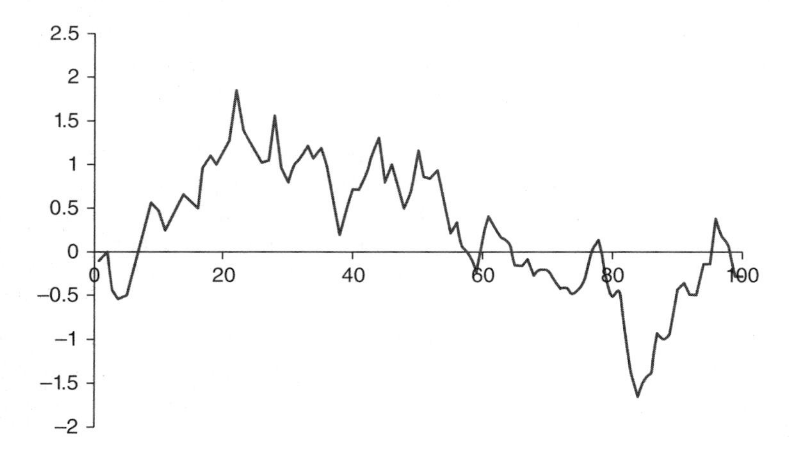

Figure 5.2 A realization of an ARMA(1, 1) drift

the fluid scales with the square of the flow rate. Accordingly, it is expected that with pressure fluctuations in an orifice meter, the differential pressure sensor signal would be proportional to the flow rate. A model for such a noise influence, n, is

$$n = \sigma\sqrt{-2\ \ln(r_1)}\sin(2\pi r_2)$$
$$y_{meas} = y_{true}(1 + n) \tag{5.14}$$

There are many other ways to implement the same phenomena.

5.5 Using the Inverse CDF to Generate *n* and *u* from UID(0, 1)

You might decide that the Gaussian distribution is not appropriate to either the independent or the drifting contribution to the disturbance. Data samplings from real experiments or fundamental analysis might reveal that alternate distributions are more appropriate. The Poisson distribution is a right fundamental model for the number of events that would occur within an interval and the log-normal distribution is one that empirically matches a natural distribution in particle sizes.

Equation 4.6 describes the Box–Muller method to generate a random variable with a Gaussian distribution. As a method to generate a random variable with an alternate distribution, consider the cumulative distribution function (CDF) plot of that distribution, as in Figure 5.3. The vertical axis is the CDF and the horizontal axis is the variable value. Use a random number generator (independent and uniform on the interval of 0 to 1), UID(0, 1), to generate a random number, r, the CDF value. Then use the inverse of the distribution to determine the variable value that results in that CDF value:

$$n = CDF^{-1}(u) \tag{5.15}$$

Although conceptually this is simple, getting a function that represents the inverse of the distribution CDF could be computational complexity.

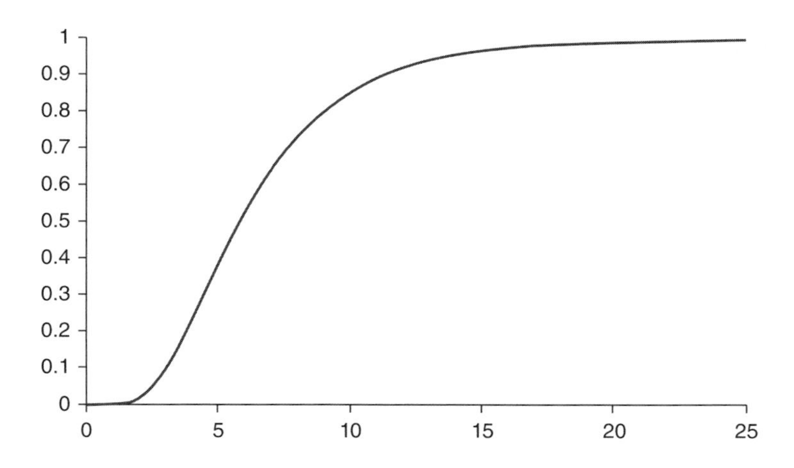

Figure 5.3 CDF used to generate random variable values

5.6 Takeaway

Validate your techniques and understanding using simulators.

Exercises

5.1 If the desired τ and σ_d for a drifting variable are 30 seconds and ± 7 units, and the simulation time interval is $\Delta t = 0.1$ seconds, what values of σ and λ should be used to simulate the d?

5.2 Use Equation 5.15 to convert a UID(0, 1) random number, r, to a uniformly distributed variable, n, on a $-a$ to $+a$ range.

5.3 Create a simulator to generate noise, n, using Equation 5.6. You choose the value of σ. Does the noise pattern agree with expectations? Does it have a mean of zero (test with a t-test)? Is the standard deviation as expected (test with an F-test)? Is the noise independent (test with an r-lag-1 test)?

5.4 Create a simulator to generate a drifting signal, d, using Equation 5.13. You choose the values of σ and λ. Does the pattern agree with expectations? Does it have a mean of zero (test with a t-test)? Is the standard deviation as expected (test with an F-test)? Is the noise independent (test with an r-lag-1 test)?

5.5 Choose a model and add a drift to the influence. Report on the simulation results.

5.6 Choose a model and add a drift to a model coefficient. Report on the simulation results.

5.7 Choose a model and add a drift and noise to the measurement. Report on the simulation results.

5.8 Combine Exercises 5.5 to 5.7. Report on the simulation results.

5.9 Choose a model (architecture and coefficient values) for a simulator and add noise and drifts as appropriate. Use the simulator to generate simulated experimental data. Use the same model architecture for regression, but adjust model coefficient values to best fit the data. If noise and drift are not used, if there is no error on the data, then regression should return exactly the same coefficient values as in the simulation model – the true values. However, with noise and drift corrupting the data, regression should provide similar, but not exactly true, coefficient values.

6

Steady and Transient State Detection

6.1 Introduction

Steady-state (SS) models should only be adjusted to fit SS data. Similarly, coefficients in models that represent the rate of change of transition between SSs should only be adjusted with transient (dynamic) data. Nominally, it is trivial to distinguish SS and transient state (TS) data. If the value changes in time it is a TS; if the value is constant in time it is an SS. However, measurement noise confounds the ideal, making the determination of SS or TS a statistical, probable determination. This chapter presents a method to determine SS and TS. Further, the steady-state identification SSID approach presented here will be shown to be useful as a scale-independent, universal, convergence criterion in nonlinear regression.

6.1.1 General Applications

During a TS something (material, energy, momentum, electrons, population individuals, boxes, etc.) is accumulating within a control volume; consequently measures of the inventory of that something (weight, composition, speed, temperature, color, voltage, etc.) are changing. By contrast, during SS, response variables (the state variables, the measures of inventory) remain constant over time because the inflow is equal to the outflow and there is no change in the inventory. However, during SS, measurement noise might make a variable value fluctuate from sample to sample.

Figure 6.1 illustrates the temperature and level in a tank of fluid in response to a hot water inflow, keeping the cold inflow the same. Initially all is at a steady state, as identified by "SS." The inflow rate and temperature are "flat-lined"; however, the level is a noisy signal that is fluctuating about a steady value.

At point "A" the inflow rate of hot water increases, affecting both level and temperature. The inflow rate is the influence (input, forcing function) and level and temperature are responses (state variables). The tank level rises because the outlet flow is dependent on the hydrostatic pressure and rises asymptotically at a rate proportional to the in–out flow

Nonlinear Regression Modeling for Engineering Applications: Modeling, Model Validation, and Enabling Design of Experiments, First Edition. R. Russell Rhinehart.

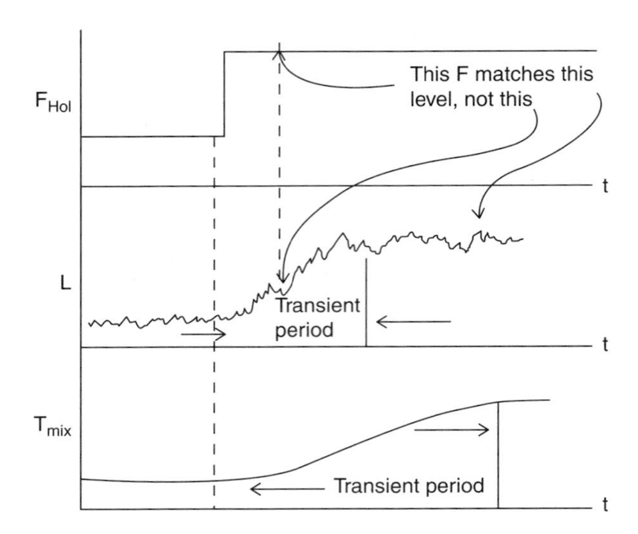

Figure 6.1 Illustration of a transient trend to the steady state

difference. The temperature rises because the fraction of hot water in the tank increases, and it rises at a rate related to the volume, tank outflow, and hot inflow. The period labeled "TS" indicates the transient state in which the responses are changing.

Note that during the TS the inflow rate in Figure 6.1 is not changing. Although I could have chosen to illustrate a gradual or progressive inflow rate change, I chose to make an ideal step-change-and-hold to illustrate that the TS is based on the responses, not the influences.

Eventually, the process settles to its new conditions. The temperature again flatlines and the level settles to a noisy steady level.

Note that the solution to first principles mathematical models of the mixing tank would reveal that the level and temperature asymptotically approach SS, but never get there. However, at some point, visually, the asymptotic trends are indistinguishable w.r.t. the thickness of the graph line, discretized pixel locations, or noise. Although a concept from the first principles models indicates that SS is never achieved, we will accept that the process is at SS when change is not detectable and in TS when change is detectable.

TS or dynamic models account for the accumulation of inventory and have time constants that relate to the speed of development of a model response toward a final value, and it may have delays that relate to the timing shift between cause and effect (input and output variables). In this example, the level and temperature both move toward asymptotic values and the response develops in time, but at separate rates. Simple material and energy balances on the tank, first principles modeling, lead to these models for tank level and temperature (key assumptions are constant density and specific heart, adiabatic, ideal turbulent flow through a fixed outlet restriction):

$$\frac{A\sqrt{h}}{k}\frac{dh}{dt} + h = \frac{\sqrt{h}}{k}(F_C + F_H) \tag{6.1}$$

$$\frac{Ah}{F_C + F_H}\frac{dT}{dt} + T = \frac{F_C T_C + F_H T_H}{F_C + F_H} \tag{6.2}$$

The model coefficients of “k” and “A” represent the discharge coefficient and effective tank surface area, and might be the model coefficients that need to be valued by experiments. The composite coefficients $A\sqrt{h}/k$ and $Ah/(F_C+F_H)$ are time constants indicating the rate of h or T rise to the final value.

By contrast, SS models do not account for the time trends in changing inventory and do not have time constants or delays. With the transient terms set to zero, the SS model versions are

$$h_{SS} = \left(\frac{F_C + F_H}{k}\right)^2 \tag{6.3}$$

$$T_{SS} = \frac{F_C T_C + F_H T_H}{F_C + F_H} \tag{6.4}$$

In Figure 6.1, the SS level is shown as a noisy (due to surface splashing) value, but a constant average value. To match SS models to data, you should only use SS data. For example, during the mid and late periods on the graph, the hot flow rate is the same but the level measurement is either low or high. Only the SS high level mechanistically matches the new flow rate. Accordingly, only data in the last SS period and first SS period should be used to determine the k coefficient value in the SS model of Equation 6.3.

As a corollary, the SS data sequence provides no useful information about the rate of change of the level or temperature, so data from the middle transient period is needed to determine the “A” coefficient value.

Since manufacturing and chemical processes are inherently nonstationary (attributes change in time), selected model parameter values need to be adjusted frequently to keep the models true to the process and functionally useful. Data from SS periods need to be isolated to adjust SS models. Data from TS periods need to be gathered to adjust TS models.

SS models are also used in process fault detection and data reconciliation.

This chapter presents an approach to determine SS and TS that I have found to be useful in chemical process data. However, other applications may have characteristics that will indicate that the approach proposed here is not the best. Chemical processes are characterized by time constants on the order of 1 second to 1 hour, they are multivariable (coupled and nonlinear), with mild and short-lived autocorrelation, and noise variance changes with operating conditions. Additionally, control computers are inexpensive and the operators have education typically at the associate degree level, both aspects requiring simplicity in algorithms. The approach presented here might not be right if mission criticality can afford powerful computers and highly educated operators or if rotating machinery creates a cyclic response as a background oscillation to the steady signal.

Identification of both SS and TS in noisy process signals is important. SS models are widely used in process control, online process analysis, and process optimization; since manufacturing and chemical processes are inherently nonstationary, selected model parameter values need to be adjusted frequently to keep the models true to the process and functionally useful. However, either the use or data-based adjustment of SS models should only be triggered when the process is at SS. Additionally, detection of SS triggers the collection of data for process fault detection, data reconciliation, neural network training, the end of an experimental trial (collect data and implement the next set of conditions), and so on.

In contrast, transient, time-dependent, or dynamic models are also used in control, forecasting, and scheduling applications. Dynamic models have coefficients representing time constants and delays, which should only be adjusted to fit data from transient conditions. Detection of TS triggers the collection of data for dynamic modeling.

6.1.2 Concepts and Issues in Detecting Steady State

If a process signal was noiseless, then SSID or transient state identification (TSID) would be trivial. At SS there is no change in data value. Alternately, if there is a change in data value the process is in a TS.

However, since process variables (PVs) are usually noisy, the identification needs to "see" through the noise and would announce probable SS or probable TS situations, as opposed to definitive SS or definitive TS situations. The method also needs to consider more than the most recent pair of samples to confidently make any statement.

Since the noise could be a consequence of autocorrelated trends (of infinite types), varying noise amplitude (including zero), individual spikes, non-Gaussian noise distributions, or spurious events, a useful technique also needs to be robust to such aspects.

6.1.3 Approaches and Issues to SSID and TSID

One straightforward implementation of a method for SSID would be a statistical test of the slope of a linear trend in the time series of a moving window of data. Here, at each sampling, use linear regression to determine the best linear trend line for the past N data points. If the process is at SS and the perturbations in the time series are independent (uncorrelated), then the trend-line slope will be, ideally, zero. However, because of the vagaries of process noise, the trend-line slope from a SS period will fluctuate with near-zero values; accordingly, a non-zero value is not a reason to reject the SS hypothesis. If the t-statistic for the slope (regression slope divided by standard error of the slope coefficient) exceeds the critical value, then there is sufficient evidence to confidently reject the SS hypothesis and claim it is probably in a TS.

Alternately, another straightforward approach is to evaluate the average value in successive data windows. Compute the average and standard deviation of the data in successive data sets of N samples and then compare the two averages with a t-test. If the process is at SS with independent perturbations, ideally the averages are equal, but noise will cause sequential averages to fluctuate. If the fluctuation is excessive relative to the inherent data variability, then the t-statistic (difference in average divided by standard error of the average) will exceed the critical t-value and the null hypothesis (process is at SS) can be confidently rejected to claim it is probably in a TS.

Note that both of these methods reject the null hypothesis, which is a useful indicator that the process is confidently in a TS. However, not rejecting SS does not permit a confident statement that the process is at SS. A legal judgment of "not guilty" is not the same as a declaration of "innocent." "Not guilty" means that there was not sufficient evidence to confidently claim "guilty" without a doubt. Accordingly, there needs to be a dual approach that can confidently reject TS to claim probably in SS, as well as rejecting SS to claim probable TS. This tutorial presents a dual approach.

Further, the conventional tests described above have a computational burden that does not make them practicable online, in real-time, within most process control computers. This chapter presents a computationally simple approach.

Further, neither approach just described would reject the null hypothesis (the linear trend is zero or the means are equal) if the data were oscillating and the peak or valley was nearly centered in the window of analysis. This chapter presents an approach that will not be confounded by oscillations.

This chapter develops, analyzes, and demonstrates a method for probable SS and probable TS identification that is robust to process events and is computationally simple. The approach has been extensively tested in lab- and pilot-scale processing units for both SSID and TSID, for the autonomous segregation of data, for model adjustment, and for the autonomous triggering of experimental plans. The method has also been applied to commercial-scale multivariable processes. Beyond its application to monitoring processes, the approach is regularly used to detect convergence in optimization and nonlinear regression.

Building on the discussion in Szela and Rhinehart (2003), there are 15 problems associated with automatically detecting SS and TS states in noisy processes. Any practicable method needs to address all of the issues.

1. Since processes produce noisy data, SSID must accommodate noisy data. When the process has settled to a final SS, the measurements do not settle at single values. Various sources of noise cause the measurements to vary, rather randomly. The average might settle in time, but the individual data keep changing, fluctuating about the time average. Statistical methods are required to identify SS or TS.
2. Statistical tests normally reject the null hypothesis. This means that normal tests would determine when there is a high probability that the process is not at SS. This leaves the other possibilities unidentified. The other possibilities are that the process is at SS, or that the process is still settling but not changing at a high enough rate to confidently claim "not at SS."
3. The process noise level can change with operating conditions. This means that the methods for statistical testing must adapt to the local/temporal process noise amplitude.
4. Processes are subject to external, uncontrolled, disturbances. Effects of these on the process response, when significant, must be identified as "not at SS."
5. Sometimes spurious, one-time and extreme, perturbations misrepresent the signal, perhaps from a missed data connection or a substituted variable due to the communication system. Effects of these must be recognized.
6. Processes are multivariable. Several variables must be monitored and the statistical tests must accommodate the increased uncertainty.
7. Related, cross-correlation between variables affects the statistical level of significance – if two variables are correlated, they are not independent, and only one needs to be included in the multivariable analysis.
8. Process state measurements do not immediately respond to the initiated change. If SS is recognized and the next set of conditions in a trial sequence is implemented, it would not be unexpected for the immediate next sampling to find the process measurement values at the same SS. This might cause the second-next conditions to be implemented too soon. A cyber operator, a virtual employee, an automaton, must wait for the effect of the new conditions to be observed, a transient, prior to looking for the subsequent SS.
9. The computational load that the method imposes on the computer system needs to be minimal, compatible with standard data acquisition and control products in use.

10. The noise distribution might not be Gaussian (normally) distributed. The method needs to be robust to the underlying noise distribution.
11. The noise might be autocorrelated. In autocorrelation, when one fluctuation is high (or low), what caused that event persists, influencing the subsequent data point to also be high (or low). Alternately, if there is a negative autocorrelation, a high data value could lead to a subsequent low value (and vice versa). Even if the data are Gaussian, they might not be independently distributed. Zero autocorrelation is a fundamental assumption in many methods and the implementation needs to ensure that the method will be insensitive or immune to any autocorrelation in the data fluctuations.
12. Process signals might be noiseless during certain periods. Perhaps the reading is zero because the valve is off, or perhaps because that value is the override to prevent an execution error, or perhaps the value is repeatedly identical because the process changes are within the discrimination ability of the transmitted signal. The method needs to cope with the extreme of zero noise.
13. Patterns in the trend may confuse certain techniques and make a not-at-SS event appear to be at SS. Consider an upward then downward trend centered in a window of analysis, fit to a linear trend line, and tested for a non-zero slope.
14. During a TS, the statistic needs to make large changes relative to its range of values at SS to clearly distinguish the two conditions. The statistic needs to be sensitive to the data pattern.
15. The method should be scale-independent so that when an operator changes variable units or setpoints the method still works.

A practicable method needs to be computationally simple, robust to the vagaries of process events, easily implemented, and easily interpreted.

6.2 Method

6.2.1 Conceptual Model

This method uses the R-statistic, a ratio of two variances, as measured on the same set of data by two methods. Figure 6.2 illustrates the concept (Huang, 2013). The dotted line represents the true trend of a process. The value starts at 15, ramps to a value of 10 at a time of 50, and then holds steady. The diamond markers about that trend represent the measured data. The dashed line, the true trend, is unknowable; only the measurements can be known and they are infected with noise-like fluctuations, masking the truth.

The method first calculates a filtered value of the process measurements, indicated by the curved line that lags behind the data. Then the variance in the data is measured by two methods. The deviation indicated by d_2 in the upper left of the figure is the difference between measurement and the filtered trend. The deviation indicated by d_1 in the lower right is the difference between sequential data measurements.

If the process is at SS, as illustrated in the 80–100 time period, X_f is almost in the middle of the data. Then a process variance, σ^2, estimated by d_2 will ideally be equal to σ^2 estimated by d_1 and the ratio of the variances, $R = {\sigma^2}_{d_2}/{\sigma^2}_{d_1}$ will be approximately equal to unity, $R = {\sigma^2}_{d_2}/{\sigma^2}_{d_1} \cong 1$. Alternately, if the process is in a TS, then X_f is not in the middle of the data, the filtered value lags behind, the variance as measured by d_2 will be much larger than

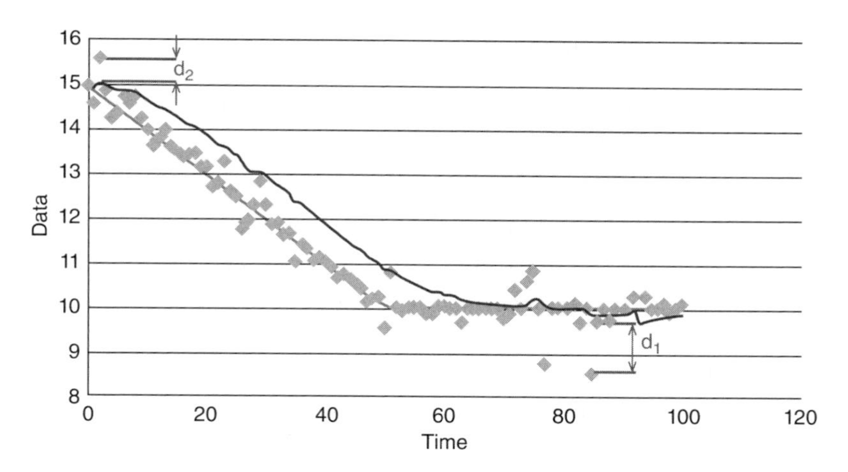

Figure 6.2 Illustration actual process (dashed line), noisy measurements (diamond markers), filtered data (solid line), and deviations

the variance as estimated by d_1, $\sigma^2_{d_2} \gg \sigma^2_{d_1}$, and the ratio will be much greater than unity, $R = \sigma^2_{d_2}/\sigma^2_{d_1} \gg 1$.

6.2.2 *Equations*

To minimize the computational burden, in this method a filtered value (not an average) provides an estimate of the data mean:

$$X_{f,i} = \lambda_1 X_i + (1 - \lambda_1) X_{f,i-1} \tag{6.5}$$

X = The process variable
X_f = Filtered value of X
λ_1 = Filter factor
i = Time sampling index

The first method to obtain a measure of the variance uses an exponentially weighted moving "variance" (another first-order filter) based on the difference between the data and the "average":

$$\upsilon^2_{f,i} = \lambda_2 (X_i - X_{f,i-1})^2 + (1 - \lambda_2) \upsilon^2_{f,i-1} \tag{6.6}$$

$\upsilon^2_{f,\,i}$ = Filtered value of a measure of variance based on differences between data and filtered values
$\upsilon^2_{f,\,i-1}$ = Previous filtered value

Equation 6.6 is a measure of the variance to be used in the numerator or the ratio statistic. The previous value of the filtered measurement is used instead of the most recently updated value to prevent autocorrelation from biasing the variance estimate, $v^2_{f,i}$, keeping the equation for the ratio simple.

The second method to obtain a measure of variance is an exponentially weighted moving "variance" (another filter) based on sequential data differences:

$$\delta^2_{f,i} = \lambda_3(X_i - X_{i-1})^2 + (1 - \lambda_3)\delta^2_{f,i-1} \tag{6.7}$$

$\delta^2_{f,i}$ = Filtered value of a measure of variance
$\delta^2_{f,i-1}$ = Previous filtered value

This will be the denominator measure of the variance.

The ratio of variances, the R-statistic, may now be computed by the following simple equation:

$$R = \frac{(2 - \lambda_1)v^2_{f,i}}{\delta^2_{f,i}} \tag{6.8}$$

The calculated value is to be compared to its critical values to determine SS or TS. Neither Equation 6.6 nor Equation 6.7 compute the variance. They compute a measure of the variance. Accordingly, the $(2 - \lambda_1)$ coefficient in Equation 6.8 is required to scale the ratio, to represent the variance ratio. Complete executable code, including initializations, is presented in sub RandomSubsetSS of Appendix B. The essential assignment statements for Equations 6.5 to 6.8 are:

```
nu2f := L2 * (measurement - xf) ^ 2 + cL2 * nu2f
xf := L1 * measurement + cL1 * xf
delta2f := L3 * (measurement - measurement_old) ^ 2 + cL3 * delta2f
measurement_old := measurement
R_Filter := (2 - L1) * nu2f / delta2f
```

The coefficients L1, L2, and L3 represent the lambda values, and the coefficients cL1, cL2, and cL3 represent the complementary values.

The five computational lines of code of this method require direct, no-logic, low storage, and low computational operation calculations. In total there are four variables and seven coefficients to be stored, ten multiplication or divisions, five additions, and two logical comparisons per observed variable.

Being a ratio of variances the statistic is scaled by the inherent noise level in the data. It is also independent of the dimensions chosen for the variable.

If the process is at SS then the R-statistic will have a distribution of values near unity. Alternately, if the process is not at SS then the filtered value will lag behind the data, making the numerator term larger than the denominator, and the ratio will be larger than unity.

6.2.3 *Coefficient, Threshold, and Sample Frequency Values*

For simplicity and for balancing speed of response with surety of decision and robustness to noiseless periods, use filter values of $\lambda_1 = \lambda_2 = 0.1$ and $\lambda_3 = 0.05$. However, other users have recommended alternate values to optimize the speed of response and Type I and Type II errors. The method works well for diverse combinations of filter coefficient values within the range of 0.05–0.2.

If R-calculated >2.5, "reject" SS with fairly high confidence and accept that the process is in a TS. Alternately, if R-calculated <0.9, "accept" that the process is at SS and reject that it is in a TS with fairly high confidence. If in-between values are given for R-calculated, hold the prior SS or TS state, because there is no confidence in changing the declaration.

The filter factors in Equations 6.1 to 6.3 can be related to the number of data (the length of the time window) effectively influencing the average or variance calculation. Based on a first-order decay, roughly the number of data in the window of observation is about $3.5/\lambda$.

Larger λ values mean that fewer data are involved in the analysis, which has a benefit of reducing the time for the identifier to catch up to a process change, reducing the average run length (ARL). However, larger λ values has an undesired impact of increasing the variability on the statistic, confounding interpretation. The reverse is true: lower λ values undesirably increase the ARL to detection, but increase precision (minimizing statistical errors).

A faster sampling frequency will speed up the recognition, but autocorrelation is an issue.

The basis for this method presumes that there is no autocorrelation in the time series process data when at SS. Autocorrelation means that if a measurement is high (or low) the subsequent measurement will be related to it. For example, if a real process event causes a temperature measurement to be a bit high and the event has persistence, then the next measurement will also be influenced by the persisting event and will also be a bit high. Autocorrelation could be related to control action, thermal inertia, noise filters in sensors, and so on. Autocorrelation would tend to make all R-statistic distributions shift to the right, requiring a reinterpretation of critical values for each PV.

It is more convenient to choose a sampling interval that eliminates autocorrelation than to model and compensate for autocorrelation in the test statistic. A plot of the current process measurement versus the previous sampling of the process measurement over a sufficiently long period of time (equaling several time constants) at SS is required to establish the presence/absence of autocorrelation. To detect autocorrelation, visually choose a segment of data that is at SS, and plot the PV value versus its prior value.

Figure 6.3 plots data with a lag of one sample (a measurement versus the prior measurement) and shows autocorrelation. Figure 6.4 plots the same data but with a lag of five samples and shows zero autocorrelation.

Generally, hydraulics (flow rates and levels) come to SS faster than gas pressure in large volumes, and faster than temperature or composition. Further, noise in flow rate and level measurements will have little persistence; however, thermal inertia of temperature sensors may extend autocorrelation. Therefore, a sampling interval for one might not be what is needed for another. The user might want to separate the process attributes into a set of hydraulic variables, another set of thermal and composition inventory (T, P, x) variables, and monitor the hydraulic state of the process and the thermal and composition states separately.

Summarizing: use $\lambda_1 = \lambda_2 = 0.1$ and $\lambda_3 = 0.05$, use R-critical = 2.5 to reject SS (accept TS), and use R-critical = 0.9 to accept SS (reject TS). Choose a sampling interval to eliminate autocorrelation from a visually determined SS period.

Figure 6.5 illustrates the method. The PV is connected to the left-hand vertical axis (log10-scale) and is graphed with respect to the sample interval. Initially it is at an SS with a value of about 5. At a sample number 200, the PV begins a first-order rise to a value of

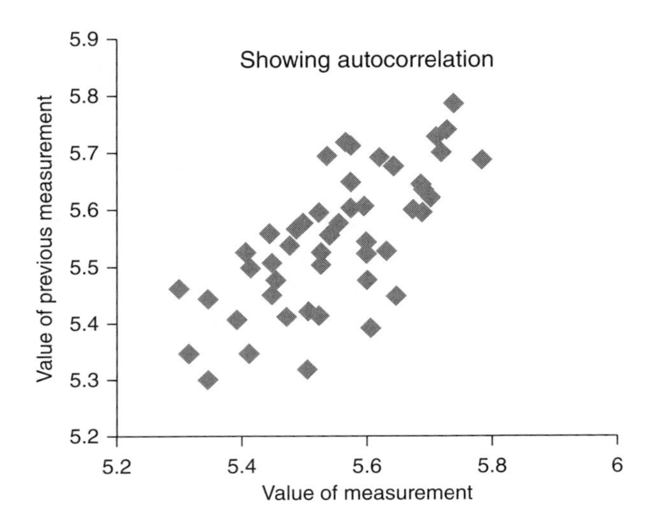

Figure 6.3 Data showing autocorrelation

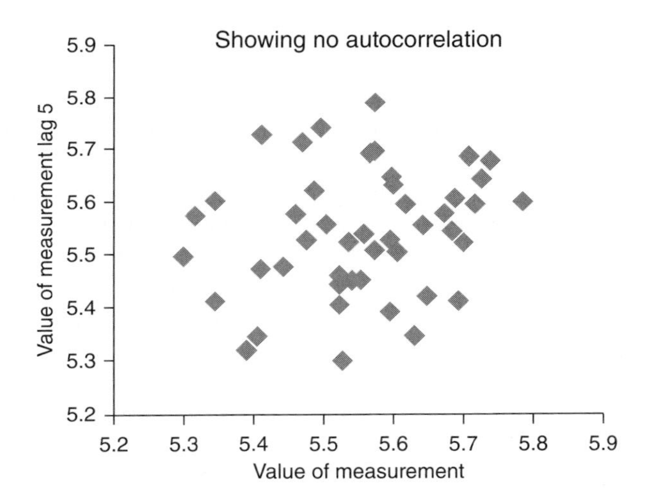

Figure 6.4 Data showing no autocorrelation when the interval is five samplings

about 36. At a sample number 700, the PV makes a step rise to a value of about 36. The ratio-statistic is attached to the same left-hand axis and shows an initial kick to a high value as variables are initialized, before relaxing to a value that wanders about the unity SS value. When the PV changes at sample 200, the R-statistic value jumps up to values ranging between 4 and 11, which relaxes back to the unity value as the trend hits a steady value at a time of 500. Then when the small PV step occurs at sample 700, the R-value jumps up to about 4

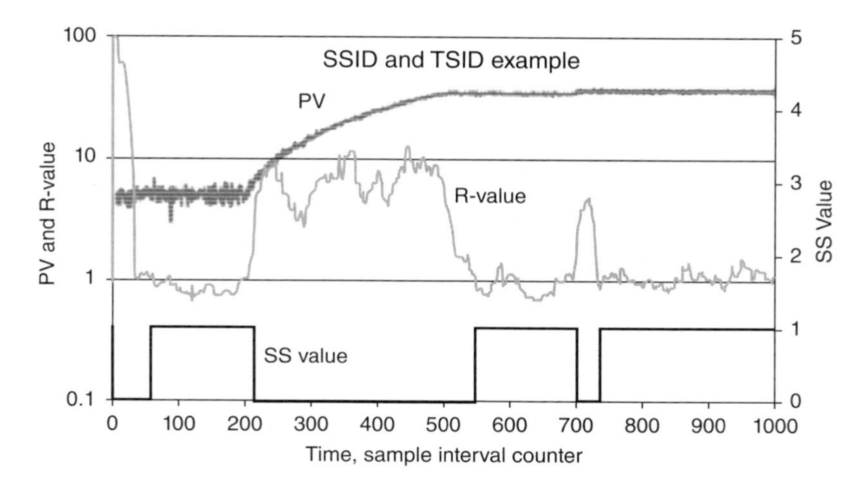

Figure 6.5 Example

and then decays back to its nominal unity range. The SS value is connected to the right-hand vertical axis and has values of either 0 or 1 that change when the R value exceeds the two limits of $R_{\beta,TS}$ and $R_{1-\alpha,SS}$.

6.2.4 Noiseless Data

As a final aspect of the method development, recognize that this statistical method only accommodates noisy data. If the data are noiseless and at SS, the data series will have exactly the same values, which would drive the denominator measure of variance to zero, eventually leading to an execution error in Equation 6.8. A noiseless signal could arise from several situations: the flow rate is off, the device is broken, the signal is being overridden, the signal is not changing as much as the discretization error of the device, and so on. Here the reported value is constant in time. To avoid this condition, add a normally and independently distributed (NID) noise signal of zero mean and small variance NID(0, σ) to the original data.

The Box–Muller method is recommended for adding Gaussian distributed independent noise to a simulated signal:

$$x_{measured,i} = x_{true,i} + w_i \tag{6.9}$$

$$w_i = \sigma\left(\sqrt{-2\ \ln(r_{1,i})}\sin(2\pi r_{2,i})\right) \tag{6.10}$$

where w_i is an NID "white noise" perturbation of zero mean and standard deviation σ variation that is added to the true value to mimic the measurement. The variables $r_{1,i}$ and $r_{2,i}$ represent independent random numbers uniformly distributed in the interval 0 to 1.

Bhat and Saraf (2004) also added artificial noise, but to mask autocorrelation.

6.3 Applications

6.3.1 Applications of the R-Statistic Approach for Process Monitoring

The original article about the technique presented in this chapter was authored by Cao and Rhinehart (1995). Subsequently, Cao and Rhinehart (1997b) and Shrowti, Vilankar, and Rhinehart (2010) presented critical values for Type I and Type II errors, respectively.

The author has been involved in many pilot-scale and commercial-scale applications. These include Brown and Rhinehart (2000) who demonstrated a multivariable version of the technique on a pilot-scale distillation process. Katterhenry and Rhinehart (2001) then used it to automatically sequence experimental transitions on the distillation process and Szela and Rhinehart (2003) demonstrated its application to automate transitions between experimental tests on a two-phase flow unit. Huang and Rhinehart (2013) reported on an application in monitoring the flow rate in a pilot-scale gas absorption unit, and Vennavelli and Resetarits (2013) reported on its application in a commercial-scale multivariable distillation unit.

Iyer and Rhinehart (2000) and Padmanabhan and Rhinehart (2005) used the approach to stop training a neural network and as the termination criterion for nonlinear regression. This will be developed in Chapter 12. The author is presently using it as the convergence criterion in optimization of stochastic models (Rhinehart, 2014). Rhinehart, Su, and Manimegalai-Sridhar (2012) used SSID as the stopping criterion for a novel optimizer.

Personal communications from industry revealed its application to triggering reactor analysis and selecting steady and transient data for neural network modeling of a process.

Other users from the journal literature include Bhat and Saraf (2004) who applied it to trigger data reconciliation on a commercial scale process unit. Jeison and van Lier (2006) used it to trigger an online analysis of transmembrane pressure in a bioreactor. Salsbury and Singhal (2006) included the technique in a patented method for evaluating control system performance. Carlberg and Feord (1997) addressed the application in model adjustment of a commercial scale reactor. Mansour and Ellis (2007) included the technique within a reactor optimization. Kuehl and Horch (2005) used the technique to detect sluggish control loops. Zhang (2001) used the technique to segregate data sections for neural network models and Ye *et al.* (2009) used it to trigger safety assessment in multimode processes.

6.3.2 Applications of the R-Statistic Approach for Determining Regression Convergence

Nonlinear, least-squares optimization is commonly used to determine model parameter values that best fit empirical data by minimizing the sum of squared deviations (SSD) of data-to-model. Nonlinear optimization proceeds in successive iterations as the search progressively seeks the optimum values for model coefficients. As the optimum is approached, the optimization procedure needs a criterion to stop the iterations. However, the current stop-optimization criteria of thresholds on either the objective function, changes in objective function, change in decision variable, or number of iterations require *a priori* knowledge of the appropriate values. They are scale-dependent, application-dependent, starting point-dependent, and optimization algorithm-dependent; correct choices require human supervision.

Iyer and Rhinehart (2000), Padmanabhan and Rhinehart (2005), Ratakonda *et al.* (2012), and Rhinehart, Su, and Manimegalai-Sridhar (2012) report on the use of this method of SSID to stop optimizer iterations when there is no statistical evidence of SSD improvement relative to the variation in the data.

An observer of an optimization procedure for empirical data will note that the SSD between the data and the model drops to an asymptotic minimum with progressive optimization iterations. The novelty is to calculate the SSD of a random subset of data (a unique, randomly selected subset at each iteration). The random subset sum of squared deviations (RS SSD) will appear as a noisy signal relaxing to its noisy "SS" value as iterations progress. At "steady state," when convergence is achieved, and there is no further improvement in the SSD, the RS SSD value is an independently distributed variable. When RS SSD is confidently at SS, optimization should be stopped. Since the test looks at the signal-to-noise ratio, it is scale-independent and "right" for any particular application.

Although the stopping criterion is based on the random and independent subset of data, the optimization procedure still uses the entire data set to direct changes in the decision variable (model parameter) values.

Figure 6.6 illustrates the procedure. The dashed line represents the root mean sum of squared deviations (rms) value of the least-squares objective function as it approaches its ultimate minimum with optimizer iterations. The dots represent the rms value of the randomly selected fraction of data (15 out of 30 data sets) and the thicker curved line that approaches the dashed line, and lags behind the data, represents the filtered random-subset rms value. The SSID procedure identified SS at iteration number 118 and stopped the optimization procedure. The graph indicates that in iterations beyond 100, the dashed line is effectively unchanging w.r.t. the variability in the data. The variability indicates both data variability and the process model mismatch. In either case, the optimizer was not making substantive improvement in the model.

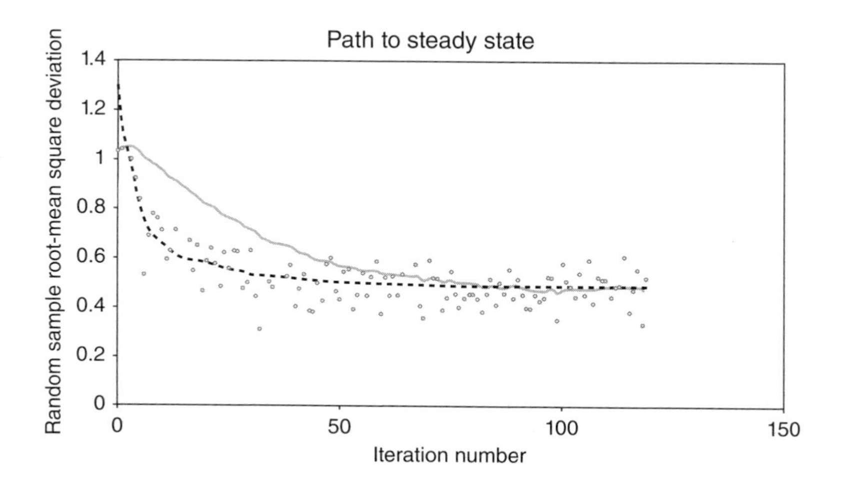

Figure 6.6 SSID application to nonlinear regression

6.4 Takeaway

This chapter presents a simple method to automate detection of probable SS and probable TS conditions. The method is scale-independent, extends to multivariable systems, and is independent of the noise level of distribution. The method has been demonstrated on a variety of process and optimization applications. The user needs to choose a sampling interval to eliminate autocorrelation.

The user needs to select variables to be monitored and criteria should be that observed variables need to be the minimum number that observes all aspects of the process.

ARL, Type I error, and Type II error need to be considered in any SSID TSID automation. With recommended lambda values of 0.1, 0.1, and 0.05, and upper and lower critical values of 2.5 and 0.9, the technique detects a TS with a signal-to-noise ratio of 3 within a few samples. Type I and Type II errors are less than about 0.5%. After a transient, the filtered values need to relax back to SS values. For large steps to a new SS value, the ARL to detect SS is about 35–60 samples.

An automated procedure to identify SS and TS removes human bias, but no automated process is 100% perfect. Perform a visual check of the data trends.

Exercises

6.1 Demonstrate the SSID, TSID approach. Perhaps use a spreadsheet. Create a column for time (or sampling interval, or sample counter). Create an adjacent column for a signal that might be observed in time. Start it with a desired value of the signal and hold that value for a desired period. Then have it change to a new value in a pattern that you would like (step, ramp, first-order, etc.) and then hold for a second steady state. A plot of signal w.r.t. time should reveal the pattern you specified. Next add noise to the signal. I recommend the Box–Muller method of Equation 6.10. Choose a value for the noise standard deviation that is about one-tenth of the change in signal so that you have a good signal-to-noise ratio for the analysis. The plot should reveal the same change in the signal, but slightly confounded by the added noise. In adjacent columns calculate the filtered signal and variance measures of Equations 6.5 to 6.7, then the R-statistic, and then a 0 or 1 signal representing confident SS or confident TS. The result should be a graph similar to that in Figure 6.5.

6.2 Repeat Exercise 1, but change the noise amplitude and signal patterns. The SSID TSID method should work for a wider variety of conditions.

6.3 Repeat Exercise 1, but change the lambda values and threshold values. The SSID TSID method should be robust to a range of values.

6.4 Repeat Exercise 1, but investigate Type I errors. Hold the signal at a noisy steady state for a long time. A Type I error is the probability that the identifier indicates TS, the number of wrong determinations over all time. You many need to have 100 000 samplings in order to have a definitive finding, so it might be simpler to run the simulation in structured code.

6.5 Repeat Exercise 1, but investigate Type II errors. After a noisy steady-state period, make a change in the signal. If the change is small relative to the noise amplitude, then the identifier might not detect the change. A Type II error is the probability that the identifier indicates SS when it should have claimed TS, the number of wrong determinations over all time. You many need to have 100 000 samplings in order to have a definitive finding, so it might be simpler to run the simulation in structured code.

6.6 Show that the ARL for the method is about $3.5/\lambda$. The first-order filter of Equation 6.5 is alternately called an autoregressive moving average (ARMA) model, or time series model, and represents a dynamic process. Start with a first-order linear ordinary differential equation (ODE) and derive its response to a step-and-hold change. Also, use Euler's explicit method (replace the differential with a forward finite difference) to convert the ODE into its ARMA model. Equate coefficients to relate lambda to tau, $\lambda = 1 - e^{-t/\tau}$. With the simulation time interval small relative to the time constant, a necessity for Euler's method to be valid, $\lambda \cong \Delta t/\tau$. Since it takes a time of about three or four time constants for a first-order ODE response to settle, then the time to recover SS from a step-and-hold signal, the settling time, the average run length, and the number of samples is about $3.5/\lambda$.

6.7 The Type I error is the probability of claiming TS when actually in an SS. What is the Type I error if applied to a multivariable process? Here SS is accepted when all variables are at SS and TS is accepted if any one variable is at TS. Show that, if the Type I error for any one variable is α, the Type I error for the multivariable process is $\alpha_N = 1 - (1 - \alpha_1)^N$.

Hint: the process is claimed at SS if variable 1 is at SS AND variable 2 is at SS AND variable 3 is at SS, AND.... In probability the AND conjunction directs multiplying individual probabilities.

Part III

Regression, Validation, Design

7

Regression Target – Objective Function

7.1 Introduction

Most engineering or science models provide a continuous-valued, deterministic response. These could either represent steady-state or transient phenomena. In contrast, models could predict a classification (nominal, class, text, or string variable) or a rank, but still a deterministic value. Alternately, Monte Carlo simulations predict a stochastic outcome, a range of possibilities, not a definitive value. There are diverse options to what you may be seeking to best fit, and you need to understand the application to choose the regression target.

Further, most often in regression, the error model, the concept for experimental variation, assumes that variation is exclusively allocated to the response (dependent variable, measurement). This is common, usually fully functional, but not fundamental or necessary.

Most of the techniques for regression are derivatives of maximum likelihood for steady-state continuous-valued, deterministic models. This chapter is grounded in those techniques, reveals how to accommodate uncertainty in the independent variables, and migrates from continuous-valued variables to reveal objective functions appropriate for classification. The fundamental concept is uncertainty.

7.2 Experimental and Measurement Uncertainty – Static and Continuous Valued

We obtain data from experiments that reveal how Nature behaves. The data represents values that are measured. Unfortunately, the measurement process can introduce both systematic and random errors. Consideration of the sources of variability leads to the derivation of the classic least-squares regression optimization objective.

Nonlinear Regression Modeling for Engineering Applications: Modeling, Model Validation, and Enabling Design of Experiments, First Edition. R. Russell Rhinehart.

This section is based on the variables of the model (both influence and response) being of the rational or real category (continuous valued) or when the discretization interval is very small relative to the range so that the variable is effectively continuous-valued. This section is also based on steady-state models.

Systematic error is often called bias or calibration error. Systematic error might be a constant offset. For example, if the pointer on a gage is slightly bent, the reading will always be off by a constant amount. Additionally, systematic error might change in time as a device warms up. Further, systematic error might change with state if a ratio, or slope, in the calibration equation is a bit off. In any case, if we know the character of the systematic error in the measurement, we can correct the measurement. The lesser the systematic error the more *accurate* is the data. Accuracy deals with the systematic, repeatable, consistent deviation from the truth, which is often measured by the average deviation from true.

In contrast, some errors are due to irreproducibility of exact experimental conditions and/or noise. For example, these could be related to electrical interference or mechanical vibration and/or uncontrolled random disturbances, which can be internal to the process (for instance, imperfect mixing, turbulence) or external from the environment (for instance, barometric pressure or electromagnetic noise on transmission). These sources of irreproducibility randomly (independently) vary from one experiment to another and lead to a distribution of measurement values, even from ostensibly the same experiment. Measurement values that have very small deviations when replicated, data that are repeatable, are termed *precise* data. Precision deals with variability and is often measured by the variance (or standard deviation) of replicate measurements.

Normally, when there are many independent, random, influences with an equivalent impact on the measurement, the resulting measurement perturbations will be normally, or Gaussian, distributed. Assuming this distribution, if there is no systematic error, the average of many repeated measurements will approach the true mean and the standard deviation of the data will approach the true population standard deviation.

Random errors, often termed noise, fluctuations, or experimental data vagaries, are those reflecting independent perturbations (deviations, errors, fluctuations) on measurements that otherwise should provide repeatable values from a process that is ostensibly at unchanging conditions (at steady state).

If the process is held at exactly reproducible experimental conditions, then the variability from repeated measurements would reflect all of the influences except for experimental control. However, experiments cannot be exactly controlled. Consider preparing a sample of 10 g of "A" and 10 g of "B" to get a 50/50 mixture as the input to a reaction. Because there is a reading error on the scales, one cannot exactly weigh something. One cannot exactly prepare a 50/50 mixture. In general, one cannot exactly, reproducibly, run experiments. One cannot replicate cooking something at an exactly known temperature, replicate driving something at an exactly known speed, and so on. As a result of variation in the input variables (the givens, the independent variable values) in repeated experiments, the measurement will reflect the variability of the experimental control (the experimental input, condition, or x value) as well as what is often called *measurement uncertainty* on the y value (the response, output, dependent variable).

Figure 5.1 revealed the concepts. The true value of x is input to the process and the true value of y is the result. However, the experimenter must attempt to set the experimental conditions,

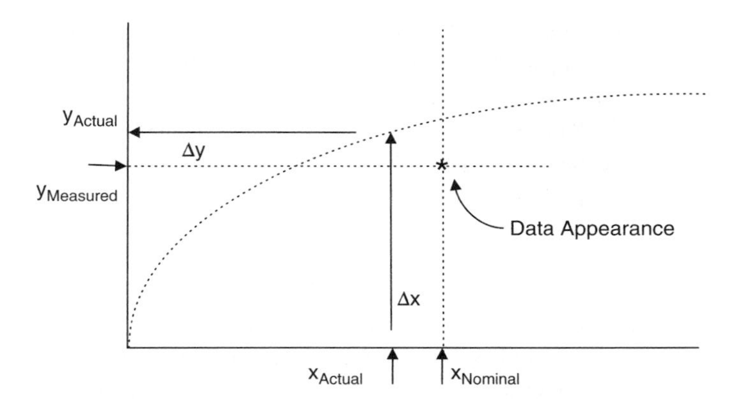

Figure 7.1 Illustration of two types of experimental uncertainty

but x-nominal, the experimenter's attempt, is not exactly what influences the process. Then the experimenter seeks to measure y, but finds y-measured, not y-true.

Figure 7.1 illustrates the impact of two sources of variability on a dependent variable, y. One source is due to uncertainty on the independent variable x (or experimental condition) and the other is y-measurement noise. The dotted curve represents the true y versus x relation, which is dotted to indicate that Nature keeps the exact relation unknowable to humans. The indication, x_{Nominal} refers to the designed (attempted and recorded) experimental input condition. The indication, Δx, represents the unknown experimental deviation from the nominal condition and x_{Actual} indicates the actual but unknowable experimental condition. The process response to the true value of x is y_{Actual}; however, measurement variability (disturbance, noise, etc.) causes the measurement to deviate as indicated by Δy, resulting in the reported value, y_{Measured}. We think y_{Measured} is the process response and, further, we think that it occurred at x_{Nominal}. We were snookered by Nature. The data point, our indication of the location of the unknowable curve, appears to be as indicated in Figure 7.1.

To quantify the impact of the two types of random uncertainty, start considering the true (but unknowable) relation, where the T subscript indicates "true":

$$y_T = f_T(x_T) \tag{7.1}$$

Since x_{True} is the result of $x_{Nominal}$ and the unknown Δx:

$$y_T = f_T(x_{Nominal} + \Delta x) \tag{7.2}$$

and because $y_{Measured}$ is influenced by the unknown Δy:

$$y_{Measured} = f_T(x_{Nominal} + \Delta x) + \Delta y \tag{7.3}$$

Propagation of variance, with x and y fluctuations independent reveals

$$\sigma^2_{y-Meas} = \left(\frac{\partial f}{\partial x}\right)^2 \sigma_x^2 + \sigma_y^2 \tag{7.4}$$

Variability on the measurement is the combination of the inherent measurement variability, σ_y^2, and the apparent impact of x-variability, the product of y-sensitivity to x times the experimental input variability, $(\partial f/\partial x)^2\sigma_x^2$.

You could measure the value of σ_{y-Meas}^2 by replicate experiments and estimate the value of σ_x^2 from analysis of the experimental procedure. With a model of $y_T = f_T(x_T)$, you could then rearrange Equation 7.4 to provide an estimate of σ_y^2:

$$\sigma_y^2 = \sigma_{y-Meas}^2 - \left(\frac{\partial f}{\partial x}\right)^2 \sigma_x^2 \tag{7.5}$$

The convention in least-squares regression, however, is to pretend that there is no uncertainty in the input variables, $\sigma_x^2 = 0$, and assign all of the experimental uncertainty to the measured value, $\sigma_{y-Meas}^2 = \sigma_y^2$. This makes the regression procedure and subsequent analysis of variance mathematically convenient. In my opinion, although this convenience is often justifiable, it is neither right nor necessary.

7.3 Likelihood

The concept of maximum likelihood provides a basis for the sum of squared deviations (SSD) objective function in regression and also reveals a method to accommodate both x- and y-variability in determining best model coefficient values.

The analysis that follows is predicated on the assumption that the model contains the functionality of Nature (that the model can be a true representation of Nature with the right coefficient values), and that the deviation from the model is normally distributed experimental noise.

Consider that a data measurement is subject to many, small, random, independent sources of deviation. In this case the distribution of values is Gaussian (normally and independently distributed) (NID) with a mean and variance of NID(μ_y, ${\sigma_y}^2$). The probability density function for a measurement y is

$$PDF(y) = \frac{1}{\sigma_y\sqrt{2\pi}} e^{-1/2((y-\mu_y)/\sigma_y)^2} \tag{7.6}$$

This Gaussian model of data variability is often a realistic approximation to the experimental data distribution, but it is not a universal truth. Variability in some processes is exponentially distributed, others are log-normal, others are chi-square.

If there is only one measurement, the best value for μ is the measured value, the average of the one data point. Then Equation 7.6 provides the probability density function representing the probability that the measurement could have some other value of y. Likelihood will be defined as the probability that the data point is within some small interval, Δy, of a particular y value.

The probability that a point could be in the interval from $y = a$ to $y = b$,

$$L(y) = p(a \le y \le b) = \int_a^b PDF(y)dy \tag{7.7}$$

For a small range, $b - a = \Delta y$, in which the PDF(y) can be assumed relatively constant, the probability can be estimated by the rectangle rule of integration

$$L(y) = p(y\ in\ interval\ y - \Delta y/2\ to\ y + \Delta y/2) = \int_{y-\Delta y/2}^{y+\Delta y/2} PDF(y)dy$$
$$\cong PDF(y)\Delta y \tag{7.8}$$

If the interval, Δy, is a constant, then $p(y$ in a Δy interval) is proportional to the value of pdf(y). This is the likelihood and, since in optimization a positive-valued constant scalar does not change the solution, the likelihood is often simply equated to the PDF: $L(y) = PDF(y)$.

Figure 7.2 illustrates the $PDF(y)$ w.r.t. y, given values for μ_y and σ_y and the likelihood as the probability a value is within a small y interval.

The situation is similar for the experimental input, x. Then the PDF of a data pair (x_1, y_1), for which deviations on x are independent of deviations on y, having a different x-value *and* a different y-value, is the product of the individual PDFs:

$$p(x\ in\ \Delta x\ and\ y\ in\ \Delta y) = L(x, y) \cong PDF(x, y) = \frac{1}{\sigma_x \sigma_y 2\pi} e^{-1/2((x-\mu_x)/\sigma_x)^2 - 1/2((y-\mu_y)/\sigma_y)^2} \tag{7.9}$$

Figure 7.3 illustrates three measured (x, y) data pairs with ellipsoids representing Equation 7.9 $PDF(x, y)$ contours. The measured data points are at the center of the nested ellipses and the ellipses represent the likelihood contours. The ellipse contours are progressively farther apart as proximity to the data point increases, reflecting the shape of the "hill" in Figure 7.3. The closer a contour is to the center, the higher is the likelihood value of the contour.

The y w.r.t. x curve in Figure 7.3 represents a steady-state model $y = f(x)$. It represents the model, not the unknowable truth about Nature, and its location depends on model coefficient values. Since the curve represents the model, it is drawn as a continuous line, not the unknowable dotted line representing the truth about Nature, a convention used throughout this textbook. The best model is the one that maximizes the likelihood that the model represents the process that would have produced data pair 1 *and* data pair 2 *and* data pair 3.

The points on the curve that maximize likelihood for each data pair are illustrated with an "X." Wherever you place the "X" on the illustration, it is on a likelihood contour. Not every contour is drawn, but any likelihood value has a contour.

Here is an argument that proves that the maximum likelihood contour is tangent to the model curve. Place your own "X" on the model curve, either to the left or right of mine. Keeping your

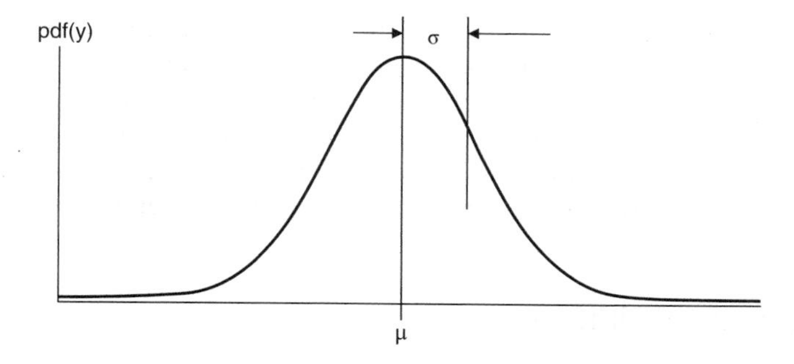

Figure 7.2 Illustration of the Gaussian distribution

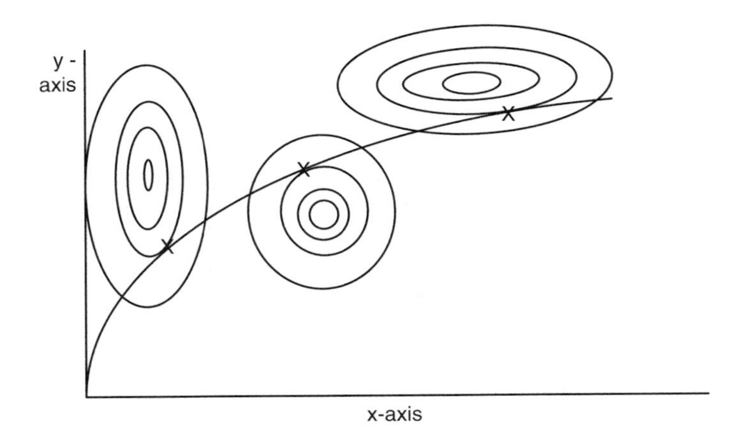

Figure 7.3 Illustration of maximum likelihood in regression

"X" on the model curve, slide your "X" so that it touches a higher-valued likelihood contour. If that contour crosses the curve, then moving your "X" a bit further places it on a contour with a higher likelihood value. At some point the tiniest slide of your "X" in either direction moves it to a lower contour. At that point the maximum likelihood contour "kisses" the model curve and the likelihood contour must be tangent to the model. If not tangent, it crosses the model. If it crosses the model, there is a direction to move your "X" that places it on a higher likelihood contour. The maximum likelihood contour is tangent to the model.

There are some key observations from Figure 7.3. First, the orientation of the nested ellipses are not the same. If $\sigma_x > \sigma_y$ the ellipse will be horizontal, as the far right data point illustrates. Alternately, if $\sigma_x < \sigma_y$ the ellipse will be vertical, as the far left data point illustrates. Variance on x and y may both change with x or y value (the experiment may be heteroscedastic).

Second, the point of maximum likelihood, indicated by the symbol "X," is not vertically below the center of the ellipses (the appearance of the x value of the data location). The distance from the data point to the curve that maximizes likelihood is not the y-distance (vertical distance) from the curve to the data location.

The model location that maximizes likelihood, the "X" locations, will be indicated by $\tilde{x}_i^*$ and $\tilde{y}_i^*$, differentiating them from the data pairs x_i and y_i and the modeled value at the ith data x value x_i and $\tilde{y}(x_i) = \tilde{y}_i$.

Likelihood can be interpreted in two equivalent ways, as illustrated in one dimension in Figure 7.4. The ith data point is located at y_i and the modeled value at $\tilde{y}_i = \tilde{y}(x_i)$. The PDFs are illustrated for each. Because the Gaussian PDF is symmetric, the likelihood value is identical for the $\tilde{y}_i$ point relative to the $PDF(y)$ centered on the ith y value and for the y_i point relative to the $PDF(\tilde{y})$ centered on $\tilde{y}$ at the ith x value. Therefore, it does not matter whether the likelihood is calculated by the data difference from the model or by the model difference from the data.

7.4 Maximum Likelihood

The likelihood of the curve representing all three data sets is the joint probability. The *and* conjunction in the prior statement, "*The best model is the one that maximizes the likelihood that*

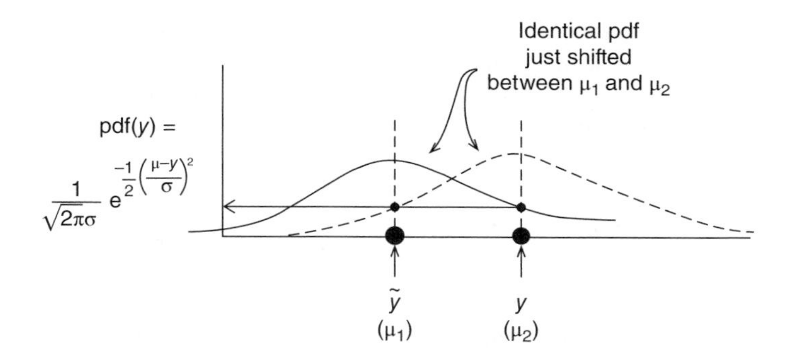

Figure 7.4 Equivalence of likelihood interpretation

it represents the process that would have produced data pair 1 and data pair 2 and data pair 3" indicates that the joint probability is the product of Equation 7.9 for each of the three data pairs (if all x and y perturbations are independent of all others and the distributions are Gaussian). We wish to find model coefficient values that maximize the likelihood, L, that the model represents the process that generated the data. Assuming that deviations are Gaussian distributed and that each data set has independent deviations, then the AND conjunctions describe a product for the total likelihood:

$$L = \prod_{i=1}^{3} PDF(x_i, y_i) \tag{7.10}$$

Since all of the coefficients (Δx, Δy, π, ...) are positive-valued constants and the exponential function is monotonic, and the sign on the argument is negative, finding the model coefficient values to maximize likelihood is the same as minimizing the exponential argument. Generalized to N (not 3) data pairs, the optimization statement is

$$\min_{\{\underline{p}\}} J = \sum_{i=1}^{N} \left[\left(\frac{\tilde{x}_i^* - x_i}{\sigma_{xi}} \right)^2 + \left(\frac{\tilde{y}_i^* - y_i}{\sigma_{yi}} \right)^2 \right] \tag{7.11}$$

where $\underline{p}$ is the vector of model coefficient values, the decision variables (DVs) in the optimization, N is the number of data sets, i is the data set index, and $\tilde{x}_i^*$ and $\tilde{y}_i^*$ represent points on the model (curve) that are closest to the ith data point (x_i, y_i) when scaled by the respective standard deviations (the X-marks-the-spot points on Figure 7.3).

Note that the deviation between the model and data point in Equation 7.11 is a scaled distance, scaled by the standard deviation of the respective variable. The ratios in parenthesis are dimensionless and describe the number of standard deviations that the experimental data point is from the closest point to it on the model curve. Further, note that the term in brackets is the sum of squared scaled distances. Accepting that variation for each variable is independent from the causes of variation on the others, the scaled distances are orthogonal and the sum of the squared distances is the square of the hypotenuse of a right triangle.

Although not explicitly apparent, the optimization objective function in Equation 7.11 directs a two-stage optimization. The outer, primary stage, searches for the DV values that represent the model coefficient values $\underline{p}$. However, for each trial solution (each new guess of

$\underline{p}$ values), there must be an inner search to find the model $(\tilde{x}_i^*, \tilde{y}_i^*)$ set, the point on the model curve that is closest to the scaled data point, (x_i, y_i), for each of the i data pairs when the distance is scaled by the standard deviation. Choose a $\underline{p}$ trial solution and then, for each data point, find the model point that is closest to the data point in terms of the variance-scaled SSD.

The inside optimization is

$$\underset{\{\tilde{x}_i^*,\, \tilde{y}_i^*\}}{Min} J_i = \left[\left(\frac{\tilde{x}_i^* - x_i}{\sigma_{xi}} \right)^2 + \left(\frac{\tilde{y}_i^* - y_i}{\sigma_{yi}} \right)^2 \right] \tag{7.12}$$

Then the model coefficients are adjusted to minimize the sum of scaled distances

$$\underset{\{\underline{p}\}}{Min} J = \sum_{i=1}^{N} J_i \tag{7.13}$$

The extension to a multivariable model, in which the response is a function of several variables is straightforward. In a multivariable regression there are multiple input, x, terms, x_1, x_2, x_3, and so on. Each of the i experimental data sets has a value for each of the j x values. Therefore, the first term in Equation 7.12 needs to be the sum of all x-direction deviations,

$$\sum_j \left(\frac{\tilde{x}_{j,i}^* - x_{j,i}}{\sigma_{xj,i}} \right)^2$$

The closest point is not the scaled x–y pair on the model curve, but the scaled $(x_1, x_2, x_3, \ldots, y)$ set on the model surface or hypersurface.

There are several issues with using the maximum likelihood method.

1. The values for the standard deviation for each input and output uncertainty need to be estimated.
2. Those values may also change with the x and y values.
3. There is a computational burden associated with the nested optimization.
4. The user needs to be prepared for the two-level optimization complexity and duality of convergence criteria.
5. The assumption that perturbations from ideal are based on the Gaussian distribution may not be valid.
6. The errors are presumed to be independent.

In my experience, one who understands the experiment can reasonably estimate variance on input and output variables. Training, practice, and software can overcome the implementation barriers. Commonly, the Gaussian distribution and error independence are reasonable approximations to data variability, but they should be verified.

Therefore, I think that maximum likelihood should receive wider acceptance. However, the implicitly and commonly accepted assumptions of constant variance and no uncertainty in the independent variables simplify maximum likelihood to "vertical" least-squares regression.

7.5 Estimating σ_x and σ_y Values

Primitively thinking, one could run replicate trials (multiple trials at the same conditions) and estimate the standard deviation on y from the multiple measurements. This would provide a value for σ_y at the one set of conditions. Then, replicates at several other conditions could be used to determine the σ_y at other conditions. Variance is chi-square distributed and if there are 10 replicates, the 95% uncertainty on sigma is between $0.69s_y$ and $1.8s_y$ where s_y is the calculated standard deviation. This represents a +80% and −31% range on the true value of the standard deviation from the data measured value. Since experimental data points are expensive (from a time, cost, resources view) it would likely require an excessive number of replicates to have a precise value for σ_y.

The difficulty in obtaining a precise value for σ_y is further confounded by the fact that propagation of variance reveals that the total variance on the y measurement is the result of both the variance on y and the variance on x:

$$\sigma^2_{y\ Total} = \sigma^2_{y\ Inherrent} + \sum \left(\frac{\partial y}{\partial x_i} \right)^2 \sigma^2_{x_i} \tag{7.14}$$

Replicate analysis of y would provide the value of $\sigma^2_{y\ Total}$, but $\sigma^2_{y\ Inherrent}$ is required for the maximum likelihood analysis. In order to determine a value for $\sigma^2_{y\ Inherrent}$ the values of $\sigma^2_{x_i}$ need to be known.

The value for σ^2_x cannot be measured. If the true value of x_i were known then you would use those values in the data. If the true value of x_i is not known, the only method to estimate σ^2_x is propagation of uncertainty based on the user's understanding of the experimental process.

The uncertainty of the estimated σ^2_x increases the +80%, −31% uncertain range on σ_y.

Accordingly, values for σ_x and σ_y need to be estimates from the experimenter's experience.

7.6 Vertical SSD – A Limiting Consideration of Variability Only in the Response Measurement

If one assumes that there is no experimental variability on the inputs, on the x values, and that all sources of variation are due to the measurement of y, then the term $(\tilde{x}^*_i - x_i)$ is eliminated, which leaves only the scaled y-deviation term in Equation 7.11. If, also, one assumes that the variance on the measurement is a constant (independent of either the time or y value, homoscedastic), then the common constant $\sigma_y{}^2$ can be factored out of the sum. Since variance is always positive and has been assumed to be a constant, then it is of no consequence to the optimization solution. Accordingly, the solution to Equation 7.11 is

$$\min_{\{\underline{p}\}} J = \sum_{i=1}^{N} (y^*{}_i - y_i)^2 = \sum_{i=1}^{N} (\tilde{y}_i - y_i)^2 = \sum_{i=1}^{N} (\tilde{y}(x_i) - y_i)^2 \tag{7.15}$$

the familiar and common minimization of the SSD.

However, this perspective reveals that the conventional least-squares objective function is a simplification for convenience, not a fundamental path toward truth.

First, it assumes constant variance; however, variance commonly changes with state measurement. Here are two common examples. As the flow rate increases, turbulence increases, and the turbulent fluctuations on an orifice differential pressure measurement increase. Since *dP* is used to determine the flow rate, as the flow rate increases orifice *dP* variance increases. This change is variance is complicated by the additional nonlinear character of the square-root relation between the flow rate and *dP*. Further, consider pH as another example. The titration curve is an "S-shaped" curve with a characteristically steep slope near pH = 7 and low slopes at high and low pH values. Concentration variability will have little influence on the pH measurement at low or high pH values, where the pH is insensitive to small perturbations in concentration. However, the same concentration variability will have a large influence on the pH measurement at medium pH values where the pH is very sensitive to small perturbations in concentration.

Second, the classic SSD objective function (OF) assumes that there is no variability in the experimental control, in the givens, in the x values of the experimental conditions.

In my experience, one can reasonably estimate variance on input and output variables. Even if the values are not exactly correct, using variance estimates in maximum likelihood regression provides better model coefficient values than the vertical SSD. Better is a judgment based on two attributes of the coefficient values – accuracy and precision. For accuracy, in a large number of independent data collection and regression operations, the average coefficient values from the maximum likelihood are closer to the true value than those from vertical SSD. For precision, in the same trials the variability in coefficient values is less for the maximum likelihood than for vertical SSD.

However, vertical SSD is easier to implement, it is commonly understood and accepted, and, for engineering purposes with ample numbers of data points, it generates models that are fully adequate. From a utility perspective, use vertical SSD. Alternately, for scientific precision, the maximum likelihood would be more defensible.

7.7 r-Square as a Measure of Fit

With minimizing vertical least squares as the regression objective, the quantity to be minimized is the SSD (Equation 7.15). However, the numerical value of SSD is not a direct measure of the goodness of fit for several reasons:

1. The SSD value depends on the number of data. If you double the number of data, effectively the SSD will also double. A study with 3 data points will have a much lower SSD than a study with 100 data points; but with 100 data, the errors will balance and the model is likely to be a better representation of the truth about Nature.
2. The SSD value is dependent on the magnitude of the noise. If two studies use many data points so that the errors balance, and the two models are good representations of nature, then the one from data with greater y-variation will have a worse SSD value, even if the models are equivalent. Since SSD represents squared deviation, if the amplitude of the noise doubles, the SSD will increase by a magnitude of four times.
3. Alternately, if the user changes the units of measurement, for instance, from Celsius to Fahrenheit, the SSD will increase by a factor of $3.2 = (9/5)^2$, even if there is no change in the model.

Accordingly, the metric usually used in reporting goodness of fit is the ratio of the variance reduction associated with the model. The variance reduction is

$$r^2 = \frac{SSD_y - SSD_m}{SSD_y} = 1 - \frac{SSD_m}{SSD_y} \quad (7.16)$$

where the symbol r^2 is the reduction in variance, SSD_y is the SSD from the average in the dependent data (the response data), and SSD_m is the SSD in the model residuals:

$$SSD_y = \sum_{i=1}^{N} (y_i - \bar{y})^2 = (N-1)\sigma_y^2 \quad (7.17)$$

$$SSD_m = \sum_{i=1}^{N} (y_i - \tilde{y}_i)^2 \quad (7.18)$$

The subscript m indicates model; later the value of m will be the number of parameters in the model, which includes the intercept, not the model order. Here, $m = 1$ indicates a one-term model, $y(x) = a$, for which the best value of the coefficient is $a = \bar{y}$. Here, $SSD_{m=1} = SSD_1 = SSD_y$.

The extreme values of r-square are 0 and 1. Zero represents the model that had no variance improvement. This might be the result if there is no underlying relation between what were presumed to be the dependent and independent variables, or it could be that the model does not contain the functionality that the data are expressing. On the other extreme, an r-square value of 1 means that the model perfectly fits the data and removed all of the residual variability. However, this does not necessarily mean that the model has the functionality expressed in the data. It could simply mean that there are an excessive number of model terms relative to the data, that the model is fitting noise and not a general mechanistic, phenomenological trend. Generally, if the user chooses $m \geq N$ then the r-square value will be perfect unity.

Because the r-square metric normalizes the goodness measure of model fit by the inherent variance in the data, it is a better indication of model fit than SSD. However, it is not a complete measure of the correctness of the model. The r-square value does not indicate that the model architecture is correct and it does not indicate that the correct independent variables are included.

As an example of how the basic model architecture could be wrong, consider that the unknowable truth is $y = \sqrt{x}$ and that noisy data are generated within the $x = 1$ to $x = 4$ region, the data trend may appear to be the upper left half of a parabola, and a quadratic model $y = a + bx + cx^2$ may seem to provide a reasonable fit, as illustrated in Figure 7.5.

An alternate metric, similar to r-square, is the ratio of the variance reduction associated with the model. This accounts for an excessive number of model coefficients. The variance reduction is

$$r_m^2 = \frac{\sigma_y^2 - \sigma_m^2}{\sigma_y^2} = 1 - \frac{\sigma_m^2}{\sigma_y^2} \quad (7.19)$$

where the symbol r_m^2 is the reduction in variance, σ_y^2 is the variance in the dependent data, response, and σ_m^2 is the variance in the residuals in a model with m number of coefficients:

$$\sigma_y^2 = \frac{1}{N-1} \sum_{i=1}^{N} (y_i - \bar{y})^2 \quad (7.20)$$

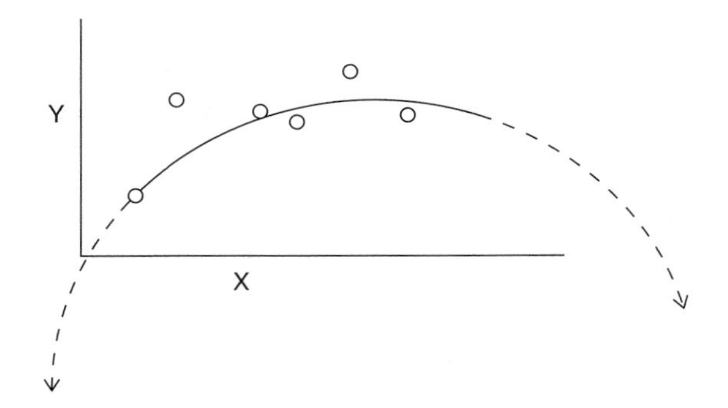

Figure 7.5 Illustration of a wrong model functionality that seems right

$$\sigma_m^2 = \frac{1}{N-m}\sum_{i=1}^{N}(y_i - \tilde{y}_i)^2 \tag{7.21}$$

The value of m is the number independent adjustable coefficients in the model, not the order or highest power on the independent variable.

In Equation 7.20, the "model" assumes that all data have the same value, the average, and measures deviations from the average. The model has one coefficient, the average value, so the denominator value $(N - 1)$ represents the degrees of freedom (DOF) in the data. In Equation 7.21 the model may have m coefficients, so the DOF for residual variance is based on $(N - m)$.

The interpretation of values of this r_m^2 metric is similar to that of r^2. It still approaches unity as $m \rightarrow N$, but not as rapidly as r^2. However, as the number of coefficients approaches the number of data, the decreasing value of the $(N - m)$ term prevents this r_m^2 from becoming unity, which prevents a naïve interpretation of improvement in that extreme case.

The relation between the two ratio-of-variance statistics is

$$r_m^2 = 1 - \frac{N-1}{N-m}(1 - r^2) = \frac{N-1}{N-m}r^2 - \frac{m-1}{N-m} \tag{7.22}$$

which forms a precursor to evaluating model designs in Chapter 13.

7.8 Normal, Total, or Perpendicular SSD

The scaled deviation terms in Equation 7.11 are scaled by a measure of variability, standard deviation. Usually an experimental plan is designed so that the range of independent and dependent variables is much, much greater than the uncertainty in the variable value. If one accepts that the variance on each variable is uniform throughout the variable range, then range can be stated as a constant multiplier, c, of the standard deviation for each variable; then an alternate simplification of the maximum likelihood method (retaining variation on both y and x) is possible.

If one uses the range of values to scale the variables, then effectively one is using the standard scaled variable notation. If z is a variable, then the scaled value, z', is defined as

$$z' = \frac{z - z_{min}}{z_{MAX} - z_{min}} = \frac{z - z_{min}}{c\sigma_z} \tag{7.23}$$

where z_{MAX} and z_{min} represent the highest and lowest values of z for the experimental data sets and the difference $(z_{MAX} - z_{min})$ is the range; σ_z represents replicate variability on z, assumed to be uniform throughout the z range.

If the variables in Equation 7.11 are scaled by the standard deviation of the variable rather than the range and c has the same value for all variables, then it can be stated as

$$\min_{\{\underline{p}\}} J = \sum_{i=1}^{N} \frac{1}{c^2}[(\tilde{x}_i^{*\prime} - x_i')^2 + (\tilde{y}_i^{*\prime} - y_i')^2] \tag{7.24}$$

Since a constant multiplier with a positive value on the objective function does not affect the optimal solution, it can be simplified as

$$\min_{\{\underline{p}\}} J = \sum_{i=1}^{N} [(\tilde{x}_i^{*\prime} - x_i')^2 + (\tilde{y}_i^{*\prime} - y_i')^2] \tag{7.25}$$

Recall that the scaled variables are dimensionless, making it possible to add x' and y' in Equation 7.25. Also, in that equation the standard deviation on the scaled variables is implicitly unity. This makes the shape of the PDF contours to be concentric circles, as illustrated in Figure 7.6, not ellipses as in Figure 7.3.

In this view (with the range proportional to the uniform standard deviation and the ratio of σ to range equal for all variables) the likelihood contours are circles. Since the maximum likelihood point on the model curve is tangent to the maximum likelihood contour and the contours are circles, then the distance from the data point (at the center of the contours) to the point of tangent is a radius of the circle. This gives the distance several properties. First, the distance between the curve and data point that maximized likelihood is the shortest distance from the curve to data. Second, the shortest distance is perpendicular to the common tangent of the model and contour, making it normal to the model curve.

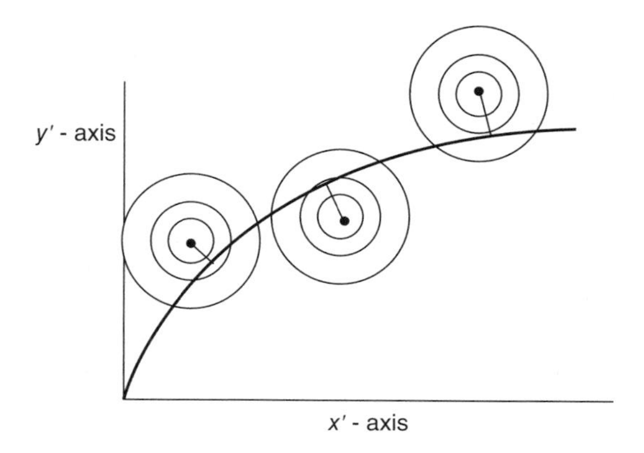

Figure 7.6 Maximum likelihood illustration in scaled variable space

Since the axis values are dimensionless, the x' and y' distances can be combined to a common distance. Then, maximizing the likelihood in this scaled variable space is the same as minimizing the distance (in scaled variables) from the data point to the curve, which also is the same as minimizing the distance (in scaled variables) from the data point along the normal to the curve (the perpendicular to the tangent). You may find this called total least squares, referring to the minimization of the hypotenuse (or perpendicular or normal) deviations from the curve.

As with Equation 7.11, Equation 7.25 requires a nested optimization. First, for each data point, find the point on the curve that minimizes distance to the data point:

$$\underset{\{\tilde{x}_i^{*\prime},\ \tilde{y}_i^{*\prime}\}}{Min} J_i = [(\tilde{x}_i^{*\prime} - x_i')^2 + (\tilde{y}_i^{*\prime} - y_i')^2] \tag{7.26}$$

Then the model coefficients are adjusted to minimize the sum of scaled distances:

$$\underset{\{\underline{p}\}}{min} J = \sum_{i=1}^{N} J_i \tag{7.27}$$

This approach removes one of the problems associated with the maximum likelihood, that of determining values for the standard deviation of each variable. However, the utility of the approach is only as right as the simplifying assumptions are right – that σ for each variable is uniform over the range, has a known value for scaling the variables, and the range-to-σ ratio for each variable is about the same.

However, the nested optimization remains a problem.

7.9 Akaho's Method

Recognizing that minimizing the perpendicular distance from the data point to the curve is equivalent to solving for the distances in Equation 7.26, we have considered Akaho's and other's methods for determining the perpendicular distance. However, these are approximations and/or iterative root-finding. However, one must question the assumptions that led us here – uniform and known σ values for all variables. In the light of the assumptions required to implement the maximum likelihood, an approximation seems fully appropriate. In my opinion, Akaho's approximation is fairly good and avoids nested optimization, providing a benefit far exceeding the detraction of its additional assumption.

The method is illustrated in Figure 7.7, which plots the σ-scaled dependent and independent variables y' and x' and focuses at the ith measurement. The data point is labeled as "A" and the $\tilde{y}(x_i)$ as "B." The true normal line intersects the model curve at point "C." Akaho's method seeks to estimate the distance from A to C. It does this by extrapolating the tangent to the model at "B" and determining the point "D," the intersection of the tangent to the perpendicular through point "A." The distance A–C is estimated by the distance A–D, which the reader may agree is a reasonable approximation. Elementary geometric analysis indicates that $d_{A,i}'^2$, the square of the Akaho distance for the ith data point in scaled variables, is

$$d_{A,i}'^2 = \frac{[y_i' - \tilde{y}'(x_i)]^2}{1 + \left(\frac{\partial \tilde{y}'}{\partial x'}_i\right)^2} \tag{7.28}$$

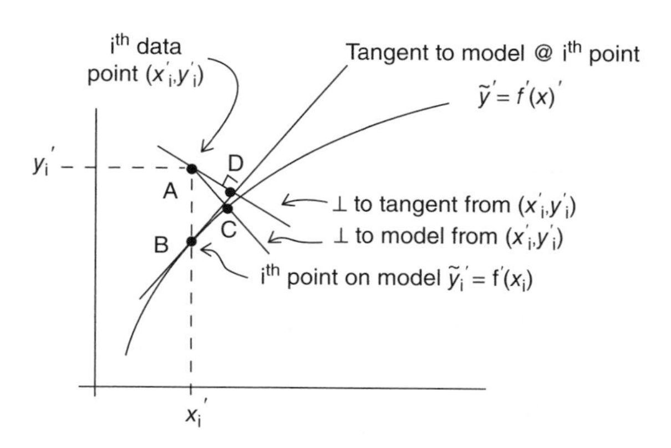

Figure 7.7 Illustration of Akaho's technique

This approach has the advantage of direct calculation of the objective function. Given the model coefficient values and the (x, y) data pairs, the deviation between the model and data, the vertical deviation, is easily obtained and is scaled by the denominator of Equation 7.28, which is easily obtained (either analytically or numerically) from the model.

This version of the normal or total least squares is an approximation to the maximum likelihood. The objective is to determine model parameters that minimize the approximated normal least squares. The approximation using the model derivative avoids the nested optimization. Solving Equation 7.29 is a single-stage optimization:

$$\underset{\{\underline{p}\}}{Min} \; J = \sum d'^2_{A,i} = \sum \frac{[y'_i - \tilde{y}'(x_i)]^2}{1 + \left(\frac{\partial \tilde{y}'}{\partial x'}_i\right)^2} \tag{7.29}$$

The multivariable version is also a direct one-step calculation based on known values:

$$d'^2_{A,i} = \frac{[y'_i - \tilde{y}'(\underline{x}_i)]^2}{1 + \sum_j \left(\frac{\partial \tilde{y}'}{\partial x_j'}_i\right)^2} \tag{7.30}$$

for which the single-stage optimization is

$$\underset{\{\underline{p}\}}{Min} \; J = \sum d'^2_{A,i} = \sum \frac{[y'_i - \tilde{y}'(\underline{x}_i)]^2}{1 + \sum_j \left(\frac{\partial \tilde{y}'}{\partial x_j'}_i\right)^2} \tag{7.31}$$

Example 7.1 The model is a simple quadratic relation

$$\tilde{y} = a + bx + cx^2$$

with one independent variable. There are three model coefficients and three decision variables {a,b,c} to be determined by regression. First, scale the dependent and independent variables with the respective standard deviations

$$y' = y/\sigma_y$$
$$x' = x/\sigma_x$$
$$\tilde{y}' = \frac{\tilde{y}}{\sigma_y} = \frac{a + bx + cx^2}{\sigma_y} = \frac{a + b\sigma_x x' + c(\sigma_x x')^2}{\sigma_y} = a' + b'x' + c'x'^2$$

Then the optimization statement (Equation 7.29) becomes

$$\underset{\{a', b', c'\}}{Min} J = \sum \frac{[y'_i - a' - b'x'_i - c'x_i'^2]^2}{1 + (b' + 2c'x'_i)^2}$$

Use optimization to solve for values of a′, b′, and c′. Then regenerate a, b, and c values:

$$a = \sigma_y a'$$
$$b = \sigma_y b'/\sigma_x$$
$$c = \sigma_y c'/\sigma_x{}^2$$

The Akaho distance estimates the true normal distance from the curve (or surface, or hypersurface if multiple independent variables). This is often a very reasonable estimate considering the other assumptions in the analysis – about the true values of σ_x and σ_y, homoscedasticity (constant variance throughout a range), the model architecture, and the limited amount of data. As the data variability decreases, the A–D estimate approaches A–C. However, where the model curvature is high (such as in pH models of strong electrolytes) this local linearization approach could misrepresent the deviations.

Chandak (2009) compared the maximum likelihood and minimizing the sum of squared normal deviations to the simpler minimization of the sum of squared vertical deviations on a pH titration response to base addition to an acidic solution. The objective was to back out acid concentration, A0, and the dissociation constant, pK_a, from simulated experimental data with uncertainty on both the measured pH and base concentration. One thousand experimental realizations provided 1000 comparisons and 1000 coefficient pairs for each technique. Figure 7.8 shows the probability of an equal or larger deviation from the true coefficient value and reveals that the Akaho (normal) distance method provides higher precision (when the uncertainty on the base concentration has an equivalent impact on pH as the uncertainty on the pH).

In the extreme case when the variability in the x value has relatively no impact on the variability in the y value, when $\sigma_x {}^{\partial y}/_{\partial x} \ll \sigma_y$, vertical SSD (conventional least squares) was as good as the maximum likelihood.

7.10 Using a Model Inverse for Regression

Desirably, models are explicit in the output variable, but sometimes the model might be an implicit form in which the answer is required to determine the answer. A common thermodynamic property model relates specific heat to temperature in a form

$$Cp = a + bT + cT^2 + dT^3 + e/T \tag{7.32}$$

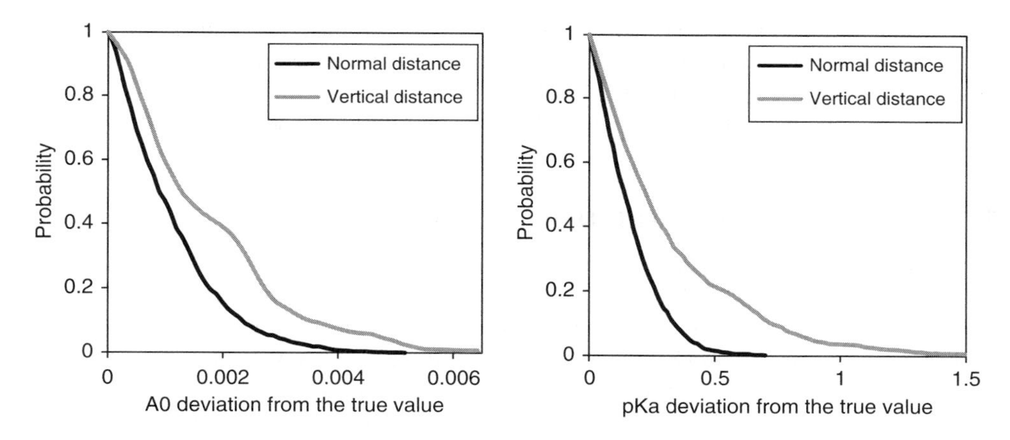

Figure 7.8 Comparison of probable error on normal and vertical deviations in nonlinear regression

However, if the question is "What temperature provides a desired Cp value?" then the value for $T = f(Cp)$ cannot be rearranged for an explicit solution. Solution of the inverse question requires using a numerical iterative procedure to solve for the dependent variable.

Determining the coefficient values in Equation 7.32 by regression is predicated on a data set containing T and Cp values. The data would have been generated by an experiment that sets T and measures Cp, where temperature is the independent variable and Cp is the dependent variable, as suggested in Equation 7.32.

However, if the model is to be used to describe the inverse, the independent and dependent variable roles are reversed. In this case, for convenience of use, the model can be explicitly rearranged as its inverse or perhaps an empirical model would be used. Define the inverse empirical model to predict the variable of interest, such as

$$T = a + bCp + cCp^2 + dCp^3 + eCp^4 \tag{7.33}$$

Then seek to minimize SSD of "x" given "y" (T from Cp) from the same data set.

Note the coefficient values $\{a, b, c, d, e\}$ in Equations 7.32 and 7.33 are not identical and the two models (one from vertical SSD and the other from horizontal SSD regression) will not be mathematically identical.

7.11 Choosing the Dependent Variable

Choose the dependent variable to reflect the in-use form of the equation. A rheology experiment might report the shear stress, σ, as a measured response to the shear rate, $\dot{\gamma}$, and the rheological model might be of a Carreau type predicting the effective viscosity, η:

$$\eta = \mu_\infty + (\mu_0 - \mu_\infty)(1 + \lambda\dot{\gamma}^2)^{(n-1)/2} = \frac{\sigma}{\dot{\gamma}} \tag{7.34}$$

The question is whether the regression objective function should be based on the calculated effective viscosity from Equation 7.34 or from the experimentally measured stress from

$$\sigma = \eta\dot{\gamma} = [\mu_\infty + (\mu_0 - \mu_\infty)(1 + \lambda\dot{\gamma}^2)^{(n-1)/2}]\dot{\gamma} \tag{7.35}$$

Log-transforming the model reveals the linear asymptotic limits, so should the SSD be based on the log-transformed model? I recommend structuring the model to predict the variable of interest for the in-use application, not for the underlying science and not for the visual affirmation of asymptotic limits. You want the model to best fit the variable for the model use.

7.12 Model Prediction with Dynamic Models

Dynamic models represent the evolution of a process with respect to time. There is not a single response value for a given input value. Place something warm in the refrigerator and that action starts a cooling process in which temperature changes in time. The time response, the dependent variable, could be influenced by several independent variables that also change in time. The salt content of the mouth of a river (where it is entering the ocean) depends on whether the tide is rising or falling and on how the upstream rain impacts the flow rate of the river.

Figure 7.9 indicates a step-and-hold trend in an input (influence) and the progressive time-dependent change in the response. Sampling intervals indicate the *i*th x and y values. Note that for the new level (a single value) of the independent variable, there are many values for the dependent variable.

By contrast, in steady-state models the dependent variable has a unique value dependent on the independent variable values. For steady-state (SS) models the sampling order in which data sets are listed does not matter. Whether it is the seventh entry or the third entry in the (y, x) list, the deviation of the data from the model is unchanged.

In transient models, however, the developing value of the dependent variable depends on the prior value, and the listing order of the data must preserve the time relation.

Use the vertical SSD of Equation 7.15 as the regression objective function. I am not aware of a convenient method to include the impact of uncertainty on either sampling time or independent variable values, when regressing models to transient data.

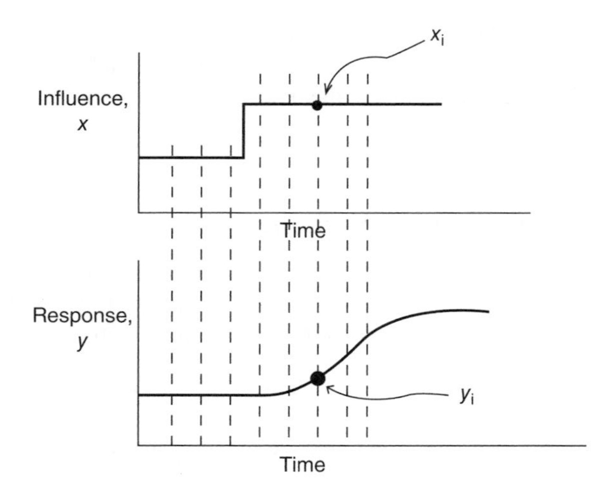

Figure 7.9 Illustration of a dynamic response

7.13 Model Prediction with Classification Models

Classification models seek to categorize an outcome such as pass/fail, 1/0, yes/no, red/white/blue, hot/warm/cool/cold, and so on. Here, the measure of goodness of the model fit would be assessed as the number (or proportion) of properly classified items and the measure of badness as the number of misclassified items. If the model must select a category, then the number of wrong classifications will be equal to the total number of trials less the number of right classifications and the objective function is simply to minimize the number of misclassifications, n:

$$\underset{\{\underline{p}\}}{Min} \; J = n \tag{7.36}$$

However, the model could be given the option to skip a classification. Something might be cold, but if the model cannot be sure of the classification, it might skip a determination. This might be perceived as better than misclassifying it as hot, which would be a mistake, but it is not properly classified as cold. A categorization of "can't tell" might not be as bad as a wrong determination. Alternately, something might be cold, which the model classifies as cool, which is better than "hot" or "can't tell." In these cases, the degree of wrongness could be used as a penalty. The user's understanding of the impact of right and wrong classifications are needed to assign a penalty for the degree of wrongness in a prediction error. If d_i is the degree of departure from ideal for an ith classification, perhaps the number of categories away from true, then d_i^2 could represent the penalty and the optimization would be

$$\underset{\{\underline{p}\}}{Min} \; J = \sum d_i^2 \tag{7.37}$$

For example, a relation is being developed to predict a human's perception of the outdoor daytime comfort. The categorizations are "It is hot" or "warm, nice, cool, chilly, cold, or freezing." This would be similar to the wind chill equivalent, which accounts for air temperature and wind speed, but the model will additionally include sun strength, humidity, and level of personal activity. Are you sitting in the shade or pulling weeds in the sun? Table 7.1 reveals some data of the human perception and the modeled characterization, and the Equation 7.37 penalty for misclassification that an optimizer would seek to minimize.

In such applications, very small adjustments to the model coefficient values would not change the discretized model characterization and consequently it would not change the sum

Table 7.1 Illustration of a classification model

Situation	Human perception	Modeled categorization	Deviation = number of classifications	Penalty = deviation squared
1	Chilly	Warm	3	9
2	Warm	Hot	1	1
3	Nice	Nice	0	0
4	Hot	Hot	0	0
5	Freezing	Cold	1	1
...	...	...	...	...

of penalties, the objective function value, for the optimizer. The optimizer, looking at how the OF surface responds to DVs (model coefficient values) would experience flat spots. This feature should lead you to choose a direct search algorithm over a gradient-based algorithm and convergence criteria that are compatible with flat spots.

Further, flat spots may make it appear that the optimizer has converged because changes in the DVs will not improve the OF value. Accordingly, multiple starts or a multiplayer algorithm should be selected to be confident that the global optimum is found.

7.14 Model Prediction with Rank Models

Some models seek to provide a rank. Exact rank pairing will have actual-to-modeled rank deviations equal to zero. One could count the number of right ranks as the measure of goodness, but this does not reveal how far off the disorder is. If the ranks are 1, 2, 3, 4, 5, then Model A in Table 7.2 is perfect. Model B (5, 2, 3, 4, 1) only has two ranks out of place but confuses the best and worst. If count is the measure of goodness, this would be as good as Model C (2, 1, 3, 4, 5), which only has the first and second shifted, or Model D (1, 2, 3, 5, 4), which has the important first three correct and misses the least important two.

Models B, C, and D each have three of the five ranks properly assigned, which makes them equivalent in a count of right rankings. However, of the three choices B, C, and D, Model D is the best choice because it gets the most important rankings correct. Therefore, we would like a metric of goodness that properly assesses the model utility.

A better measure than count-of-number-of-right-answers is the sum of absolute rank differences. A still better method is to weight large deviations stronger by using the SSD. Finally, if the top ranks are more important to the application than the lower ranks, then one can scale the deviation by dividing by the rank. These options are illustrated in the table.

Regression can be used to determine model coefficients that predict rank. Again, the user needs to consider a measure of badness to be minimized to be sure that the choice is consistent with the in-use situation of the model.

As with categorization predictions, the optimizer will experience flat spots on the OF response to DVs and the user is encouraged to use a direct search approach over a gradient-based procedure, to choose a compatible convergence criterion, and to use a multiplayer algorithm to ensure confidence that the global optimum will be found.

Table 7.2 Illustration of rank assessments

Metric	True rank	Model A	Model B	Model C	Model D	Model E
	1	1	5	2	1	3
	2	2	2	1	2	2
	3	3	3	3	3	1
	4	4	4	4	5	5
	5	5	1	5	4	4
Count right		5	3	3	3	1
Sum Abs(error)		0	8	2	2	6
Sum e^2		0	32	2	2	10
Sum e^2/r		0	19.2	1.5	0.45	5.78

7.15 Probabilistic Models

There are a variety of models that describe distributions. Some provide probability values or fractions, which could be either an event probability of a discrete event or an interval probability – the probability of a value between two values or less than a particular value. In all cases these produce a continuous valued number and can be compared in a least-squares sense to the data, with an OF such as Equation 7.15.

The model might predict the value of a distribution coefficient, such as the standard deviation. Again these are continuous valued numbers that can be tested against data in a least-squares sense.

7.16 Stochastic Models

Contrasting models that deterministically describe a distribution, Monte Carlo simulations, provide the simulated distribution, the histogram of expected outcomes. They do not provide identical values for a given input, but provide a realization of possible outcomes. To adjust model parameters so that the simulation best matches the experimental data requires a best fit in all sections of the cumulative distribution function (CDF). Use either the Kolmogrov–Smirnov maximum deviation or the chi-square contingency statistic as the regression OF. Using the KS deviation, the optimizer will seek to minimize the single maximum deviation and the OF response to the DV may likely be discontinuous as the point of worst deviation changes with distribution parameters. A direct search algorithm is preferred over a gradient-based algorithm.

The chi-square statistic looks at all of the data. It normalizes the square of the deviation by the expected number, which is equivalent to scaling the deviations by the standard deviation so that a difference of 30 in a bin that is expected to have 847 is of similar impact as a difference of 5 in a bin expected to have 21. However, you need to choose the bin intervals. Have at least 10 bins to be able to critically see features of the histogram and plan the number of experiments so that there are no fewer than five counts in each bin. The KS statistic is simpler to implement and removes human choices from the OF, reducing possible human objections to the outcomes.

However, for either approach the OF will be a stochastic response to the DVs. If the optimizer replicates the DV values, the Monte Carlo simulation will produce a different realization, which will have a different OF, even for the same experimental data. Accordingly, not only does the optimizer need to deal with discontinuities but it needs to accommodate a stochastic surface. In addition, the criteria for convergence needs to be consistent with the local noise on the surface. My preference is to use leapfrogging (a multiplayer direct search) with a steady-state convergence criterion.

7.17 Takeaway

In regression the purpose is to maximize likelihood (or one of its variations such as to minimize the sum of squared residuals). In the terminology of optimization, the purpose is called the objective and the rule (mathematical formula) that calculates the value of the objective is termed the *objective function*. In regression, the model coefficients are to be adjusted to minimize the OF. However, in optimization the adjustable variables, the model coefficients, are termed the *decision variables*. In regression terminology the statement, "Adjust the model coefficients to get the best model," would be translated to "Adjust the DVs to minimize the OF" in optimization speak.

First, use standard SSD with "vertical" deviations. Usually this will be fully adequate. It is also simple, familiar to your audience, and easy for them to accept.

Only for steady-state models, when there is fairly precisely known uncertainty in the independent variable value, and when seeking the highest precision in a coefficient value (such as for science purposes) is the maximum likelihood either needed or its effort is defensible. The effort includes: checking if the distribution of random errors on each variable is Gaussian, testing for independence, assigning reasonably accurate σ values for each variable, and implementing the nested optimization.

If there is a need to account for random uncertainty in the independent variable, then Akaho's method is a useful compromise. It provides effectively identical results to maximum likelihood (given assumptions) for much less computational effort.

Define the objective function so that it is a proper measure of the in-use application of the model outcome. Understand the context to devise an appropriate OF for regression.

Models that predict rank or categorization usually lead to flat spots in the OF response to DVs and need an optimizer, convergence criterion, and a number of initializations that are appropriate to the flat spots and confidence in finding the global.

Monte Carlo simulations lead to stochastic OF surfaces and need both an optimizer and an appropriate convergence criterion.

Exercises

7.1 Here are three (x, y) points: (1, 1), (3, 5), and (4, 6). Graph these and fit with a best linear relation. Now reconsider the linear regression fit if the certainty of the y location of the (1, 1) point is very low.

7.2 Demonstrate for a least-squares objective function that it is equivalent to consider the model as true and the point a deviation from it or to consider the point as true and the model a deviation from it.

7.3 Derive Equations 7.9 and 7.10 from Equation 7.7.

7.4 Describe a procedure to determine the point $(\tilde{x}_i^*, \tilde{y}_i^*)$ as indicated in Equation 7.12.

7.5 If the model is a linear relation $y = a + bx$, with constant x and y variance, show that the maximum likelihood OF is equivalent to the vertical least-squares OF weighted by $1/(1 + b^2)$.

7.6 Derive Equation 7.15 from Equation 7.11 with the assumptions of no x-uncertainty and constant y-variance.

7.7 What limiting conditions would permit vertical least squares to provide just as good a model as the maximum likelihood?

8

Constraints

8.1 Introduction

Typically, on a generic regression application such as $y = a + bx + cx^2$ there are no constraints on the optimization. The coefficients a, b, and c could have either positive or negative values. However, in phenomenological models coefficients represent phenomena and their values are constrained. For example, a delay and a time constant must both have non-negative values. Further, in regression of a model to fit an engineering application there are many other variables to consider. One application could be to fit a distillation column tray-to-tray model to data by adjusting coefficients representing tray efficiency and ambient heat losses. Here, there are many other constraint considerations than just those on the decision variables (DVs), such as whether tray efficiency should be non-negative. Constraints would include internal model variables that relate to physical limits (order, equilibrium, composition) and asymptotic limits (equilibrium, steady-state values, idealization limit). Constraints could be single limits (non-negative), bounds (0–100%), combinations (reflux must be less than boil-up), or discretization (integer values for a delay counter).

Nonlinear regression is a nonlinear optimization application and should address constraints.

8.2 Constraint Types

There are many ways to classify constraints. One classification contrasts *hard* as opposed to *soft*. The *hard constraint* may not be violated. An example is that the mole fractions in a material must sum to unity, a "law" of Nature. Another is to keep a concentration lower than its explosion limit. The other constraint type is a *soft constraint*, which desirably is not violated, but if violated there is a penalty proportional to the magnitude and duration of the violation. An example is, "Keep the product within specification." Violating the specification limit is physically possible and perhaps generation of waste is the penalty. Another is, "Don't cross the border." Risk is the penalty – the probability of being caught times the consequence of being caught. Both soft constraint examples were laws of Man. Although the Nature–Man

Nonlinear Regression Modeling for Engineering Applications: Modeling, Model Validation, and Enabling Design of Experiments, First Edition. R. Russell Rhinehart.

idealization often segregates hard and soft, the reality is that one situation may make a particular constraint hard (the boss might say, "Do that again and you're fired") and another situation may make the same constraint soft. *Hard* and *Soft* constraints are accommodated differently in the optimization algorithm that determines best model coefficient values.

Constraints may be imposed on the DVs (the model coefficient values). For example, in transient process modeling, time constants and delays must have non-negative values. Constraints may be imposed on model outputs. For instance, a consequential temperature must be above a freezing point or, another, the tank cannot be totally drained in the future. Constraints may be on intermediate calculations or relations. For instance, concentrations must be non-negative, the trial solution cannot violate the second law, or a divide-by-zero cannot be executed.

Each of the constraint types (hard and soft) may be imposed on each of the variable types (model coefficient values, model output values, intermediate variables, or computer code calculation operations).

You need to understand your application and the situational context to be able to define constraints and to decide whether they are hard or soft. Since constraints can have a significant impact on the optimization outcome, you need to carefully identify them. Omission of constraints, improperly stated constraints, or constraints that are mistranslated into a mathematical form can lead to wrong optimization outcomes.

There are additional ways to classify constraints. They could be constraints on value or on the rate of change. They could be equality or inequality constraints. They may be on a single variable or on combinations of variables. They may be on the current property or a future property.

Constraints on value, level, or magnitude of a single variable may be *inequality* or *equality* or both. The user may specify that noise level must be less than 35 dB or that tank level must be above 1 ft and below 19 ft. The sign indicates that the posted speed limit is 55 mph. To keep water a liquid at atmospheric pressure, the temperature must be above 32 °F and less than 212 °F. The exit flow rate and cash flow must be non-negative. The argument of the log function must be greater than zero. The valve position must be between 0 and 100%.

Constraints on the rate of change of a single variable may also be equality or inequality types. Acceleration must be less than 5 Gs. The rate of change of the main steam valve must be less than 5% per minute. The radius of curvature on a road must be less than 10 degrees of arc per 100 m. The second derivative of the function at the extremes of the data range must be zero.

Constraints may be on combinations of variables, again equality or inequality. The sum of all mole fractions must equal unity. Vapor and liquid compositions must be in thermodynamic equilibrium. Material and energy balances must have zero deviation. dQ/T must be positive. The autoregressive moving average (ARMA) model of Equation 2.24 must be open loop stable. The average slope should not exceed 500 ft per mile.

8.3 Expressing Hard Constraints in the Optimization Statement

As an indication of constraints (but with no indication of how they are to be treated) simply add the statement "subject to" (ST) and list the constraints with the conventional optimization statement. The following example is stated as: *find the model coefficient values to minimize the conventional "vertical" sum of squares deviation of data to model subject to three constraints*:

$$\underset{\{\underline{p}\}}{Min} \; J = \sum_{i=1}^{N} (y_i - \tilde{y}_i)^2 \tag{8.1}$$

ST

$$p_4 > 0.5$$

$$x_1 + x_2 = 1$$

$$\left| \frac{\partial y}{\partial z} \right| \leq 3.3\overline{33}$$

$$p_2 \textit{ is a non-negative integer}$$

Here the variable p_4 in the first constraint represents the fourth coefficient in the model. This is a decision variable, a value the optimizer changes to minimize the objective function (OF). In the second constraint, the variables x_1 and x_2 are not explicitly revealed in either the DVs or the OF. These represent intermediate model state variables, such as component fractions. The second constraint is shown as a equality constraint, as opposed to the limit of a value. The third constraint is a rate of change constraint. It would pertain for all values of the variable z, not just at one point. If there is an analytical expression for $y(z)$, then you could take the partial derivative and have an expression that must be true for all z values. Alternately, the derivative could be numerically evaluated at a sufficient number of z increments. Finally, the fourth constraint indicates that a coefficient value must be an integer, such as a model coefficient that represents a delay for an ARMA model.

When constraint violation is impermissible, use hard constraints. Example cases include: phase equilibrium constrains P and T, composition constrains mole fractions to sum to unity, values for time constants must be non-negative, at steady state the reflux rate cannot be larger than the vapor product rate.

Here the optimizer is told, "I know your algorithm led you to calculate that set of DV values for a trial solution. However, those are not permissible. I'm not going to calculate a new OF value. Deal with it."

Direct search optimizers can handle "pass–fail" as an OF as well as a numerical value. In Chapter 10 this textbook will recommend two direct searches: a cyclic method with heuristic expansion and contraction factors for the search and leapfrogging. However, there are many other well-accepted direct search approaches including Hooke–Jeeves, Nelder–Mead (a version of the Spendly, Hext, and Himsworth simplex method), particle swarm, differential evolution, and so on.

8.4 Expressing Soft Constraints in the Optimization Statement

If constraints are *soft*, then it is desired not to violate them, but a small violation is tolerable. For soft constraints we add a penalty to the OF for a violation. Usually, the penalty is based on the magnitude and/or duration of the violation of the constraint. For the inequality constraint on p_4 the magnitude of the violation, the measure of badness, b_1, would be stated as

$$b_1 = 0 \qquad if\ p_4 > 0.5$$
$$b_1 = p_4 - 0.5 \qquad if\ p_4 \leq 0.5 \tag{8.2}$$

The badness, undesirability, or penalty for a violation usually increases faster than the magnitude of the violation increases. Consider the experience some acquaintance might have had with the penalty associated with speed over the speed limit, or the number of sequential times a child is caught misbehaving, or the number of missed shipments to a customer. The first penalty might be forgiven, the second equivalent violation gets a small penalty, and the third identical violation gets a large penalty. Conventionally, the penalty is proportional to the square of the violation. I am not aware of a fundamental aspect of Nature that leads to a mathematical derivation of the quadratic function for the penalty for a deviation from desired, but I think it adequately represents the human "feel" for undesirability. The square of the deviation arises in the Gaussian distribution, Taguchi analysis, and it is a smooth function with continuous differentials, making it familiar and convenient, as well as heuristically consistent. Therefore, define the penalty as the square of the measure of badness:

$$P_1 = {b_1}^2 \tag{8.3}$$

Nominally, the penalty is added to the OF, but first the relative human importance of the goodness-of-model fit to the data, the sum of squared deviations (SSD), and the extent of soft constraint violation need to be balanced. This scaling factor, λ, is usually termed a Lagrange multiplier. The addition of a penalty for coefficient p_4 violation of its 0.5 limit is illustrated as

$$\underset{\{\underline{p}\}}{Min}\ J = \sum_{i=1}^{N} (y_i - \tilde{y}_i)^2 + \lambda_1 {b_1}^2 \tag{8.4}$$

$$\text{ST}$$

$$x_1 + x_2 = 1$$

$$\left|\frac{\partial y}{\partial z}\right| \leq 3.\overline{333}$$

Note that the p_4 constraint condition is now not shown in the ST list because the constraint has been taken care of in the OF calculation.

Also note that the Lagrange multiplier has several functions. It must equate the impact of units on the y and p variables in the first and second terms of the OF, as well as balance the relative importance of the two aspects (primary least-squares objective and penalty). It is also scale-dependent. For instance, if y is measured in degrees Celsius and subsequently switched to degrees Fahrenheit, the numerical value of the squared y-deviation increases by a factor of 3.24 ($=1.8^2$), then the value of λ should also increase by a factor of 3.24 to keep the magnitude of the constraint in the same balance with the model fit. Further, if the number of data points in the sum, N, were to increase, the SSD would be larger, and λ would have to be proportionally larger. These scaling changes for λ are relatively straightforward. However, the evaluation for human concern is neither absolute nor equation-based. If, on another day, the new boss is very concerned about his/her leadership image that could result from a constraint violation, then the magnitude of λ should increase to reflect this new balance of concerns. While the value of λ does depend on scale and N primarily, the value of λ is based on the human concern, which changes in time and situational context. The value of λ is situation-dependent.

The remaining constraints can also be expressed as soft penalty functions. The equality constraint is violated for any deviation from unity so the measure of violation is a simple statement:

$$b_2 = x_1 + x_2 - 1 \tag{8.5}$$

The remaining inequality constraint would be evaluated for all z values, from the first ($k = 1$) to the last ($k = M$):

$$\begin{aligned} b_{3k} &= 0 \quad if \left|\frac{\partial y}{\partial z}\right| \leq 3.3\overline{33} \\ b_{3k} &= \left|\frac{\partial y}{\partial z}\right| - 3.3\overline{33} \quad if \left|\frac{\partial y}{\partial z}\right| > 3.3\overline{33} \\ b_3 &= \sum_{k=1}^{M} \varepsilon_{3k} \end{aligned} \tag{8.6}$$

With penalties defined for the soft constraints, the optimization statement becomes

$$\underset{\{\underline{p}\}}{Min}\ J = \sum_{i=1}^{N} (y_i - \widetilde{y}_i)^2 + \lambda_1 {b_1}^2 + \lambda_2 {b_2}^2 + \lambda_3 {b_3}^2 \tag{8.7}$$

There are, now, three Lagrange multipliers in the OF of Equation 8.7, adding complexity to determining their values.

An alternate approach that provides structure in determining the relative weighting factor values was revealed to me by process control engineers at [DMC] Corporation. It is called the *equal concern* (EC) method for weighting. *Concern* cannot be quantified, but it is felt, intuitively understood, as a level of anxiety about the collective consequences of something. The consequences may be financial, risk, personal image, safety, utility of use, marketing competitiveness, others' opinions, ethics, and so on, or any combination of those aspects. To determine equal concern factors, follow this procedure. First, define a magnitude of a deviation from ideality ($D_{primary}$) for the main OF, think about the myriad of consequences resulting from the deviation, and feel the level of concern or anxiety that deviation creates. Next, take each penalty factor, b_j, one at a time. For each, consider deviations from ideality, feel the associated concern, and state a deviation (D_j) that results in the same level of concern as that associated with $D_{primary}$. Third, normalize the terms in the OF by the equal concern factors. Figure 8.1 illustrates the concept.

The left-hand graph indicates how concern increases with the deviation from ideal for the primary OF quantity. The horizontal axis will have D values and units matching the OF deviation from ideal. However, since concern is a confluence of multiple values, the vertical axis cannot be quantified; it can only be qualitatively felt. The right-hand graph reveals concern relative to one of the penalty function deviations from the constraint threshold. Again, the numerical value or the units on concern cannot be assigned, but there is an intuitive feel for the value of D_j that produces a concern equal to that produced by $D_{Primary}$.

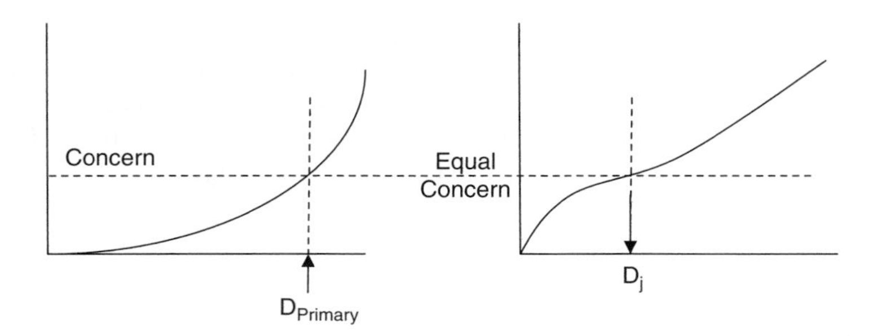

Figure 8.1 Illustration of equal concern deviations

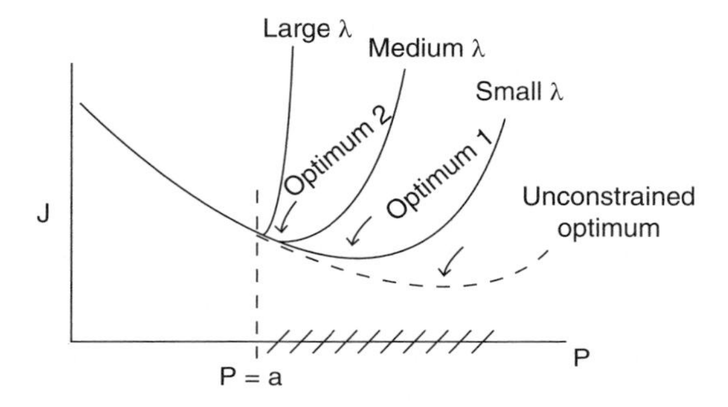

Figure 8.2 Illustration that a soft constraint permits a small violation

Equation 8.8 shows how the EC factors can be used to normalize the OF. Here, the deviation of the primary OF is based on one data point, so the 1/*N* factor in front of the first sum normalizes it. Each of the individual penalties are represented in the second sum:

$$\underset{\{\underline{p}\}}{Min}\ J = \frac{1}{N}\sum_{i=1}^{N}\left(\frac{y_i - \tilde{y}_i}{D_{Primary}}\right)^2 + \sum\left(\frac{b_j}{D_j}\right)^2 \tag{8.8}$$

The term *soft* for the constraints is an acknowledgement that constraint violations might not be eliminated. The optimization statement of Equation 8.8 balances the penalty for constraint violations with an attempt to minimize model data deviations. However, it often permits some constraint violation, depending on the problem structure.

Figure 8.2 illustrates how a penalty function, treating a constraint as soft, permits a constraint violation. Here the decision variable, the model coefficient being adjusted, is p and the OF value is J. There is a constraint on p that it should be less than a. To the left of $p = a$ there is no penalty, but to the right of $p = a$ a quadratic penalty is added to J. The dashed line represents the value of the OF with no penalty. The solid line represents the value with the penalty. At the value of $x = a$ the OF is still dropping rapidly, but at x values just greater than a, the penalty for the constraint violation is very small. Therefore, just to the right of the constraint, the total OF value continues to decrease. However, further to the right, the deviation squared begins to add a large value to the OF, dominating the trend. The net result is that the optimum is slightly into the constraint region. The impact of adjusting the penalty weighting factor is also illustrated. Higher penalty factors reduce the constraint violation, but cannot remove it. Excessively high penalty factors effectively create a discontinuity in the OF.

Often the problem structure is such that as a DV value changes to decrease the primary OF, its influence increases a penalty. If the Lagrange multiplier (or the equal concern factor) is such that the increase in penalty is less than the decrease in the primary OF, then the optimizer will move in that direction. If constraints are active, if $b_j \neq 0$, decreasing the EC factor, D_j, increases the penalty for the jth constraint violation and shifts the DV choices to lessen the constraint violation, but there will be a point at which the change in penalty is less than the change in benefit of the primary OF. Although soft penalty functions often cannot eliminate the violation, this soft constraint behavior is acceptable in many instances.

There is not much to be done about this situation. The only way the soft penalty will prevent a constraint violation is if the primary OF is not decreasing at the constraint point. Larger lambda values will move the optimum closer to the constraint, but still permit a violation. Extremely large lambda values make the penalty seem like a discontinuity to the optimizer, which can lead to optimization difficulty.

Conceptually, you could use absolute values of deviations to represent the constraint penalty or you could use a fixed penalty if the constraint is violated, instead of the quadratic penalty in Equation 8.8. However, either of these approaches creates a slope or a level discontinuity in the OF surface, which is a difficulty for many optimizers.

An advantage of the soft constraint, quadratic penalty function approach is that the smooth and quadratic functionality of the OF makes optimization relatively easy. Since there are no surface discontinuities on the OF or forbidden regions for the DVs, gradient-based optimizers (Cauchy, Newton–Raphson, Levenberg–Marquardt, Rosenberg, conjugate gradient, etc.) all work. There are no constraints on the optimizer choice of DV values. The optimizer is always told, "Yes, you may choose that set of DV values for the trial solution. Since you chose that set of DV values, here is the OF value. Continue."

However, when a constraint is absolutely not permissible, we use *hard* constraints. This changes the objective statement and affects the choices of optimizers.

8.5 Equality Constraints

Equality constraints may actually be a benefit to optimization. Often, when they are on the DVs, they can be used to reduce the number of DVs. This is in contrast to what was presented in Equation 8.5, where the deviation from the balance (or equality constraint) is the measure of badness and its penalty was added to the OF as a soft constraint.

Often the equality constraint can be rearranged with other constitutive relations to explicitly reveal the DVs in the equality constraint, and this equation rearranged to solve for one DV value given the others. Where this is possible there is no need to search for all M number of the DV values independently. Since a particular set of $M - 1$ DVs and the equality constraint permits calculation of the remaining DV, the optimizer only has to search for $M - 1$ DV values. This reduces the DV search space dimension from M to $M - 1$. Since many optimizers have a computational burden that has a quadratic dependence on the DV dimension, using equality constraints to reduce the DV dimension can yield a significant reduction in computational burden.

8.6 Takeaway

Seek to use equality constraints to reduce the optimization DV dimension. Soft constraints are compatible with gradient-based optimizers, but require user-defined equal concern factors and permit some constraint deviations. Hard constraints create a discontinuity that is best accommodated by direct search optimizers. Constraints are critical to the solution, so develop them with as much care as you invest in developing the model or experimental plan.

Exercises

8.1 Describe and state constraints associated with a first-order plus delay model.

8.2 Describe and state constraints associated with an ARMA(2, 1) model that represents an open-loop stable process.

8.3 Describe and state constraints associated with an ARMA(1, 1) model that represents an open-loop unstable process with a delay.

8.4 Choose a situation for model regression with one constraint that can be presented in the form of Equation 8.4. Decide on the number of data, units for the modeled result, and units for the constrained model coefficient. Discuss issues related with the appropriate value of lambda and choose a value.

9

The Distortion of Linearizing Transforms

9.1 Linearizing Coefficient Expression in Nonlinear Functions

Unconstrained linear regression, see Equations 1.1 to 1.8, has some desired attributes over nonlinear optimization. Linear regression is guaranteed to find a solution, to find a unique solution, and to find a solution within a time interval that can be specified based on problem dimension. First learned, frequently reinforced, and less complex in concepts, linear regression is better understood than nonlinear regression. More commonly used, it is more familiar to both user and the user's audience. As a result, we often seek to linearize coefficient expression in models that are nonlinear in the coefficients. Most commonly log, reciprocal, square root, and similar functional transformations are used.

This use of the term linearization is different from the substitution of a truncated Taylor series expansion of nonlinear terms, and then arguing that small deviations from a nominal value permit truncation of higher order terms, leaving only the linear terms as an approximation. *Functional transformation* is mathematically exact, it is not an approximation, and it does not require limiting a variable range to keep a truncated approximation reasonably true.

As examples of linearizing transforms, of functional transformation of the nonlinear problem to a linear form, here are three common models. First, is the power law model relating flow rate to orifice pressure drop, where the coefficient b appears nonlinearly:

$$F = a(\Delta P)^b \tag{9.1}$$

Log transformation (taking the natural logarithm of both sides) yields:

$$\ln(F) = \ln(a) + b \ \ \ln(\Delta P) = c + b \ \ \ln(\Delta P) \tag{9.2}$$

In this form, since values for F are known, values for $\ln(F)$ are known. Similarly, values for $\ln(\Delta P)$ are known. Since a is a constant, $\ln(a)$ is a constant, explicitly indicated as c. The

Nonlinear Regression Modeling for Engineering Applications: Modeling, Model Validation, and Enabling Design of Experiments, First Edition. R. Russell Rhinehart.

log-transformed model is linear in coefficients c and b, which can be more easily seen if it is represented by more familiar variables:

$$y = c + bx \tag{9.3}$$

The second common example is a dimensionless group correlation, in which coefficients b and c both appear nonlinearly:

$$Nu = a\, Re^{b} Pr^{c} \tag{9.4}$$

Log transformation linearizes it:

$$\ln(Nu) = \alpha + b\ln(Re) + c\ \ln(Pr) \tag{9.5}$$

Again, this is now linear in the coefficients, $\alpha = \ln(a)$, b, and c.

Third is the logistic function, sometimes called a mono-polar sigmoidal function, often used as a reasonable approximation to the Gaussian (or normal) cumulative distribution function. Frequently, the probability of an event occurring, p, is dependent on the value of a particular situation, x, with c representing the center (or x-value of a 50% probability event) and s represents a scale factor:

$$p(x) = \frac{1}{1 + e^{s(x-c)}} \tag{9.6}$$

This function can be linearized in the s and c coefficients by several stages. Take the reciprocal of both sides, subtract unity, log transform, and then combine coefficients to obtain

$$y = \ln\left(\frac{1}{p(x)} - 1\right) = sx - sc = sx + b \tag{9.7}$$

Fourth is the elementary form of the Arrhenius reaction rate temperature dependence, in which the activation energy, E, appears nonlinearly.

$$r = k_0 e^{-E/RT}[c] \tag{9.8}$$

Divide by concentration and then log-transform:

$$y = \ln\left(\frac{r}{[c]}\right) = \ln(k) - E\left(\frac{1}{RT}\right) = a + bx \tag{9.9}$$

This last example uses three transformation techniques: rearrange the equation, take the log of both sides, and use the reciprocal of R times T to convert it into the form that is linear in coefficients. The activation energy value is directly determined by the negative of the slope and the value of $k = e^{a}$.

There are many other possible equation rearrangements, variable substitutions, or functions that could be appropriate to linearize coefficient expression in model equations that have alternate forms. Mathematically, these transformations are permissible and defensible.

By contrast, again, note that these are not linearizing approximations, as would be approximating the nonlinear term with a truncated Taylor series expansion. In a linearizing approximation the response variable remains the same. By contrast in a functional transformation to linearize the regression coefficients, the response variable has a different functionality and different units.

9.2 The Associated Distortion

Because transformations change the response variable (from r to $\ln(r/[c])$ for example), they distort the relationship between the coefficient and the response variable. Consider the log-transformation of the variable y, subject to uncertainty, ε_y. Figure 9.1 plots $\ln(y)$ w.r.t. y and illustrates the distortion. Note that equal increments of y on the horizontal axis map unequal increments on the $\ln(y)$ axis. There are two aspects of this distortion.

First, consider what might arise from experimental data. Figure 9.2 illustrates experimental data from a typical power law relation, such as flow rate and orifice-reported transmission signal milliamperes reading. Note that experimental design has placed the flow rate data on fairly uniform increments through the entire experimental range. Also note that the experimental error, the deviation from the dashed line, the unknowable truth about Nature, is equivalent for both high and low flow rates.

However, when log-transformed, Figure 9.3 shows how the low F values have an amplified leverage, relative to the centroid, which amplifies their importance in the least-squares objective. Note that most of the log-transformed data are clustered to the upper right of Figure 9.3, and the low value is far from the data centroid. This is one aspect of the distortion that gives the low data point excessive weighting on the location of the best line. The second aspect of the distortion is that experimental errors for low values are also amplified. Again, this is illustrated

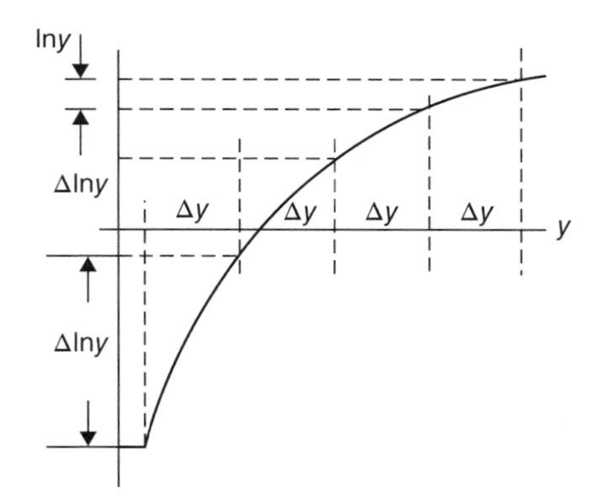

Figure 9.1 Nonlinearity of the log transformation

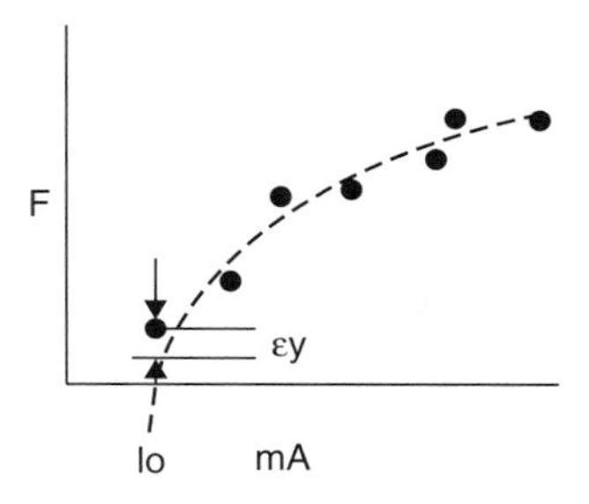

Figure 9.2 Illustration of experimental data

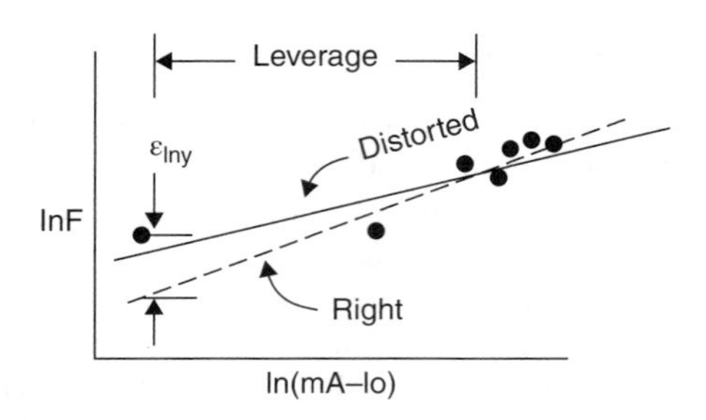

Figure 9.3 Log-transformed model and data

in Figure 9.3, where the dashed line is the unknowable truth. See how far above the line is the leftmost data point compared to the rightmost one. The large deviation and faraway leverage of the leftmost data lead to a least-squares fit (solid line in Figure 9.3) that is substantially different from the truth. Another data set might have a negative error on the leftmost point and lead to a bad model in the other extreme.

Log-transformed variables effectively lead to a weighted least squares that magnifies the uncertainty on lower values and amplifies their leverage on the line, emphasizing them while dismissing higher values. Each of the two aspects of the effect can be mathematically quantified.

First, consider the weighting or leveraging aspect. A weighted sum of squares objective function is represented by

$$J = \sum \omega_i (y_i - \tilde{y}_i)^2 \tag{9.10}$$

where $(y_i - \tilde{y}_i)$ is the residual, the customary vertical deviation, and ω_i is the weighting placed on the importance of the ith data set.

If the objective is to minimize the sum of squared log-transformed deviations,

$$J = \sum (\ln(y_i) - \ln(\tilde{y}_i))^2 = \sum \ln^2(y_i/\tilde{y}_i) \tag{9.11}$$

then the implied weighting factor for Equation 9.10 is

$$\omega_i = \left(\frac{\ln(y_i/\tilde{y}_i)}{y_i - \tilde{y}_i}\right)^2 = \left(\frac{\ln((\tilde{y}_i + \varepsilon_i)/\tilde{y}_i)}{\varepsilon_i}\right)^2 \tag{9.12}$$

where ε_i represents the residual, the process to model mismatch.

The reader should be able to visualize the impact of both the y value and experimental error on the weighting factor.

Accordingly, if using Equation 9.11 for the least-squares regression, the weighting required to make $\omega_i = 1$ in Equation 9.10 is the reciprocal of Equation 9.12.

One could apply the reciprocal of the weighting to the log-transformed deviations and seek to determine model coefficients that minimize

$$J = \sum \omega_i^{-1} (\ln(y_i) - \ln(\tilde{y}_i))^2 \tag{9.13}$$

However, the reader can affirm that this reduces to the original untransformed objective:

$$J = \sum (y_i - \tilde{y}_i)^2 \tag{9.14}$$

Second, propagation of error reveals that the uncertainty on $\ln(y)$ is

$$\varepsilon_{\ln(y)} = \left|\frac{\partial \ln(y)}{\partial y}\right| \epsilon_y = \frac{1}{y}\epsilon_y \tag{9.15}$$

Consider a variable, y, that has a uniform uncertainty, ε_y, throughout its range. Equation 9.15 indicates that when the value of y is large, the uncertainty on the log-transformed value, $\varepsilon_{\ln(y)}$, is small, and vice versa. This results in small y values having an amplified $\varepsilon_{\ln(y)}$ (vertical displacement), which disproportionately increases the dominance of data in the range of low y values and exacerbates the weighting importance due to distance leveraging.

Propagation of uncertainty in the transformed logistic model of Equation 9.7 results in

$$\varepsilon_y = \left|\frac{\partial y}{\partial p}\right| \epsilon_p = \frac{1}{p(1-p)}\epsilon_p \tag{9.16}$$

which indicates that when p is either nearly unity or zero, the displacement in the y value is greatly amplified over that for mid-range p values.

When experimental variability is very small, or there are many data points appropriately placed to average the sensitive areas, or all of the data are in the same range (with equivalent transformation relocation), the impact of the distortion by functional linearization will be small, and functional linearization is often defensible when balancing perfection with sufficiency. However, the larger is the variability in measured values, or the larger is the range on the measured values, or the fewer are the number of data points, the greater is the impact of the distortion on the resulting coefficient values.

Functional linearization has several drawbacks. Consider that the original desire is to minimize deviations in the reaction rate in Equation 9.8. The linearized Equation 9.9 will direct the optimizer to minimize the log-of-reaction-rate-divided-by-concentration, which leads to several undesirable attributes. (i) If you understood the concern about deviations in the reaction rate, how do you relate concerns to the log of the reaction rate scaled by concentration? (ii) The log transform distorts the importance of deviations, weighting small values much greater than large values, leading to a model that attempts to fit the noise on small values and disregard the importance of large values. (iii) The new response variable may not reflect the desired objective. Even the units on sum of squared deviations (SSD) and standard error of estimate from linear regression will be irrelevant to the assessment of uncertainty on the modeled output. (iv) Even if ε_y is uniform throughout the y range, the nonuniform $\varepsilon_{\ln(y)}$ violates the idealized basis in analysis of variance in linear regression. (v) Sometimes when real data (with noise) is entered into linearizing transforms, it can lead to undefined operations such as divide-by-zero, log-of-a-negative-number, and so on, which would not be encountered in the untransformed data.

However, functional linearization of the model permits linear regression techniques which are guaranteed to have a single solution, are well analyzed, familiar, provide derivative properties (standard error of the coefficients), can be used for screening and initial coefficient range evaluation, and are conventionally accepted. Functional linearization permits data presentation that should easily be identified as linear or not (to test model and data), and provides reasonable first estimates of the model coefficient values. Therefore, sometimes you might want to start with linear regression on functionally transformed models.

As an aside, if you use Excel Charts to provide a power law fit to data, and if you independently log-transform the data and ask the Excel Chart to provide a linear trend line, you will find that the coefficient values are exactly the same. This is because Excel Chart trend line calculations use linear techniques and log-transform the power law model to use linear regression. You could use Excel Solver, instead, to independently find power law coefficients and to find the linear model coefficients on the log transformed data. If you do, you will find Solver returns the same coefficient values on the log-transformed data as did the Excel Chart trend lines, and that the nonlinear power law coefficients are slightly different. The greater the variability (noise) on the data the greater will be the difference.

9.3 Sequential Coefficient Evaluation

There seems to be a common view that a helpful "trick" in nonlinear regression is related to determining coefficients in a model sequentially. I have observed this view in work related to chemical reaction kinetics, equilibrium thermodynamics, and viscoelastic modeling. Therefore, I suspect it is widespread. The "trick" is to simplify the complicated model, determine simplified model coefficient values, then add complexity to the model (more functionality, more coefficients), and, keeping the coefficient values of the simple model, determine the added coefficient values of the complex model that best fits the data.

This could be an appropriate procedure if the steady-state and transient portions of a model are independently fitted to the data. First, take steady-state data and fit the model to it. Then, using that steady-state model, adjust the dynamic model parameters to best fit the transient data. However, this presumes that the steady-state coefficients and transient coefficients are separable – that the values of the transient coefficients have no impact on the steady-state portion of the model.

Here is a simple illustration of the problem with sequentially determining coefficient values. The data reflect a simple truth about nature, $y = x^2$, the relation with noise on y. The concept is to use a power series (Taylor's series) expansion of a model $y = a + bx + cx^2$. However, the "clever" modeler may erroneously determine the coefficients sequentially. First $y = a$. This is graphed as Model 1 in Figure 9.4. Then with that value of a fixed, $y = a + bx$, Model 2. Then

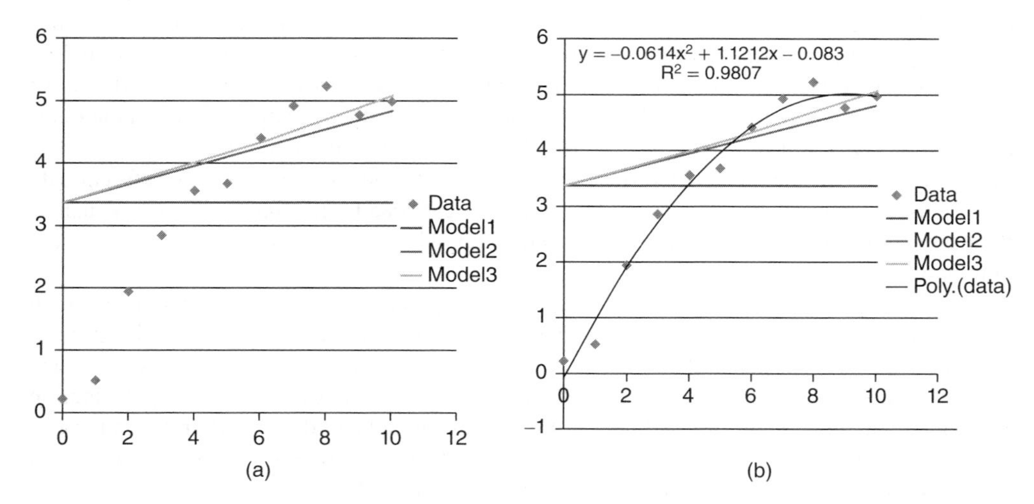

Figure 9.4 Sequential coefficient evaluations: (a) the findings and (b) comparison to simultaneous regression

with those values of a and b fixed, $y = a + bx + cx^2$, Model 3. It should be obvious that Model 3 is not a good fit to the data. The data are curved downward, but Model 3 is actually curved upward, and Model 3 is far from the data in the $0 < x < 4$ range.

By contrast, a quadratic trend line, in which all coefficients are evaluated simultaneously, provides a much better fit.

9.4 Takeaway

Linearizing transformations distort data impact (leverage and weighting, and interpretation of concern). As a result, I recommend using nonlinear regression on the original model function, not a linearized version. However, transforming the model to obtain linear trends may be useful as a first estimate of a nonlinear trial solution or to see if the trends are as expected.

Do not evaluate coefficients sequentially. In some instances it is good to evaluate coefficients in a simplified model to provide initial guesses for their values in the more complex model. However, if you progressively add complexity, let the optimizer adjust all coefficient values simultaneously.

Exercises

9.1 Choose a model that is nonlinear in coefficients, for example, Equations 9.1, 9.4, 9.6, or 9.8. Choose coefficient values. The model and coefficient values would represent the truth about Nature. Choose about eight values for the independent variable over a wide range and use the model to generate dependent variable data. This would represent the truth about Nature's response, but you cannot know it exactly. Add random noise to the dependent variable data to represent what might actually be measured from an experiment. This data set is termed a realization, one particular possible outcome. Now do regression in two ways. For one, use nonlinear optimization on the untransformed data to minimize the SSD of the dependent variable. For the other, use linearizing transform(s) on the model to make coefficients linear and then use linear regression to determine coefficient values. Compare the obtained coefficient values to the true values that generated the model. Plot the resulting models with the "unknowable truth." The better approach will compute coefficients that are closer to the true values and result in a model curve closer to the true curve. For nearly every realization, you should find that the nonlinear regression approach provides better results. In generating data, be true to reality. Choose a nominal x value, but perturb it a bit before calculating the y value. Then add noise to the modeled y value to represent the measurement. Your data representing your experimental knowledge of the process should be the nominal-x and y-measurement pairs. You cannot know either the actual (perturbed) x value or the true y value.

9.2 Explore several aspects in comparing nonlinear regression to regression of the transformed data and model. You choose the number, location, and noise aspects of the data so that your comparison of the two regression techniques reflects the number, placement, and noise on the data.

(a) If your data has zero noise, then both regression approaches should give identical coefficient values, equaling the true (but unknowable) values; however, the convergence criteria on the numerical algorithm will limit precision.

(b) If the noise is too high, neither method will give good values.
(c) If there are many data points, then, in any one region, "+" and "−" deviations will balance the "pull," and the linearizing transform approach should not be biased by the weighting distortion.
(d) However, if there are too few (just two) data points, the two approaches should return identical results, but bad, wrong, coefficient values.

9.3 Repeat Exercises 1 and 2 for the following:
(a) A translated reciprocal relation (the regression parameters are a and b)

$$y = \frac{a}{x - b}$$

(b) Chick's law for a microbe count in relation to the chlorine dose and time of exposure (the regression parameters are k and n):

$$N = N_o e^{-kc^n t}$$

(c) A mixing rule for two components (the regression parameter is p):

$$y_{mixture} = y_1 x_1{}^p + y_2(1 - x_1)^p$$

(d) A packed bed reaction model (A + B = C + D, excess B so that the reverse reaction can be ignored, B assumed constant, and the reaction first-order in A, using Hougan–Watson or Langmuir–Hinshelwood kinetic model and uniform flow at steady state, the regression parameters are k_0, k_1, and E):

$$\ln\left(\frac{c_o}{c_i}\right) + k_1(c_o - c_i) = \frac{Lk_0}{v} e^{-E/RT}$$

(e) The Swamee–Jain explicit approximation to the Colebrook–White expression for the Moody–Darcy friction factor in flow through full circular pipes (the regression parameters are a, b, c, and d):

$$f = a\left[\ln\left(\frac{\varepsilon}{bD} + \frac{c}{Re^d}\right)\right]^{-2}$$

10

Optimization Algorithms

10.1 Introduction

For the most part, you will use pre-written optimizers to solve for the model coefficient values. In optimization terminology these are the optimal DV (decision variable) values, noted by the symbol DV*. However, you will need to make many choices related to the algorithm and its parameter choices, to the convergence criterion and its thresholds, to the number of starts, to the start locations, and so on. Accordingly, you should understand optimization.

10.2 Optimization Concepts

In regression, the objective is to find the best model coefficient values, meaning those coefficient values that minimize the objective function (OF), as presented in Chapter 7. Alternately, the OF is called the cost function or fitness function. The act of finding best values is termed optimization. For models that are linear in the coefficients, optimization is a sequence of one-step procedures in linear algebra, which ends at the optimum at the end of one pass through the procedure. However, for nonlinear optimization, the search is an iterative procedure that progressively moves the coefficient values toward their best. The DVs in regression are the coefficient values and a trial solution (TS) is a set of coefficient values.

A common tool for conceptualization of optimization is to imagine a contour map of a region of the Earth's surface, and relate the OF to the land elevation, and the DVs to the East–West and North–South directions. The optimization objective is to find the latitude (East–West) and longitude (North–South) values that indicate the point of lowest elevation. Often the surface is illustrated as a net in a three-dimensional isometric illustration with the vertical z axis representing elevation (the OF value) and the horizontal x–y plane representing the latitude and longitude (the DV values). Alternately, the surface is represented as a contour map. Figure 10.1 illustrates these two viewpoints.

A *trial solution* (TS) is a guess of what might be the best x–y (DV) values. The initial guess might be chosen by the user or it might be randomly generated. The job of the optimizer is to sequentially find new trial solutions (values for the DV set) that move the trial solution downhill to the lowest point (to the minimum OF value). The optimum DV and corresponding OF value are noted as DV* and OF*.

Nonlinear Regression Modeling for Engineering Applications: Modeling, Model Validation, and Enabling Design of Experiments, First Edition. R. Russell Rhinehart.

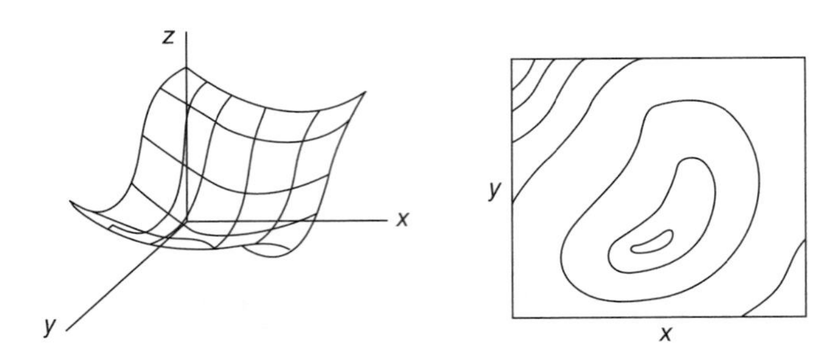

Figure 10.1 Illustration of a 2-DV optimization, 3-D surface net, and 2-D contour map

Moving downhill seems simple. Balls roll downhill. Water flows downhill. Inanimate objects can do it. If you were standing on a hill somewhere and looking around, you could see the downhill direction and you could scan the visible surface for the lowest point. Then you could jump to it in one move or walk directly toward it in sequential downhill steps. Recognize, however, that when you look you are seeing the entire surface elevation (OF value), you are seeing the OF value at every location, and can immediately point to the minimum. You can instantly see every OF value, the result of every DV TS. In Figure 10.1, for example, you can see the entire surface.

By contrast, the optimizer only knows the OF value at the trial solution point and at the prior points visited. It has not explored all DV values, so it does not know all OF values. It cannot "see" the entire surface. It is like standing on a surface in a dense fog. You can only know your elevation, the local slope (which you can sense by the angle of your feet), and the elevations of where you had been (past trial solutions). However, an optimizer is at the trial solution, a point. It is like being on a pogo stick, in the fog, with no "feel" for the downhill direction. The optimizer, in a sense, is blind and must test the local contour with tentative, small steps before knowing which way is downhill, in order to determine what direction to move the trial solution and how far to move it.

Although this introduction presented the concepts by describing a surface in three-dimensional space, with a familiar context for human visualization, the aspects are scalable to higher dimensions for which we cannot visualize the N-dimensional "space."

Although there are many ways to classify optimization algorithms, there are two general contrasting classes: gradient-based and direct search. *Gradient-based* optimizers use the local slope of the surface to point the direction downhill and to calculate how far to jump to the next trial solution. The jump-to trial solution is an estimate of the location of the surface minimum based on local surface information, and usually a quadratic model assumption. However, it is likely not to be the exact minimum; and at the new TS, the slopes are re-evaluated and the next best estimate of the minimum is calculated. The procedure is iterative – analyze slopes, estimate a new trial solution, evaluate the OF at the new TS, and repeat. Names for classic gradient-based approaches include Cauchy, incremental steepest descent (ISD), Newton–Raphson, and Levenberg–Marquardt (LM).

Successive quadratic (SQ) is a method that uses local samples of the surface to generate a quadratic model of the surface, a surrogate model, and then jumps to the TS at the minimum of the surrogate surface. Although SQ does not explicitly use gradients, it presumes that the entire surface is an extension of a local quadratic model and has the same properties as gradient-based techniques.

If the surface is not well-behaved, not quadratic-like, then the gradient-based methods can send the trial solution in strange directions. This happens when constraints and model discontinuities are active, and happens when the TS is on the other side of an inflection from the DV*. Further, time or space discretization in models can create local surface undulations that confound gradient estimates. Finally, the gradient value can be strongly impacted by the user's choice of forward, backward, or central difference, and the size of the DV step used in a numerical estimate of the derivative.

Direct search methods are in contrast to gradient-based methods. They do not presume a surface structure or hope to jump to the optimum in one move. They use heuristic rules to direct incremental changes for the next trial solution. Names for direct search methods are cyclic heuristic, Hooke–Jeeves, Nelder–Mead, particle swarm, leapfrogging, and genetic algorithms.

When the quadratic surface assumption is nearly valid, the gradient-based techniques will jump to the proximity of the solution and then jump very near to optimal in a few iterations. This would seem to make them more efficient. However, at each new trial solution, the optimizer must apply the effort to assess all first and second derivatives and whether analytically or numerically this can be a substantial computational load. Direct search algorithms only use the single TS value. Further, direct search methods are more robust to surface aberrations and do not presume the nonlinear surface is quadratic. They often get to the optimum in fewer function evaluations (with less computational work).

This chapter will provide an overview of the mechanics of important optimization procedures and their issues related to nonlinear regression.

10.3 Gradient-Based Optimization

Gradient-based optimizers are powerful and common, but have undesirable attributes as well.

Gradient-based optimizers evaluate the local slope of the surface to determine what direction is downhill. The steepest downhill direction may not be aligned with a DV axis. Accordingly, there will be a slope for each axis. The slope, of course, is the partial derivative of the OF value with respect to each DV:

$$\text{Slope}_i = \frac{\partial OF}{\partial DV_i} \tag{10.1}$$

In Equation 10.1 the subscript i represents the DV number. For an M dimension application (one with M DVs) there are M partial derivatives to define the slope.

If one could generate an equation to represent the derivative, the estimate of the slope, solving the equation is a computer procedure. If there are M number of DVs then there are $M + 1$ number of functions to be evaluated, one for the OF and M for the slopes. However, the function might not provide a convenient or tractable form for solving the derivative analytically. Numerical estimates are usually used.

10.3.1 Numerical Derivative Evaluation

Conceptually, the slope could be calculated from an analytical expression, but, more generically and hence commonly, it is evaluated numerically as a finite difference approximation. This means that after the trial solution OF value was evaluated; the OF value must be evaluated at least M additional times (either analytically or numerically). If a forward difference or backward difference approximation to the derivative is used, the partial derivatives would be approximated with the finite differences:

$$\text{Slope}_i \cong \frac{OF(\underline{DV}) - OF(\underline{DV}_i^-)}{\Delta DV_i} \tag{10.2}$$

$$\text{Slope}_i \cong \frac{OF(\underline{DV}_i^+) - OF(\underline{DV})}{\Delta DV_i} \tag{10.3}$$

Equations 10.2 and 10.3, respectively, represent the backward and forward finite difference, where the subscript i represents the ith DV and the superscripted "+" and "−" symbols mean that the OF is evaluated at the trial solution, but with the ith DV increased or decreased by ΔDV_i. Either of these is a one-way look at the local slope. If ΔDV_i has the same sign as the general direction of the trial solution movement, then the forward difference is a better estimate of where the trial solution is likely to be moving. However, the size of the forward difference increment may move the $\underline{DV}_i^+$ location across a constraint. The backward difference increment may have a higher probability of representing a feasible slope estimate, but it characterizes search history, past direction not future.

In either case, each slope estimate requires one additional OF evaluation for each DV. The number of function evaluations, $M + 1$, is the same whether using an analytical function or the numerical approach.

Figure 10.2 illustrates forward, backward, and central difference approximations. The central difference is based on the two adjacent points. As illustrated, this provides a better surface characterization of the true slope at the base point. However, the central difference approximation requires $2M$ additional OF evaluations:

$$\text{Slope}_i \cong \frac{OF(\underline{DV}_i^+) - OF(\underline{DV}_i^-)}{2\Delta DV_i} \tag{10.4}$$

In either case (forward, central, or backward), someone must choose the value for ΔDV_i. It could be a fixed value. However, if it is too big for the end game (near the optimum) it may not provide a valid estimate of the local slope. This is illustrated in Figure 10.3, where the trial solution (x_b) is to the left of the optimum point (x^*), but the numerical central difference slope indicates that the subsequent trial solution should move further toward the left, the wrong

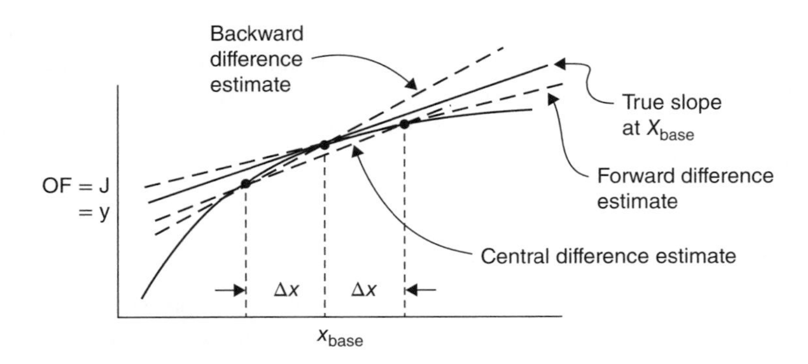

Figure 10.2 Illustration of numerical approximations to the slope

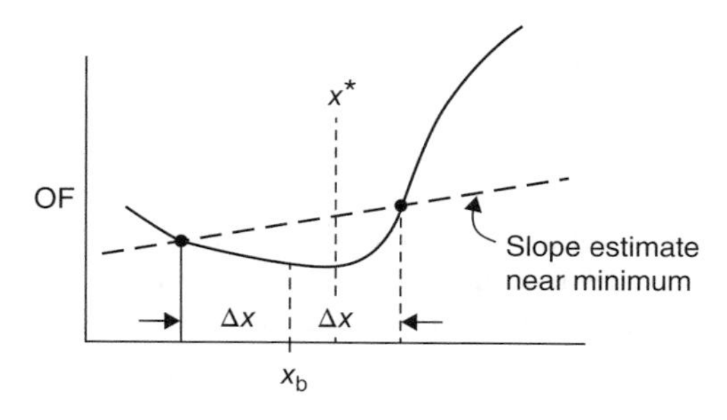

Figure 10.3 Illustration of possible misdirection by numerical slope estimates

direction due to numerical approximation, and so would the forward difference slope in this illustration.

Alternately, ΔDV_i could be calculated as proportional to the last trial solution change in the DV, which would provide a range for evaluating the slope that is locally appropriate. However, if ΔDV_i is too small, the digital truncation error can distort the slope estimate. Proper choices require user input, which usually needs *a priori* knowledge of the surface features.

10.3.2 Steepest Descent – The Gradient

Viewed from a high elevation, and looking down on the surface as a two-dimensional contour map, the steepest descent direction is a straight line from the trial solution in a direction that is perpendicular to the contour. This is illustrated in Figure 10.4. It is relatively easy to mathematically prove that the direction of steepest descent is normal to the contour (perpendicular to the tangent to the contour at the trial solution point). The set of equations that would describe the initial direction of motion of a ball placed on the surface and beginning to roll downhill due to gravity represents the steepest descent direction.

The line of steepest descent does not necessarily point to the minimum, as illustrated in Figure 10.4.

Mathematically, the line of steepest descent can be defined in vector notation as a deviation from the local point. In two dimensions, x and y with the trial solution at (x_0, y_0), the (x, y) pairs that define the line are defined as

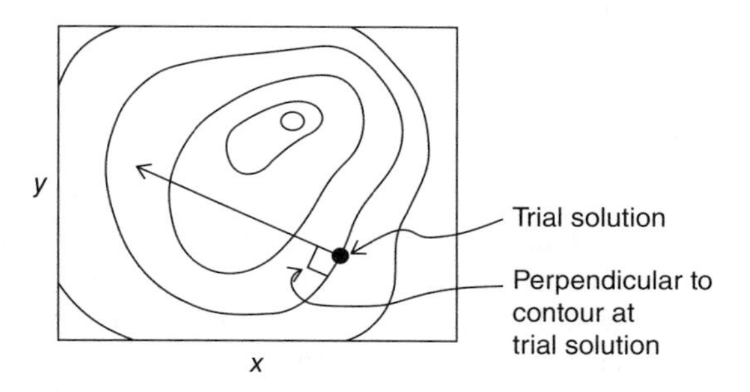

Figure 10.4 Illustration of steepest descent in a 2-DV contour map

$$x = x_0 - \alpha \frac{\partial OF}{\partial x}$$
$$y = y_0 - \alpha \frac{\partial OF}{\partial y} \tag{10.5}$$

where α is a common scale factor for both dimensions. This is often termed a parametric model in which the x and y values are dependent on the value of the parameter α. The value of α is related to the distance along the line from the TS point. If the distance is the Pythagorean distance

$$S = \sqrt{(x - x_0)^2 + (y - y_0)^2} \tag{10.6}$$

then

$$S = \alpha \sqrt{\left(\frac{\partial OF}{\partial x}\right)^2 + \left(\frac{\partial OF}{\partial y}\right)^2} \tag{10.7}$$

or, if the distance, S, along the line were to be specified, then the α value required for the (x, y) pair to move that distance is

$$\alpha = \frac{S}{\sqrt{\left(\frac{\partial OF}{\partial x}\right)^2 + \left(\frac{\partial OF}{\partial y}\right)^2}} \tag{10.8}$$

This mathematical analysis is commonly found in introductory optimization texts. Equation 10.5 can be shown to indicate the direction of steepest local descent and the analysis is scalable to M dimensions. In vector notation, Equations 10.5 and 10.8 become

$$\underline{\Delta x} = -\alpha \underline{\nabla} f \tag{10.9}$$

$$\alpha = \frac{S}{\sqrt{\underline{\nabla} f \cdot \underline{\nabla} f}} \tag{10.10}$$

in which $\Delta x = x - x_0$.

The analysis presumes that the units on x and y are identical. If the model coefficients represent different model attributes, then it is likely that the units are not identical. Practically, this is not an issue in optimization and mathematically the equations are valid. However, the unit mismatch can be discomforting to an engineer. If so, use scaled variables in which the DV would be scaled by its expected range. Then DVs are dimensionless and there is no dimensional conflict in Equations 10.5 to 10.10.

10.3.3 Cauchy's Method

Cauchy's method searches along the line of steepest descent. When it gets to the minimum along that line, it again evaluates the direction of steepest descent and begins a new line search. Advantages of this approach are the simplicity of evaluating the steepest descent, followed by a simple one-dimensional line search (regardless of the number of dimensions, M).

An issue with this sequential line search is that function evaluations are wasted when finding the minimum along the initial lines that are still not near the optimum. Initially, there is no need to find the true minimum along the line of steepest descent.

10.3.4 Incremental Steepest Descent (ISD)

Figure 10.4 also illustrated a hill surface in which the steepest descent path curves as the TS progresses along it toward the minimum. Move a bit from the TS along the steepest descent line and see how the direction of steepest descent changes. Water trickling from overwatered plants on a driveway follows the steepest descent path, re-evaluated at each step downhill, as illustrated in Figure 10.5.

Also, the set of equations that would describe the motion of an inertialess ball rolling downhill due to gravity represents an optimizer that moves incrementally in the steepest descent direction.

In an ISD algorithm, take a step of length S along the direction of the steepest descent and re-evaluate the direction of steepest descent. Use Equation 10.10 to calculate α from S and the gradient and then Equation 10.9 to calculate the new TS DV values. If the step was a success, if the OF value at the new TS is better than at the old, then make S larger and repeat. Alternately, if the new TS is either worse or violates a constraint, return to the old TS value and try again with a smaller S value. I find that an S expansion of 1.1 and contraction of 0.5 provide robust and efficient acceleration and contraction values, and are heuristically comfortable.

10.3.5 Newton–Raphson (NR)

In an ISD algorithm, the search takes cautious incremental steps along the direction of the local steepest descent. By contrast, if an optimizer could characterize the curvature of the downhill

Figure 10.5 An example of incremental steepest descent

path, then it could jump to a trial solution where it seems to end (at the minimum) and would find the minimum faster.

This higher-order surface characterization could be obtained by calculating second-order derivatives. The Newton–Raphson method is the basic approach. Illustrated for convenience in a 2-DV situation with DVs represented by x and y, the NR algorithm can be derived starting with the desire that the derivative of the OF, f, with respect to each DV, x and y, is zero at x^* and y^*. Using g to represent the derivative and the subscripts to represent w.r.t. which DV for x, the notation desiring the derivative to be zero is

$$g_x = \left.\frac{\partial f}{\partial x}\right|_* = 0 \tag{10.11}$$

Expanding $g(x, y)$ in a Taylor series at the local function value at x^* and y^*, and truncating to linear terms gives

$$g_x(x^*, y^*) = 0 = g_x(x_0, y_0) + \left.\frac{\partial g_x}{\partial x}\right|_o (x^* - x_0) + \left.\frac{\partial g_x}{\partial y}\right|_o (y^* - y_0) \tag{10.12}$$

$$g_y(x^*, y^*) = 0 = g_y(x_0, y_0) + \left.\frac{\partial g_y}{\partial x}\right|_o (x^* - x_0) + \left.\frac{\partial g_y}{\partial y}\right|_o (y^* - y_0) \tag{10.13}$$

Combining Equations 10.12 and 10.13 to solve for DV*,

$$\begin{bmatrix} \left.\frac{\partial^2 f}{\partial x^2}\right|_0 & \left.\frac{\partial^2 f}{\partial x \partial y}\right|_0 \\ \left.\frac{\partial^2 f}{\partial x \partial y}\right|_0 & \left.\frac{\partial^2 f}{\partial y^2}\right|_0 \end{bmatrix} \begin{bmatrix} \Delta x \\ \Delta y \end{bmatrix} = -\begin{bmatrix} \left.\frac{\partial f}{\partial x}\right|_0 \\ \left.\frac{\partial f}{\partial y}\right|_0 \end{bmatrix} \tag{10.14}$$

Note that the right-hand side (RHS) of Equation 10.14 is the negative gradient, $-\underline{\nabla} f$, exactly as in Equation 10.9. The left-hand side (LHS) matrix of second derivatives is often termed the Hessian and the RHS vector is often termed the Jacobean. In vector–matrix notation, Equation 10.14 is

$$\underline{\underline{H}} \cdot \underline{\Delta x} = -\underline{\nabla} f \tag{10.15}$$

which can be rearranged in the same form as Equation 10.9, except that the inverse Hessian replaces α, providing step factor values for each dimension that are specific to that dimension:

$$\underline{\Delta x} = -\underline{\underline{H}}^{-1} \cdot \underline{\nabla} f \tag{10.16}$$

This is the Newton–Raphson method. It does not have the dimensional inconsistency problem of Equation 10.9 and does not need scaled variables for engineering sensibility.

While derived for a 2-DV case, Equations 10.15 and 10.16 are valid for M DVs.

The Hessian is a matrix of second derivatives. Along the main diagonal, the second derivatives are w.r.t. the same DV. There are M of these homogeneous second derivatives, one for each DV. To numerically evaluate the homogeneous second derivative use

$$\frac{\partial^2 f}{\partial DV_i^2} = \frac{f(\underline{DV}_i^+) - 2f(\underline{DV}) + f(\underline{DV}_i^-)}{\Delta DV_i^2} \tag{10.17}$$

Each of the M main diagonal elements requires one more function evaluation over that needed for the forward or backward difference gradient estimates, or no more if the gradient is estimated using the central difference approximation.

There are M^2 elements in the Hessian; M of them along the main diagonal are derivatives of a common DV. The remaining $M^2 - M$ off-diagonal elements are mixed derivatives. If the surface is continuous (no cliffs) and has continuous derivatives (no slope discontinuities), then the i–j derivative is the same as the j–i derivative:

$$\frac{\partial^2 f}{\partial x \partial y} = \frac{\partial^2 f}{\partial y \partial x} \tag{10.18}$$

so that only $(M^2 - M)/2$ number of off-diagonal terms in the Hessian matrix need to be evaluated. Numerically, the mixed second derivative is

$$\frac{\partial^2 f}{\partial DV_i \partial DV_j} = \frac{f(\underline{DV}^{++}) - f(\underline{DV}^{+-}) - f(\underline{DV}^{-+}) + f(\underline{DV}^{--})}{\Delta DV_i \Delta DV_j} \tag{10.19}$$

where the "+" and "−" superscripts on the DV vector indicate the deviations in the ith and jth DV values. Each of these off-diagonal elements requires four new function evaluations.

Therefore, in addition to the OF evaluation and the $2M$ evaluations needed for the central difference estimates of the gradient, $4(M^2 - M)/2$ additional function evaluations are required for the Hessian. Summing, each iteration requires $1 + 2M^2$ number of function evaluations to generate the next TS value. For many applications the additional work adequately provides the "intelligence" to jump the TS directly to a close proximity to the optimum.

There are alternative approaches that reduce the number of function evaluations. In some, the Hessian elements are updated once every few iterations. In quasi-Newton approaches the Hessian elements are estimated from past changes in the gradient.

Although the incremental changes to the DV to generate the next TS are expressed as one-line Equations 10.15 and 10.16, they represent a set of linear relations. Solving the set requires linear algebra numerical procedures such as Gaussian elimination. This adds a computational burden.

However, the major objection to this approach is that it seeks the point where the derivatives are zero, which might be a maximum, saddle, or minimum. We only want the optimizer to seek a minimum. The NR method presumes that the truncated Taylor series of Equation 10.12 is a legitimate model of the surface (if the gradient is linear, then the OF is quadratic), which is often an untenable assumption, and if the derivatives are numerical, they are only estimates. Therefore, in the proximity of the optimum with good derivative estimates, NR is a powerful, efficient, jump-to-the-optimum procedure, but it can be misdirected by surface features.

10.3.6 Levenberg–Marquardt (LM)

If the function is nearly quadratic and the estimates of the first and second derivatives are valid, then NR directs the TS jump to near the minimum. However, when not in the proximity of the minimum, local surface shapes can send the jump-to point far from the minimum. NR needs a quadratic-like surface. NR could also seek a maximum or a saddle point.

Various enhancements solve these NR issues by beginning the search with the steepest descent rule (with first-order derivative calculations), then progressively shift calculation of sequential trial solution values with the second-order derivative information. The methods of Marquardt, Levenberg-Marquardt, and Rosenbrock are most commonly used. To obtain an LM rule, combine Equations 10.9 and 10.15:

$$\left(\underline{\underline{H}} + \lambda \underline{\underline{I}}\right) \cdot \underline{\Delta x} = -\underline{\nabla} f \tag{10.20}$$

where $\lambda = 1/\alpha$ and α is the ISD step-size factor in Equation 10.9. Start with a large λ value, classically $\lambda = 10^4$. This should indicate that the λ values in the matrix on the LHS of the equation dominate the Hessian values. Then the rule of Equation 10.20 is essentially an ISD search with a small step size – safe, cautious, small steps downhill. As iterations progress, reduce the λ value. The LM technique halves λ at each iteration. Eventually, the λ value will be small relative to the Hessian elements and Equation 10.20 represents the NR method – a bold jump to the anticipated minimum.

10.3.7 Modified LM

As a personal preference, I do not like the uncertainty in choosing an appropriate value of the initial λ value so that it dominates all of the Hessian terms, or the rule to halve it at each iteration regardless of the result. Further, if the initial λ value is too large, the initial step sizes may be so small that the sequential TS values appear to have met convergence criteria. As an alternative, I multiply Equation 10.9 by λ and Equation 10.15 by the complementary value of $(1 - \lambda)$. Here $0 \leq \lambda \leq 1$ and λ indicates a weighting of ISD and NR; it is not the LM reciprocal of the step size coefficient. Summing results in

$$\left((1-\lambda)\underline{\underline{H}} + \frac{\lambda}{\alpha}\underline{\underline{I}}\right) \cdot \underline{\Delta x} = -\underline{\nabla} f \tag{10.21}$$

In this modified LM method, start with $\lambda = 1$, indicating the safe downhill ISD method, and select a step size, S, that is appropriate for the situation based on *a priori* experience. This should be a reasonable estimate, larger than the convergence criterion threshold, but smaller than the expected DV range. Determine the value for α from Equation 10.10. If the step is a success (no constraint violation and a better OF value) accept the new TS as the base case and incrementally decrease λ to permit more NR influence for the next move. If the step is not a success, then return to the prior successful TS and take two actions. First, incrementally increase λ to return to more influence from the safe ISD. Second, use ISD alone from the old TS. If ISD gives a better value, take it as the new TS and increase the step size. If not a success, reduce the step size. Then return to Equation 10.21 to calculate the new TS from the new λ and S values. A wide variety of reasonable rules work for adjusting λ and S values. Here is what I do. To expand S, multiply it by 1.1. To contract S, multiply it by 0.5. To increase λ, rapidly to return to the safe ISD, use

$$\lambda := 0.5(1) + 0.5\lambda \tag{10.22}$$

To reduce λ cautiously, but not too slowly, to shift to NR, use

$$\lambda := 0.8\lambda \tag{10.23}$$

10.3.8 Generalized Reduced Gradient (GRG)

If the downhill direction leads across a constraint, the above optimizers will aim downhill, meet the constraint, and stop there. The generalized reduced gradient (GRG) will follow along a constraint, when encountered, and come off it to continue the gradient-based search when the surface permits. To do so, there has to be logic in the optimizer that converts an active inequality constraint to an equality constraint, which reduces the search dimension. Which DVs remain in the reduced set? What to do if a second constraint becomes active? The programmer will have chosen the precedence.

10.3.9 Work Assessment

Recognize that, for an M dimension DV exercise, there are M first-order derivatives in the Jacobean and M^2 second-order derivatives in the Hessian. If the surface does not have discontinuities, the i–j derivative has the same value as the j–i derivative. This means that each trial solution, each optimizer iteration, must evaluate the OF $1 + 2M + 4(M^2 - M)/2 = 1 + 2M^2$ times. The computational work increases as the square of the number of DVs. For example, an 8-DV application will have about 1.8 times more function evaluations per iteration (for each progressive TS) than a 6-DV application. This explosion in computational burden, or experimental cost if experiments are providing the OF values, is often called the “bane of dimensionality.” Again, the correct values for ΔDV_i that are required for numerical estimation require *a priori* knowledge.

10.3.10 Successive Quadratic (SQ)

Another approach to characterizing the surface so that the optimizer can jump to the minimum is SQ.programming. Here, model the entire surface with a quadratic surrogate. With two DVs, x and y, the OF model is

$$f = a + bx + cx^2 + dy + ey^2 + fxy \tag{10.24}$$

There are six model coefficients in this estimate of the OF response to the DVs, which would require six function evaluations at each TS to determine the surrogate model coefficients. Then linear algebra techniques use the six sets of OF and DV values to determine the model coefficient values. This would be followed by the linear algebra solution of the two normal equations to determine x^* and y^* for the surrogate model.

In M dimensions, there is one intercept coefficient, two coefficients for each DV, and $M(M - 1)/2$ for the cross-product terms, for a total of $(2 + 3M + M^2)/2$ coefficients in the surrogate model requiring that number of OF evaluations. Then one linear algebra set would have $(2 + 3M + M^2)/2$ simultaneous equations and the other would have M simultaneous equations.

Although this is not explicitly gradient-based, there are strong equivalences. In SQ, explore the local surface area with $(2 + 3M + M^2)/2$ trial solution evaluations (for large M this requires about one-fourth of the number as with the Newton–Raphson type approaches). Use this information to create a quadratic model of the local surface, pretend that it is a reasonable approximation to the global surface, use linear optimization to determine the minimum of the surrogate modeled surface, and jump to that point. Defining the location of the exploration points also requires user-specified ΔDV_i values. If too large, they cannot make local sense of the surface, which is especially important at end-game convergence. If too small, they lead to truncation error in characterizing the surface and calculating the jump-to spot.

SQ also seeks minimum, maximum, or saddle points. Like NR, the results can be very dependent on the initial TS value.

10.3.11 Perspective

If surfaces were quadratic, and with ideal ΔDV_i values used to either define the surrogate surface model or to estimate the derivatives, SQ and any of the Newton–Raphson techniques will jump to the exact minimum after one iteration. However, surfaces are usually more complicated than a simple quadratic and you will likely not have chosen ideal values for the ΔDV_i. For this reason, a succession of trial solutions progressively moving toward the minimum is required. Depending on surface complexity, optimizers require several, to many, iterations.

There are many enhancements to gradient-based techniques, to enhance robustness, and to reduce the number of function evaluations. They can use higher level surface intelligence in the optimizer determination of the next trial solution. Excel Solver uses a GRG search, which searches along constraints when they are encountered. Quasi-Newton methods do not evaluate the Hessian elements at each iteration, but use the progression of Jacobean elements to estimate the second derivatives.

When a surface is well-behaved (nearly quadratic) or when the trial solution is near to the optimum (where the nearly quadratic approximation is a valid approximation), and when there are no constraints, and when there is a unique optimum, then gradient-based searches are very efficient. They find the minimum with a low number of function evaluations. However, gradient-based searches can become wholly confounded with surfaces that have discontinuities (in either slope or level, such as ridges and cliffs), flat spots, inflections, stochastic responses, or experimental noise. They also cannot analytically accommodate hard constraints. The user must specify the ΔDV_i values used in estimating the numerical derivatives. Because of this, my preference is to use direct search techniques, which do not use gradient information.

10.4 Direct Search Optimizers

In a *direct search*, the optimizer only uses the OF value. It does not use gradient information. Traditional direct search techniques are the *Nelder–Mead Simplex search* (Nelder and Mead, 1965; Spendly, Hext, and Himsworth, 1962) and the *Hooke–Jeeves pattern search* (Hooke and Jeeves, 1961). Each of these use an exploratory pattern to determine the local surface nature and then incrementally moves either away from the worst (NM) or in the direction indicated by the best (HJ). The size of the incremental move at each iteration changes as experience on the surface develops. It increases with TS success and contracts when the jump-to trial solution is not as good as the present trial solution. With these algorithms the user must specify an initial size of the Simplex or the pattern, which has the same too large or too small issues as those associated with choosing the ΔDV_i values in gradient-based methods. Although these optimizers make incremental progress downhill, which would require more iterations than gradient-based optimizers to get to the optimum, each iteration requires fewer OF evaluations. In my experience with nonlinear optimization, the Hooke–Jeeves search finds the optimum with fewer OF evaluations than the gradient-based methods. Although I do not have extensive experience with NM, I believe HJ and NM are equivalent in metrics to assess algorithms (the number of function evaluations to converge, the probability of finding the global optimum, the ability to handle constraints, the code complexity).

A simple cyclic direct search used to be my favorite nonlinear optimizer. It is very effective and extraordinarily simple to understand and code. Although not quite as fast at finding the optimum as the *Hooke–Jeeves* or *Nelder–Mead* methods, it is nearly as fast, far simpler to explain, to understand, and to program.

10.4.1 Cyclic Heuristic Direct Search

In this optimization search, start with a single feasible trial solution and evaluate the OF value. This is the base point. Then increment the first DV by ΔDV_1, keeping the other DVs at the base value, and evaluate the OF. If the new trial point is both feasible and has a better OF value than the base point, then you are moving in the right direction. Therefore, (i) make the new trial point the base point and (ii) increase the ΔDV_1 value so that the next time you explore that direction, you make a bigger step. Alternately, if the new trial point is either infeasible or does not have a better OF, then you have either moved too far or moved in the wrong direction. Therefore, (i) retain the prior base point and (ii) reverse sign and contract the ΔDV_1 value for the next turn of that DV. Repeat this for each DV, one-by-one. When each has been explored, repeat the cycle. Each cycle through all DVs is an iteration.

I recommend values of 1.05–1.2 for the expansion factor. My preference is 1.1. If moving in the right direction, each iteration moves 5–20% further than the past. In my experience, larger expansion values make the exploration point too aggressively moved after too few successes. Smaller values increase the number of iterations to get to the optimum.

I recommend values of 0.5–0.8 for the contraction factor. My preference is 0.5. These heuristic values also arise from qualitative arguments and personal experience. If the new trial solution is not a success, then reversing the search direction sends the search back in the direction it came from. Since the current point is better than a prior point, the reverse ΔDV should be smaller than the prior step. Therefore, the contraction factor should be less that the reciprocal of the expansion factor.

The user must choose the initial ΔDV_i values. I recommend starting with 10% of the DV range. If the wrong size or sign is chosen, contraction or expansion will quickly adjust the ΔDV_i values.

The algorithm is robust to surface aberrations (cliffs, slope discontinuities, flat spots, inflections), effective, and relatively efficient, and the code is very simple. The procedure can be easily implemented by a person guiding experimental trial-and-error choices.

Here is the active section of code in which comment lines are underscored:

```
For i = 1 to Number_of_DVs
  u(i) = ub(i) + du(i)          'Increment ith DV to search.
  CALL ConstraintTest           'Test for hard constraints violations.
  IF Constraint = "PASS" THEN   'No hard constraints are violated.
     Call ObjectiveFunction     'Determine the OF value.
     IF OF < OFb AND Constraint = "PASS" THEN
                                'If better OF and no violation, then …
        ub(i) = u(i)            '… accept the new DV value, …
        OFb = OF                '… update the SSD value, and …
```

```
      du(i) = 1.2 * du(i)       '... make the next change larger.
    ELSE                        'Equal or worse OF, then ...
      u(i) = ub(i)              '... reset DV to the base case, and ...
      du(i) = -0.5 * du(i)      '... reverse and contract search.
    END IF
  ELSE                          'Constraint was violated, TS was bad.
    u(i) = ub(i)                'Reset the DV to the base case, and ...
    du(i) = -0.5 * du(i)        '... reverse and contract search.
  END IF
NEXT i
```

When I compare either HJ or the cyclic heuristic approach to SQ or LM, on a variety of applications I find that the direct search approaches find the solution fewer function evaluations, especially when the regression challenge is increased with nonlinear coefficients or discontinuities from partitioned models.

10.4.2 Multiplayer Direct Search Algorithms

Classic optimizers start at a single trial solution and proceed downhill from that spot. However, if the initial trial solution is near to a local minima, then the optimizer will advance to that local optima, not the global optima. Recently, due to the power of computers, optimizers have been developed with multiple searches working in "parallel" and sharing information. The multiple searches are from TSs independently scattered throughout the entire DV space. Each TS is a player. Sometimes these are called multi-individual, multiparticle, multiplayer searches. Whether from true parallel computation or serially computed, the concept is that all players are simultaneously exploring their region, while observing the outcomes of other players. With multiple players scattered throughout the feasible DV space, there is a higher likelihood that one of them is near the global optimum and will attract other players toward it, or that the path that others take toward the best might uncover a better place yet.

One of the earliest multiplayer search algorithms, remaining very popular, is *Particle Swarm Optimization* (PSO). This direct search technique is intended to mimic the way that birds or gnats swarm toward a target (Kennedy and Eberhart, 1995). In it, randomly place N ($\approx 10M$) players (particles, individuals, trial solutions) simultaneously on the surface (or hypersurface). Evaluate the OF for each. Identify the particle (DV trial solution) with the best OF value. In each iteration, each particle is driven to a new spot by several forces. One is random motion, another is the draw to return to the best spot that an individual had found so far, a third is momentum of motion that will tend to keep the particle moving in the past direction, and the fourth is the attraction toward the best spot found in the swarm (cluster, team) history. The draw of each particle to the global best is tempered, so as iterations progress in time, the particles seem to buzz about their local spot with a progressive movement toward the global best. With a large number of particles, PSO acquires broad surface characterization as the particles migrate toward the best and sweep across regions. This makes PSO likely to identify a global minimum.

However, for most of the particles, the random local exploration at each iteration provides no value, yet it costs OF evaluation in either computational burden or experimental cost. In addition, although the probability of detecting the global minima is high, it is neither perfect nor guaranteed.

Other multiparticle optimizers are variously termed simulated annealing, ant colony, differential evolution (DE), bee colony, and genetic algorithms. There are many more. Their names refer to a life process that we observe which seems to intelligently take a species evolution or daily activity of a population toward an optimal location.

Leapfrogging is now my preference for a direct search. It is a bit more complicated than the cyclic heuristic direct search, but still much simpler than either gradient-based techniques or most multiplayer direct techniques, and extensive tests reveal that it requires much fewer OF evaluations. We tested it on nonlinear optimization problems with 2–70 DVs (Rhinehart, Su, and Manimegalai-Sridhar, 2012) and applied it to problems of over 200 DVs.

10.4.3 Leapfrogging

Leapfrogging is named after the children's playtime game. The terminology is also used to describe movement of military units or law enforcement personnel as they progressively approach a target, moving one unit at a time, covered by the others that remain stationary.

Start with N players (N individual trial solutions, $N \approx 10M$) randomly placed in the feasible DV space. Identify the TS with the worst OF and the player with the best OF. Leap the worst over the best. Only the player with the worst OF is relocated. All other players remain in place. Only one new OF is evaluated. In the leapover, the one player in the team with the worst OF value is relocated to a random position within its DV space reflection on the other side of the player with the best OF value. This is the leapfrogging aspect, and the leap-to position is defined by

$$x(i)_{new} = x(i)_{current\_best} - r_i * (x(i)_{current\_worst} - x(i)_{current\_best}) \tag{10.25}$$

where i = the ith dimension of the DV space, $x(i)$ = value of the ith DV dimension, *current_best* refers to the player with the best OF-value, *current_worst* refers to the player with the worst OF-value, *new* refers to the leap-to position for the former worst, and r is a uniformly distributed random number [0,1] independent for each dimension and iteration.

If the DV does not have continuous values, the reported value of the jump-to location is rounded to the nearest discrete value, which permits the optimizer to handle mixed real-integer problems. Note that when the player leaps, the *current_worst* solution is replaced by the *new* solution.

Figure 10.6 shows the leapfrogging concept for a two-dimensional function. The best and worst players have OF values of 8.8 (the circle) and 10.4 (the triangle), respectively. The jump-to-area is indicated by the dashed rectangle, the reflection of the solid rectangle, and the jump-to location could be any random point within the jump-to area. The player does not jump along the straight line, but lands in a random spot within the reflected window. As

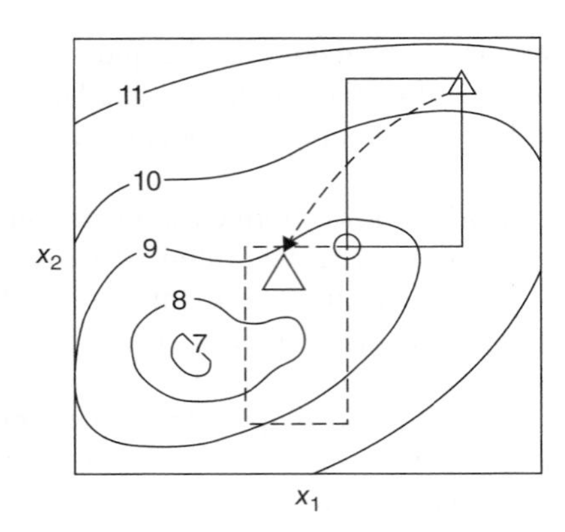

Figure 10.6 Leapfrogging in a two-dimensional function

illustrated with the dashed line and arrow, the player with the worst OF will jump to a new location, vacating the old site.

With Equation 10.25, each leap-over places the former worst to a position averaging half the distance away from, and on the "other side" of, the best player. If the new position is feasible, the move is complete, regardless of the OF value at the leap-to position. If the new position is infeasible, then the iteration continues with a leap of the same player from the infeasible location back over the same *current_best*, using Equation 10.25, again reflecting and cutting the distance in half, on average. Unless the leap-to location is infeasible, each leap-over requires one new function evaluation. Desirably, the new leap-to trial solution is in the direction of the optimum and the new leap-to position becomes the new best. However, the new position may not be a new best for the team (for all players).

After the leap-over is complete, the move of the next player nominally begins with a search for the worst and best. However, computational effort can be reduced recognizing: (i) there is no need to search for the best, because the leap-to spot is either better than or not better than the prior best, so the best is known and (ii) there is no need to search for the worst if the leap-to spot is either infeasible, equal to, or worse, than the previous worst.

An iteration in the optimization procedure is nominally one complete cycle through the procedure. In SQ it would represent the $(2 + 3M + M^2)/2$ function evaluations, linear algebra computation of the surrogate model coefficient values, and linear algebra evaluation of the surrogate model minimum to determine the next TS. In LM-type searches, an iteration would be comprised of the $(1 + 2M^2)$ function evaluations to determine the derivatives and then the linear algebra solution for the next TS. An iteration for the cyclic heuristic search would require M function evaluations, one for each DV. On average, HJ requires $(1 + 1.5M)$ function evaluations per iteration. These examples reveal that an iteration requires diverse numbers of function evaluations. In LF the iteration could be defined as the single leap-over, with one function evaluation per iteration. However, one function evaluation does not provide as much knowledge as the M, or $(1 + 1.5M)$, or $(1 + 2M^2)$ function evaluations per iteration associated with the other optimization algorithms. When progress is observed by iteration or when convergence is tested after an iteration, LF will not seem to make much progress. Accordingly, I define an LF iteration as M (number of DVs) feasible leap-overs.

Most optimizers seek to move the best trial solution to a better spot. However, leapfrogging seeks to relocate the worst, which is similar to the NM Simplex approach, which reflects the worst through the opposite face of the simplex with only one OF evaluation per reflection. However, unlike the Simplex approach, which characterizes the local spot on DV-space with $M + 1$ locally placed trial solutions, leapfrogging can use any number of players randomly dispersed in feasible space to provide an immediate, coarse global characterization. The leapfrogging use of multiple players, attraction of players toward the global best position, and randomization of new location is similar to particle swarm techniques (Kennedy and Eberhart, 1995). However, in leapfrogging each individual does not continue a local search at each iteration, which is usually a wasted effort by most particles in the swarm. This greatly reduces the number of function evaluations in leapfrogging relative to particle swarm.

Leapfrogging also has similarities to differential evolution. DE also uses multiple individuals, players, or particles, but they are represented as vectors from the origin. Differences between the vectors are new vectors. DE randomly adds the difference vectors to existing vectors to create new trial vectors. It reviews the trial and old vectors and keeps the best, in a manner similar to that of genetic or evolutionary algorithms that select the fittest individuals for survival.

When the OF has a stochastic nature, the best individual in a search iteration may have been defined by a fortuitous OF evaluation, and it would be undesirable for an algorithm to follow an improbably fortuitous path of chance "best" OF values into a region of probable penalties or shortfalls. To handle stochastic OF situations in leapfrogging, the OF for the seemingly best player in the team is re-evaluated. If it remains the best, use it. If a re-evaluation of the seemingly best returns a worse OF value, assign that new value to the player and re-search for a new best player in the team. In this method, only one player needs to be re-evaluated, saving experimental or computational effort in evaluating the OF and avoiding probable penalty areas. For a convergence criterion, observe the OF of the worst player with each iteration. This will be a stochastic variable that progressively improves (transitions to a lower value) until it reaches a noisy steady state. Claim convergence when the OF value of the worst player reaches steady state (Rhinehart, 2014).

Our studies reveal that leapfrogging can handle hard and soft constraints, surface aberrations (including discontinuities and flat spots), and mixed discrete-continuous variables. For a large range of test problems, leapfrogging finds the optimum with computationally less burden (as measured by the number of function evaluations) than gradient-based and other direct search algorithms.

Use a minimum of 20 players, but at least 5–10 players for each DV dimension. A fewer number of players can increase the probability that the team will miss surface features or will converge on a gradual slope toward the minimum. A higher number increases the computational work.

Figure 10.7 provides a flow chart of the LF procedure.

10.5 Takeaway

Use direct search optimizers to avoid problems of surface aberrations, hard constraints, nonlinearities, and the issues of deciding on appropriate numerical methods to calculate derivatives. The cyclic direct search is simple and robust and easily implemented in a human

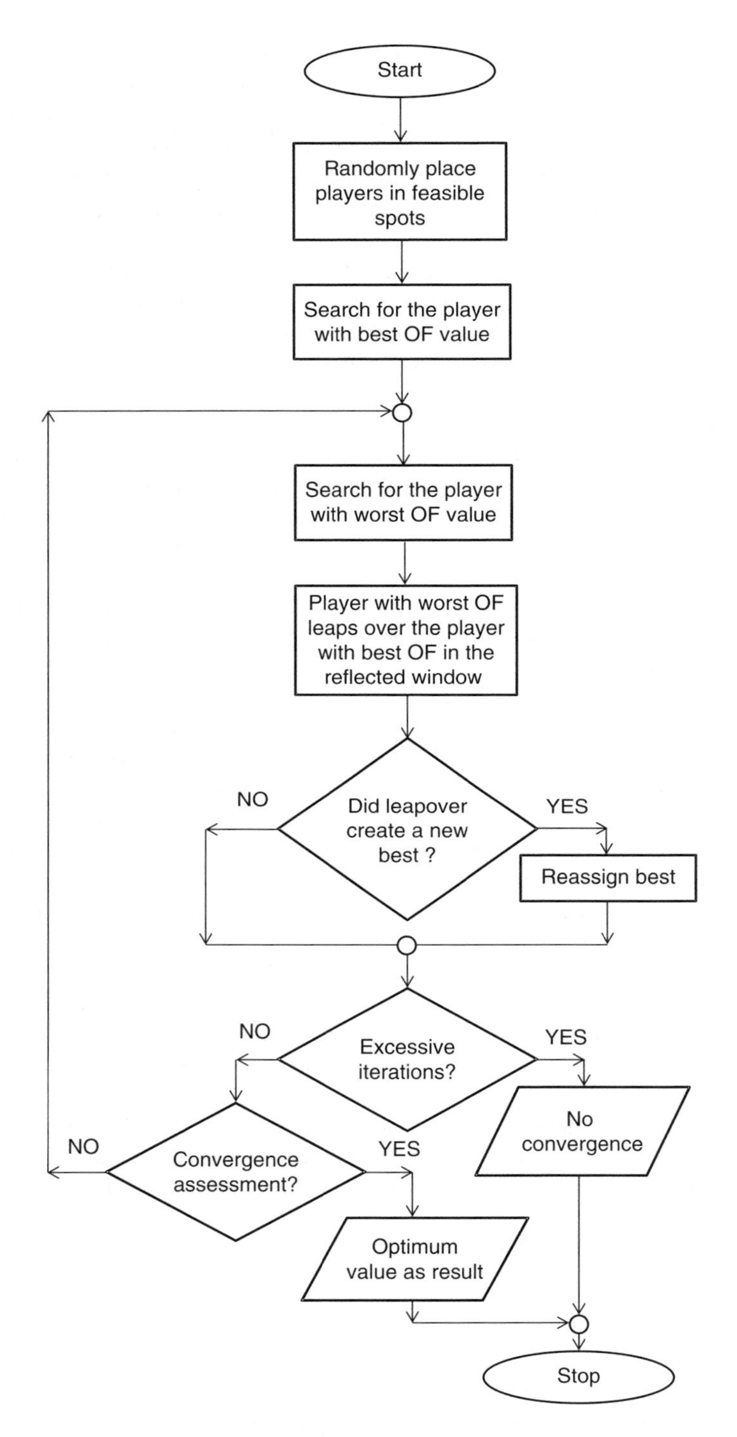

Figure 10.7 Leapfrogging procedure flow chart

trial-and-error search. In my opinion, leapfrogging is best in class for global optimization, robustness, understandability, and computational burden. It can also cope with stochastic and mixed discrete-continuous situations. Appendix B provides codes for regression using the LF optimizer and SS convergence criterion.

11

Multiple Optima

11.1 Introduction

It is not uncommon for a nonlinear function to have multiple minima. Figure 11.1 illustrates two examples of this with a possible plot of objective function (OF) w.r.t. decision variable (DV). The objective of the optimizer is to find the point DV* representing the best of the several optima.

Each graph in Figure 11.1 illustrates three local minima. The left-hand graph might illustrate a model that has discrete levels or a compartmental model that shifts equations with something related to DV magnitude. This example reveals local optima with discontinuities. The right-hand graph shows an OF as a continuous function with multiple optima. Both graphs reveal an optima on the DV boundary.

In each case, one optimum is clearly the global best (lowest of all). This is called the *global optimum*. The others are called *local optima*. We want our optimizer to find the global, overall best.

The two-dimensional contour map of Figure 11.2 shows three local optima at OF values of +1, +3, and 0.

One could interpret Figure 11.2 as a population density map around cities. If the best for a location to live means in a large city, and you start near Tulsa, Oklahoma, the direction of immediate improvement takes you toward Tulsa (the area has about 1 000 000 people). However, near the Texas Metroplex (of Dallas, Fort Worth, and Arlington), the best local direction leads you there (about 7 000 000 people). At any of these proximities, moving away from the city center moves you toward open country, low population density, toward a worse local OF value. If a large city population within the United States is "best", then NYC wins (at about 8 500 000), and Atlanta, Baltimore, Chicago, … , San Francisco, Tulsa … are all local optima.

Consider the illustration in Figure 11.3. If a search is a steepest descent search, then starting with an initial trial solution in Region (1) leads you to the OF_1 optimum. Starting in Region (2) you find OF_2, and in Region (3) you find OF_3.

Consider that the optimizer is a steepest descent search. Starting with an initial trial solution anywhere in Region (2), the downhill direction will lead to convergence at the OF_2 value. If Region (1) has 25% of the DV range (or area if two DVs, or volume if three DVs, or hypervolume if more), Region (2) 55%, and Region (3) 20%, and if the starting

Nonlinear Regression Modeling for Engineering Applications: Modeling, Model Validation, and Enabling Design of Experiments, First Edition. R. Russell Rhinehart.

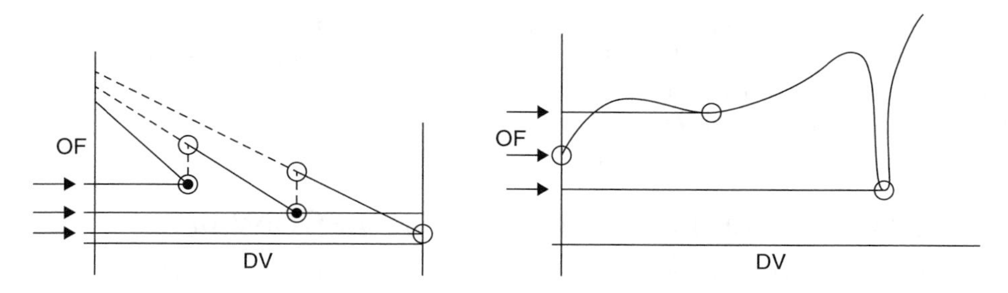

Figure 11.1 Illustrations of multiple optima for a single DV model

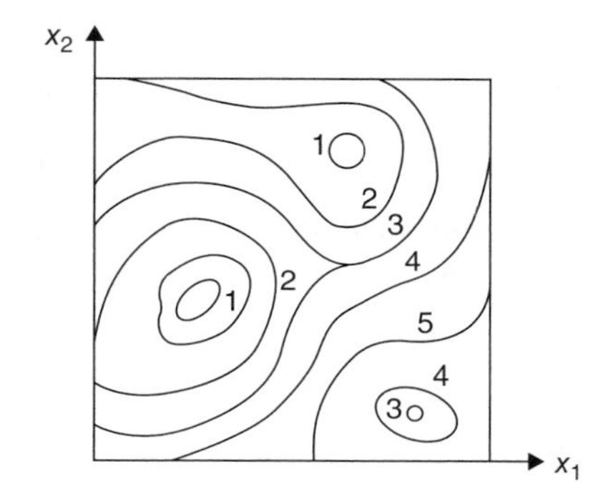

Figure 11.2 Illustrations of multiple optima for a two DV model – contour plot

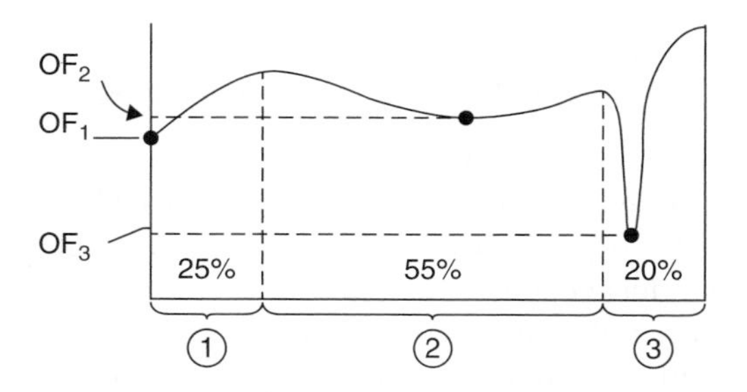

Figure 11.3 Quantifying multiple optima

point $(x_{10}, x_{20}, x_{30}, \ldots)$ was randomly selected from the permissible DV range, then, from Figure 11.3, there is a 25% chance of finding OF_1, a 55% chance of finding OF_2, and a 20% chance of finding OF_3. Using the symbol P_i to indicate the probability than an optimizer finds the ith optimum:

$$P_1 = \text{probability of finding } \text{OF}_1 (= 25\% \text{ in the example})$$

$$P_2 = \text{probability of finding } \text{OF}_2 (= 55\%)$$

$$P_3 = \text{probability of finding } \text{OF}_3 (= 20\%)$$

The problem with a single optimization trial from one random initial DV is that you might discover OF_1, or OF_2, but not the global best optimum, OF_3. As a cure, to increase the probability of finding the global optimum, start the optimizer many times (trials) from random locations and take the best of N trials.

11.2 Quantifying the Probability of Finding the Global Best

To answer the question, "How many trials?" start with an analysis of the probability of finding OF_3 in N random starts. The analysis is simplified by computing the NOT complement and expanding the NOT to each event:

$$\begin{aligned} &P(\text{finding } OF_3 \text{ in at least } 1 \text{ of } N \textit{ random starts}) \\ &\quad = 1 - P(\text{not finding } OF_3 \text{ in } \textit{any} \text{ of } N \textit{ random starts}) \\ &\quad = 1 - P(\text{not in } 1^{st} \ \underline{\text{and}} \text{ not in } 2^{nd} \ \underline{\text{and}} \ldots \underline{\text{and}} \text{ not in } N^{th}) \end{aligned} \tag{11.1}$$

Since P (not finding OF_3) is purely random and independent of trial number the AND conjunction is computed as the product of individual probabilities. Again, using the NOT complement:

$$\begin{aligned} &P(\text{finding } OF_3 \text{ in at least } 1 \text{ of } N \textit{ random starts}) \\ &\quad = 1 - P(\text{not in } 1^{st})\, P(\text{not in } 2^{nd}) \ldots P(\text{not in } N^{th}) \\ &\quad = 1 - [1 - P(\text{finding } OF_3 \text{ in } 1^{st}] \cdot [1 - P(\text{finding in } 2^{nd})] \ldots [1 - P(\textit{in } N^{th})] \end{aligned} \tag{11.2}$$

Since the probability of a trial leading to a success is independent of trial number, the result is

$$P(OF_3 \text{ in at least } 1 \text{ of } N) = 1 - [1 - P(OF_3 \text{ in any randomly initialized trial})]^N \tag{11.3}$$

If $P(OF_3 \text{ in any one random trial}) = 0.20$, as illustrated in Figure 11.3, then $P(OF_3$ in at least 1 of $N)$ calculated from Equation 11.3 is indicated in Table 11.1. The numbers reveal that increasing the number of trials from independent initial trial solutions increases the probability of finding the global optimum.

If you have no knowledge of where the optimum is, but the P (finding it in any one trial) is 20%, then with five starts from random initial trial solutions, the probability of finding the global is 67%. With $N = 20$ randomized starts it is 99%.

If you want to be 99% confident of finding the global in the above example, you need to run 20 independent trials. There is a balance between excessive computational work and solution confidence. Considerations have led to many methods seeking to improve the probability of finding the global without requiring an excessive number of trials.

Table 11.1 The impact of repeating optimization trials from random locations (if the probability of finding the global in any one trial is 0.20)

N	1	2	3	4	5	10	20
$P(OF_3 \text{ in } N)$	0.20	0.36	0.49	0.59	0.67	0.89	0.99

11.3 Approaches to Find the Global Optimum

There are many approaches that will, hopefully, find the global optimum:

1. Some people run an optimizer 10 times from independent initial trial solutions and take the best model from the 10 optimizer trials. The number 10 is an intuitive choice for a sufficient number of trials to get a solution close enough to the true optimum. It may be good enough usually, but it is not defensible. As Table 11.1 indicates, if the probability of finding the global on any one trial is 20%, then 10 randomized trials only have an 89% chance of finding the global.
2. Particle swarm optimization, differential evolution, leapfrogging, and other "global optimizers" start with a number of individuals on the surface and use a combination of random and heuristic rules to provide broad surface exploration, which results in a high probability that it will find the global. The higher the number of players on the surface, the greater the probability of finding the global. However, there is no guarantee that it will find the global minimum on one trial, but certainly the computational burden per trial is increased with a high number of players.
3. When the region has been explored in prior applications, and prior information indicates likely values for the DV*, if you start with those DV values, you will be likely to find the global optimum. This reduces the computational burden, but requires *a priori* knowledge. If the knowledge is available, use it!
4. If a person is observing the resulting OF from one random start, then another, then another, … , the person might come to realize that all starts in one particular region lead to the same optimum, and the person might begin to override purely random initialization and shape where future initial trial solutions start. You could use fuzzy logic (or something similar) to automate heuristic rules that would preferentially explore other starting places. This would override initializations based purely on random numbers. Start in places where there were no prior starts. Do not start in places that seem to draw the search to the same, repeated, optimum.
5. Additionally, you could observe the path of trial solutions from each random initialization to the converged optimum and use this developing information to characterize the surface. Then place new starts in the locations that have not yet been characterized.
6. Start searching with one trial solution, when near a minimum, and start adding random perturbations to the next guess, which are intended to have the individual jump out of the hole or away from a constraint. Heuristic rules can guide the frequency and extent of the one-last-look-around-before-going-in searches (Li and Rhinehart, 1998).
7. Start searching with one guess and randomly add jumps to other positions. Start at one place and optimize. Randomly start at another and optimize …. Stop when Baysean *posteriori* probability of finding the best is achieved (Snyman and Fatti, 1987; Snyman and Kok, 2008).
8. At each iteration, add a random placed trial solution in feasible space. If better (than best or worst of individuals), replace that individual with it. Otherwise ignore the random search.
9. Start N times and take the best of N solutions. Use the equations in Section 11.4 to determine the value of N that meets your desired probability of success.
10. Devise a two-stage search. Choose 100 (or some large number) of random starting places and evaluate OF at each. This initial stage permits a wide initial surface exploration and is likely to be the best of the large number of initial trial solutions in the vicinity of

the global optimum. The second stage is to optimize from the best one only. The fine tuning only generates the optimization work of moving one trial solution to its local optimum.

11. Devise another type of two-stage coarse fine search. For the coarse search, start the optimizer *N* times from independently initialized trial solutions, but use a coarse convergence criterion that claims convergence in the general vicinity of the optimum. This will provide *N* OF results. None will be very precise, but none require excessive computational work. Return to the best of *N* and use that DV* as the initial trial solution and a tight convergence criterion to fine-tune that search for the desired precision. If the coarse search provided multiple DV* values with equivalent OF* values, then there is a high confidence that the vicinity of the global OF will have been found. If the *N* coarse searches provide one isolated best, then do more coarse searches to be confident that the vicinity of the global best has been discovered. The probability of finding the global on any one trial can be estimated from the number of times the best was found in the coarse search stage, and this information can be used to determine *N*.

Unfortunately, there is no universal guarantee of finding the global optimum. In some restricted situations of a particular class of problems with particular topographical properties, there are proofs that a particular optimizer will find the global or the vicinity of the global. However, my experience indicates that these certain cases rarely arise in practical problems. Increasing *N* (number of independent trials or number of players in a multiplayer optimizer) increases the likelihood of finding one of the best few optima.

11.4 Best-of-*N* Rule for Regression Starts

Consider a histogram of OF values from 1000 (or some large number of) optimization trials, each from a random starting trial solution.

The illustration on the left of Figure 11.4 represents an optimization application and the histogram on the right indicates expected results of finding each local optimum. In 1000 independent trials, we expect the optimizer to find OF_3 20% of the time (in 200 of the 1000 trials).

Because the convergence criteria stops short of perfection (which would require an infinite number of iterations), the stops nominally at OF_3 will not exactly stop at the same place. Looking closely at the histogram in Figure 11.5, one sees the "spread" in OF values.

Normalizing the histogram (divide each bin population by 1000) and smoothing the curve (it actually might require millions of trials with random starts to make it appear smooth) produces a probability density function as illustrated in Figure 11.6.

Unless we can see the entire OF surface, we do not know what the histogram of OF values at convergence looks like. Further, one optimizer that uses a steepest descent algorithm would provide the distribution of OF values as indicated, but an optimizer that jumps around would generate a different distribution of OF values at convergence. Further, for any particular optimizer, the distribution of OF* values at convergence will depend on the user's choice of acceleration factors, convergence criterion, convergence thresholds, forward or backward finite differences, and so on.

Regardless of the details in the OF* histogram, there is always a best 10% of all possible OF values for a particular optimizer with particular coefficient choices on a particular problem.

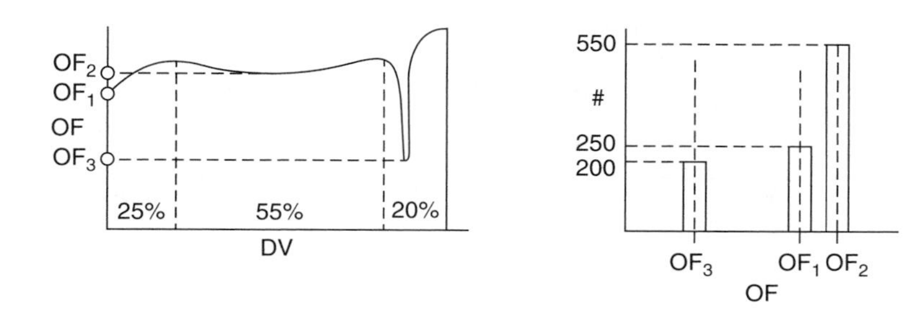

Figure 11.4 Histogram of OF results

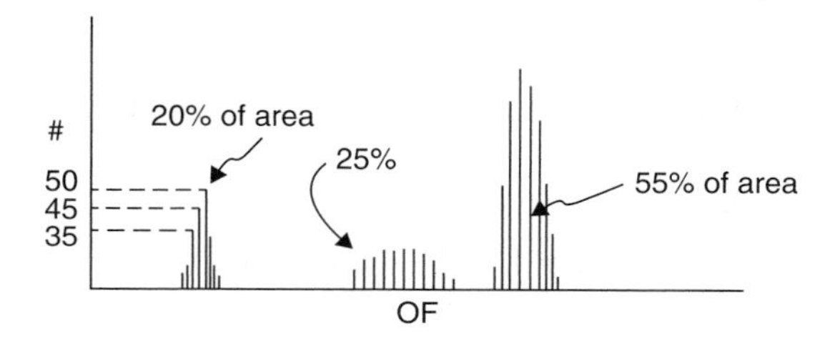

Figure 11.5 Histogram of OF results – greater detail

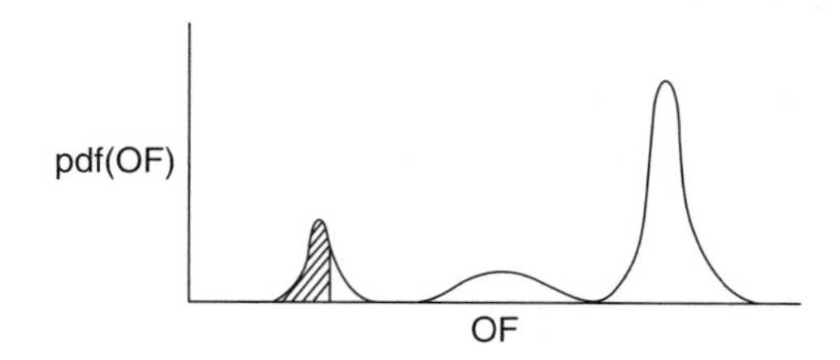

Figure 11.6 Probability density function of OF results

This is indicated as the shaded area in the left portion of Figure 11.6. The upper row of figures in Figure 11.7 illustrates several PDF(OF*) shapes. As well, there is always a best 2% and a best 27.324%. The lower set of graphs in the figure reveal the CDF(OF*) that corresponds to the PDF graphs.

The best 10% does not mean OF values less than 1.10 OF*. It means the OF values that demark the first 10% of the area on a PDF(OF) graph or the CDF = 0.10 value on a CDF(OF) graph.

From Equation 11.3,

$$P(\text{finding best 10\% of all possible solutions in at least 1 of } N \text{ random starts}) = 1 - [1 - 0.1]^N \tag{11.4}$$

Equation 11.3 can be re-stated, then rearranged to solve for N given the user-specified desired probability, $P_{Success}$, of finding one of the best fractions of OF values in at least 1

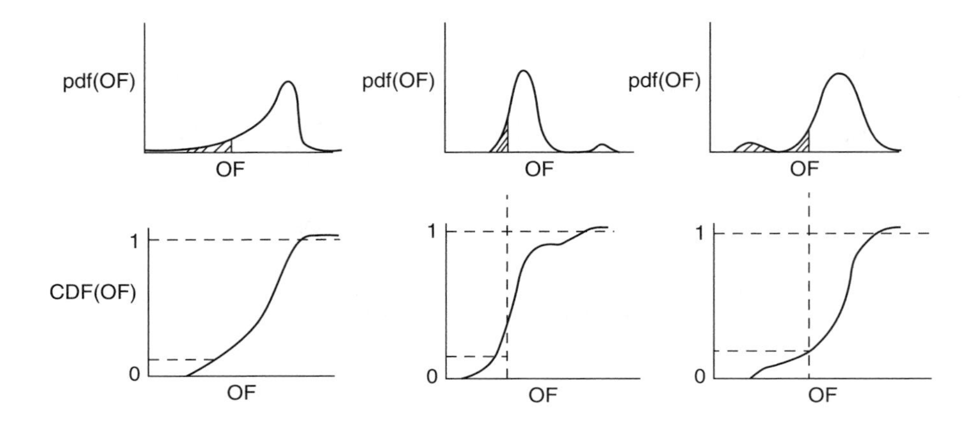

Figure 11.7 Varieties of PDF(OF) results and associated CDF(OF) graphs

of N independent trials:

$$P_{success} = 1 - [1 - \text{best fraction}]^N \tag{11.5}$$

Solving for N,

$$N = \frac{\ln(1 - P_{success})}{\ln(1 - \text{best fraction})} = \frac{\ln(1 - p)}{\ln(1 - f)} \tag{11.6}$$

where f represents the fraction of best solutions that you desire to be found and p represents the confidence (%/100) that the best of N trials will find such a solution. Choose the desired best fraction of possible OF values that you want to find in at least 1 of N trials, say the best 5%. Choose the probability that you want the N trials to find one (or more) from that best fraction, say 99%. Then calculate N:

$$N = \frac{\ln(1 - 0.99)}{\ln(1 - 0.05)} = 89.78113496\ldots \tag{11.7}$$

Round to an integer. Ninety trials are required (90 randomized starts) to be able to find one of the best 5% of all possible OF values at least once, with a 99% confidence. The final equation is

$$N = \text{INT}\left(\frac{\ln(1 - p)}{\ln(1 - f)} + 0.5\right) \tag{11.8}$$

where p is the desired probability that at least 1 of N optimizer trials from random DV starts will find an OF that is from the best fraction, f, of the best of all possible OF values.

If interested in details and validation, see Iyer and Rhinehart (1999), but Equation 11.8 is all you need.

11.5 Interpreting the CDF

As you progressively run optimization trials to find the DV* solution, the developing CDF provides information to choose the value of f in Equation 11.8. Figure 11.8 provides three examples after $N = 20$ trials. In the left illustration it seems obvious that there are two optima

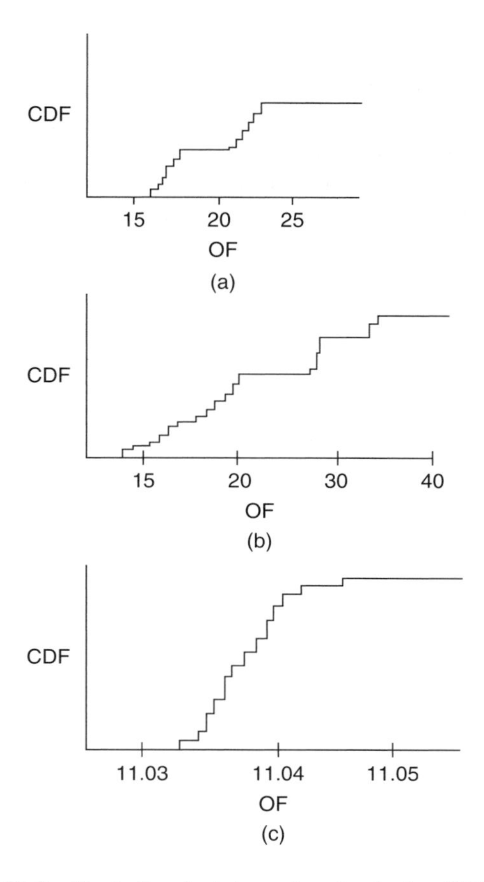

Figure 11.8 Illustration for interpreting developing CDF graphs

and that the probability of finding the best is about 35%. There is no sense in striving to find the best 1% if there is a 35% chance of finding the global. Therefore, use $f = 0.35$ in Equation 11.8.

By contrast, the center figure does not have a sharp CDF rise, indicating that we are not confident that the vicinity of the best has been found. It also reveals a second and third local optima with fairly narrow spread, indicating that local optima can be precisely identified. This pattern could indicate that the global optimum is in a gently sloping narrow valley. Depending on your needs, you might choose a desired best fraction of 1%, $f = 0.01$, in Equation 11.8, or tighten the convergence threshold so that the optimizer does not give up so early.

A third situation is illustrated in the right-hand side of Figure 11.8. It also indicates a tail to the beginning of the CDF graph, but note that all OF values are similar. There is no indication of other local optima. This indicates that each search from the randomized initializations are finding the vicinity of the same global best. Here there is no need for more searches, but if greater precision is desired, the optimizer convergence criterion could be tightened.

11.6 Takeaway

Start N times and take the best of N solutions. Use Equation 11.8 to determine N with initial choices for p and f. As the number of trials progresses, interpret the developing CDF to reshape your choices of confidence and best fraction. To minimize the computational burden, do a coarse initial search. Only use the final precision-desired convergence criterion on the final fine search starting at the DV* from the best coarse search.

12

Regression Convergence Criteria

12.1 Introduction

Optimizers iteratively move the trial solution toward the optimum. It is likely that no iteration will jump exactly on the optimum, but, hopefully, each iteration gets closer to the optimum. Once close enough to the optimum there is no sense in trying to get closer. The optimizer should stop when the transient state (TS) is close enough to the optimum. Seeking infinite digit perfection is usually not justified.

As a familiar concept, consider how many digits you need when using pi. They have computed over 4 million. Using 3.14 is often fully adequate. Using 3.14159265358979 (rounded to the 15th digit is closer to perfection, yet not perfection and probably excessive for any engineering application.

More than perfection not being necessary, in regression fitting of models to data, the model itself is an imperfect representation, and the data are corrupted with various forms of experimental uncertainty. Accordingly, the precise optimum DV* (decision variable) value will be infected with the experimental vagaries and will not represent the truth about Nature. There is therefore no justification for excessive precision.

The optimizer should stop when the TS is close enough to the optimum. A convergence criterion needs to be defined, to stop the optimizer when the proximity to the perfect value is close enough.

12.2 Convergence versus Stopping

Convergence means that the optimizer has located the close proximity of an optimum and that continued iterations to get excessive precision are not justified. We want our optimizer sequence of trial solutions to converge on one spot. Convergence means success. This is different from terminating the optimizer because of a failure or non-success. Use *stopping criteria* to stop an optimizer when it is headed for non-success.

Sometimes the optimization must be stopped prior to convergence. Here are three reasons. (i) If the search leads to an internal variable value that would cause an execution error (divide by zero, log of a negative number, subscript out of range, etc.), use error trapping to terminate the

Nonlinear Regression Modeling for Engineering Applications: Modeling, Model Validation, and Enabling Design of Experiments, First Edition. R. Russell Rhinehart.

search. (ii) Stop if the search leads to a region of infeasibility that violates a material or energy balance or a composition that is less than zero or greater than 100%. (iii) If the search begins to oscillate, encircle, or diverge with no improvement in the objective function (OF) value after excessive iterations, stop the program. *Stopping criteria* are usually based on excessive iterations, excessive computational time, or error trapping. Stopping for such reasons means that the optimum was not confidently identified. Do not accept results when the optimizer stopped due to such stopping criteria. In such cases, the optimizer should explicitly state that it stopped prior to meeting convergence criteria.

The next sections discuss choices for convergence criteria, not stopping criteria.

12.3 Traditional Criteria for Claiming Convergence

There are several commonly used criteria for claiming convergence, which are easy to understand and evaluate. One is to claim convergence, stop the optimizer, and report the found answer, when the OF value is low enough. As illustrated in Figure 12.1, the user must decide on the threshold value, which raises several issues If the threshold value is too high, you will stop not near the optimum. If the threshold value is too low you will never stop. Setting the right threshold value requires *a priori* knowledge of the OF versus DV trend. However, if you are searching for the optimum on a new application, you probably do not have such knowledge.

A second common approach is better – stop when the incremental improvement in OF is small, as illustrated in Figure 12.2. The associated problems are: too large a value of the threshold on Δf could stop far from the optimum, while too small a value could take excessive iterations. Again, *a priori* knowledge is needed for the right threshold choices. A nice aspect of this is that the OF value is probably understood by the user, who can set a sensible threshold value on the change in OF. For instance, if the standard deviation of replicate

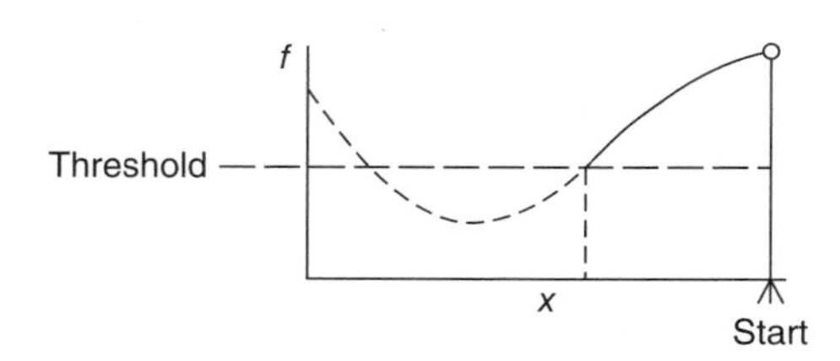

Figure 12.1 Illustration of the problem with threshold on OF as convergence criterion

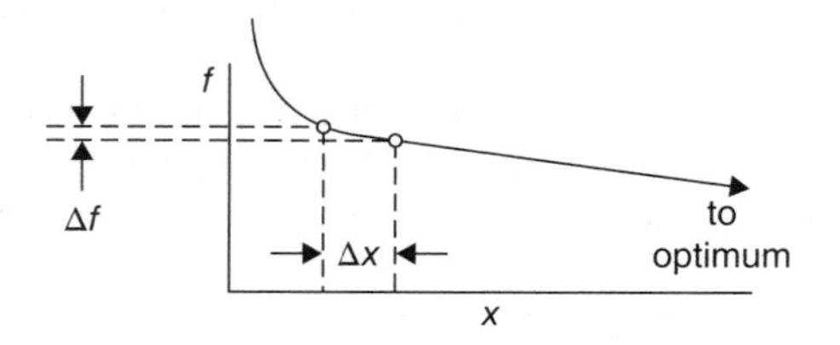

Figure 12.2 Illustration of the problem with threshold on ΔOF as convergence criterion

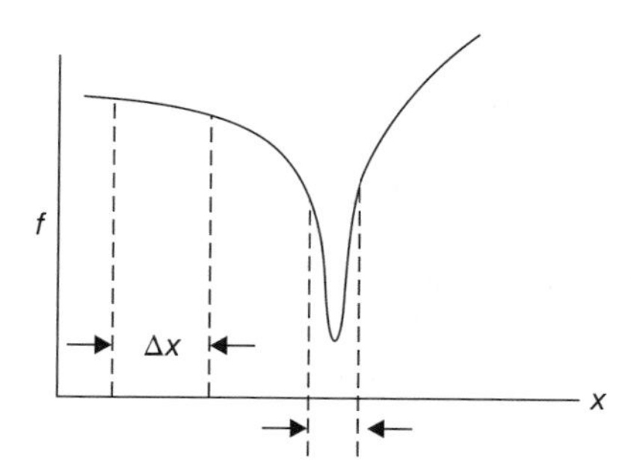

Figure 12.3 Illustration of the problem with threshold on ΔDV as convergence criterion

experimental values is known, then when optimizer improvements in the root-mean sum of squared deviations (rms) between the model and data is 1% of that, the incremental improvement is 2 orders of magnitude smaller than the inherent data variability. That would justify claiming convergence in the proximity of the optimum.

Alternately, and probably the most common approach is to stop when ΔDV is small, as illustrated in Figure 12.3. Associated problems are: a steep OF response could mean that the trial solution is still far from optimum if it is very sensitive to the DV value as illustrated and a low slope could mean that you stop prior to locating the optimum.

Further, if multivariable, each of the M ΔDV thresholds needs to be determined. This is complicated by the likely situation that each DV has its distinct units and range, requiring independent ΔDV thresholds for each DV. Stop only if each is less than its threshold. Unfortunately, many optimizers that use this criterion use a common threshold for all DVs, regardless of units or scale. Accordingly, using scaled DVs would permit a common threshold value.

This is also complicated by the sensitivity of the OF to the DV. A DV with a large value might have a significant impact on the OF, while one with a small value might have an inconsequential impact. Against intuition, in this case, the large-valued DV might justify a much smaller ΔDV threshold than the small-valued DV.

A solution to the scale issue is to use relative changes. Stop when the relative change in DV, $\Delta x/x$, is small. Determine all DV thresholds, but in scaled terms. Stop only if each $\Delta x/x$ is less than threshold. Still, this approach has the problem of the choice of threshold. Too large, and it misses optimum; too small, and it causes excessive iterations. In the chance that x could have a value near zero, use $|\Delta x| < x * threshold$ as the stopping criteria to prevent an overflow computer execution error.

Usually, stopping criteria are on the ΔDV thresholds because these are simple to set. However, they are difficult to relate to the more meaningful impact on the OF or to the uncertainty in the model prediction.

To include OF performance use a relative threshold to overcome lack of knowledge about the ultimate OF value. To eliminate a possible divide by zero or overflow issues stop when the relative change in the OF is small, $|\Delta f| < f * threshold$.

The only way to judge whether the stopping criteria are too lenient, or too severe, is to observe the results. For instance, plot OF w.r.t. iteration and each DV w.r.t. iteration to observe progress, to visualize whether the approach has settled to one point. However, this requires either *a priori* knowledge or active human monitoring.

Further, selection of the right values for any of the threshold or limits depends on scale (2 °F is the same interval as 1 °C), features of the function topography, and the utility or user-need for the result (propagation of uncertainty defines tolerance of OF and DV intervals). Selection of right values requires user *a priori* knowledge, but often in new applications, values and scale for coefficients are not known and it takes several trials to be comfortable that your choice of thresholds is consistent with the application.

Considering these issues, following are three alternate approaches to identify convergence for regression, which are not dependent on either user knowledge or on scale-dependent thresholds.

12.4 Combining DV Influence on OF

The OF often has a user-meaning, an interpretability related to desirability. By contrast, a change in the DV does not. Further, changes in the DV could have compensating effects, so it is difficult to say what change in DV is sufficiently small to meet desired convergence or not having any meaningful improvement in the OF w.r.t. iteration. Therefore, use a combined measure by propagation of maximum impact of DV change on the OF as the sum of the absolute values of the partial derivative of OF w.r.t. DV times the change in DV that the optimizer is implementing. This is a propagation maximum error concept.

Consider that the OF is a function of the DVs:

$$OF = f(DV_1, DV_2, DV_3, \ldots) \tag{12.1}$$

and consider that the current optimizer change in DV represents a delta-DV. Then the propagation of maximum error indicates the possible collective impact of the changes in the DVs on the OF:

$$\Delta OF_{max} = \sum \left| \frac{\partial OF}{\partial DV_i} \right| |\Delta DV_i| \tag{12.2}$$

It will probably require a numerical approach to define the sensitivity of OF to DV, the partial derivative of OF w.r.t. DV. This means that the equation is approximate. For useful derivative estimates, the delta-DV used to assess the partial derivative should be small relative to the iteration-to-iteration trial solution change, to ensure that you are assessing the local surface feature. Finally, if the optimizer is creeping upon the optimum, one would want to keep it going until the maximum collective change is perhaps an order of magnitude, or two, smaller than a meaningful change in the OF. Therefore, claim convergence when

$$\sum \left| \frac{\partial OF}{\partial DV_i} \right| |\Delta DV_i| \leq 0.01 \Delta OF_{meaningful} \tag{12.3}$$

In this manner, claim convergence if the propagation of maximum change in OF due to iteration-to-iteration changes in the DV is less than some amount that is inconsequential to the situation. This is a bit more complicated than convergence criterion on the OF or on the DVs, but relates the incremental DV changes to a metric that the user can understand and limits of that metric that the user can define. Since the OF is likely to be a sum of squared deviation (SSD) of model from data, or its rms value, and the user can have a sense for the inherent variability in replicate data, the user can define the value for $\Delta OF_{meaningful}$ from any of several perspectives. One approach could evaluate $\Delta OF_{meaningful}$ based on the inherent variability in the data, with another based on the human impact of the OF.

12.5 Use Relative Impact as Convergence Criterion

For regression, you are creating a model for a particular use. Even if the model coefficient values are perfect, the inputs to the model will have uncertainty, making the model-calculated outputs uncertain. If the model-calculated values are uncertain, then perfection in model coefficients is excessive. My friends tell me to teach students to be able to balance perfection with sufficiency. With this perspective, there are a variety of choices for the convergence criterion. In one, choose the convergence criterion based on sensitivity of the OF value due to coefficient precision relative to uncertainty of the OF value to the in-use input variables of the model.

This stopping criterion is also based on propagation of maximum uncertainty and is developed as follows.

A model depends on both coefficient values (which the optimizer is seeking) and the input variables. Accordingly the OF is also dependent on both:

$$OF = f(x_1, x_2, x_3, \ldots, DV_1, DV_2, DV_3, \ldots) \tag{12.4}$$

Propagation of maximum uncertainty reveals that uncertainty from both DVs (the model coefficient values that are adjusted in regression) and in-use variable values (represented by x) lead to uncertainty on the OF value:

$$\Delta OF_{max} = \sum \left| \frac{\partial OF}{\partial DV_i} \right| |\Delta DV_i| + \sum \left| \frac{\partial OF}{\partial x_i} \right| |\varepsilon_{x_i}| \tag{12.5}$$

Claim optimizer convergence when the maximum impact of the change in the DV values is much less than the maximum impact of the uncertainty of the in-use values of x. Perhaps "much less" can be set as less than one-tenth: claim convergence if

$$\sum \left| \frac{\partial OF}{\partial DV_i} \right| |\Delta DV_i| < 0.1 \sum \left| \frac{\partial OF}{\partial x_i} \right| |\varepsilon_{x_i}| \tag{12.6}$$

This approach does not need *a priori* interpretation about the tolerable limits on the OF. It uses experimental uncertainty on the control variables in the physical world to define what is a rational limit on the OF and hence the DVs.

This stopping criterion will change as the optimizer evolves model coefficient values.

In a similar approach, consider the in-use model-calculated output. Choose the convergence criterion based on sensitivity of the calculated output due to coefficient precision relative to uncertainty of the model-calculated value to the in-use input variables of the model. The analysis parallels that above. The model is used to calculate y:

$$\tilde{y} = f(x_1, x_2, x_3, \ldots, DV_1, DV_2, DV_3, \ldots) \tag{12.7}$$

Propagation of maximum uncertainty reveals that uncertainty from both DVs and in-use variable values leads to uncertainty on the y value:

$$\Delta \tilde{y}_{max} = \sum \left| \frac{\partial \tilde{y}}{\partial DV_i} \right| |\Delta DV_i| + \sum \left| \frac{\partial \tilde{y}}{\partial x_i} \right| |\varepsilon_{x_i}| \tag{12.8}$$

Claim optimizer convergence when the maximum impact of the change in the DV values is much less than the maximum impact of the uncertainty of the in-use values of x. Perhaps much less can be set as less than one-tenth: claim convergence if

$$\sum \left| \frac{\partial \tilde{y}}{\partial DV_i} \right| |\Delta DV_i| < 0.1 \sum \left| \frac{\partial \tilde{y}}{\partial x_i} \right| |\varepsilon_{x_i}| \tag{12.9}$$

This approach also does not need *a priori* interpretation about the tolerable limits on the y value. It uses experimental uncertainty on the control variables in the physical world to define what is a rational limit on the model-calculated y value and hence the DVs.

This stopping criterion will change as the optimizer evolves model coefficient values.

One could choose a variety of options on either approach. For instance, propagation of maximum error due to uncertainty on the coefficient values relative to propagation of 95% probable error on the model input values:

$$\sum \left| \frac{\partial OF}{\partial DV_i} \right| |\Delta DV_i| < 0.1(1.96)\sqrt{\sum \left(\frac{\partial OF}{\partial x_i} \right)^2 (\sigma_{x_i})^2} \tag{12.10}$$

12.6 Steady-State Convergence Criterion

Consider these aspects of nonlinear regression from experimental data. Nonlinear regression is an iterative procedure in which an optimizer seeks the best values of model coefficients. "Best" could be in terms of conventional least squares, total least squares, or maximum likelihood. Since experimental data contain noise (random deviations from the true value, often termed experimental random error), a model (even one with a perfect structure) cannot exactly match the data. Six points are important:

- First, with each iteration of the optimizer, the coefficient values progressively improve, but as they approach the best values for the particular data set, the progressive improvements in coefficient values reduce. Upon nearing the best possible set of model coefficients, the optimizer progressively fine tunes the coefficient values, nearing the true minimum, but never quite reaching perfection.
- Second, even if the optimizer was using the true model and could find the true optimum, it cannot eliminate the residual variations in the experimental data from affecting the model coefficient values.
- Third, a different set of data, representing a different realization of the random experimental errors, would lead the optimization to slightly different coefficient values. Perfection, a true optimum, for one particular data set is not universal perfection.
- Fourth, the user's choice of the OF (vertical SSD, total sum of squares, minimize maximum deviation, maximize likelihood, and underlying distribution or choices about data variance) has different data weighting, which impacts the resulting optimum.
- Fifth, the user-defined model architecture (functional relations, choice of inputs, etc.) does not represent the elusive truth about Nature.
- Sixth, some criteria must be specified to define when the optimization iterations can be terminated and when the answer is close enough to perfection with that particular data set. The user also chooses this.

Regression is infected with user choices and data vagaries. Expecting to determine the exact perfect truth is unreasonable. This steady-state convergence criterion is another method that seeks to balance perfection with sufficiency.

Usual convergence criteria are based on the magnitude of the change in the OF or the magnitude of the changes in the model coefficients. However, the user must specify such threshold

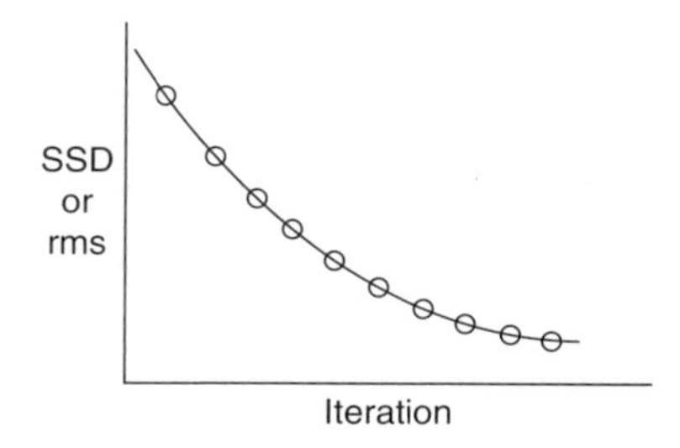

Figure 12.4 Characteristic OF approach to optimum w.r.t. iteration

values, and because they are scale-dependent, dependent on the data variance, and dependent on the sensitivity of the model prediction to the coefficient values, the proper selection requires post-trial knowledge.

An alternate concept for identifying regression optimization convergence is that, when model improvement is inconsequential to inherent variability in residuals (data from model), then stop the iterations. To understand this alternate relative indication of convergence, first observe how the OF generally improves with optimization stages in Figure 12.4. The OF could be vertical SSD, rms value, total SSD, likelihood, number of right categories, or any desired measure of closeness to fit. As iterations progress, the OF value progressively drops, asymptotically approaching the minimum.

SSD and rms are defined as

$$SSD = \sum_{i=1}^{N} (y_i - \tilde{y}_i)^2 \tag{12.11}$$

$$rms = \sqrt{\frac{1}{N}\sum_{i=1}^{N} (y_i - \tilde{y}_i)^2} \tag{12.12}$$

Instead of observing the SSD or rms that the optimizer is using to guide a sequential trial solution, at each iteration measure SSD or rms on a random sampling of about one-third of the data:

$$rms_{random} = \sqrt{\frac{3}{N}\sum_{i=1,\ random}^{\frac{N}{3}} (y_i - \tilde{y}_i)^2} \tag{12.13}$$

As a side note, when I implement this case, random means without replacement. Once a data point is included in the random set, it is excluded for the full set of data and thus it cannot be duplicated in the random set. This is in contrast to bootstrapping, which samples with replacement and permits a data point to be duplicated. I have not explored this stopping criterion with replacement. It might work just as well and be simpler to code.

At each iteration, a new random subset of the data is selected for convergence testing. (The OF for regression optimization continues to use all of the data.) Observing the value of rms_{random} with iteration, there will be a general trend downward, asymptotically approaching a minimum, as in Figure 12.4. However, the trend will have random perturbations, making

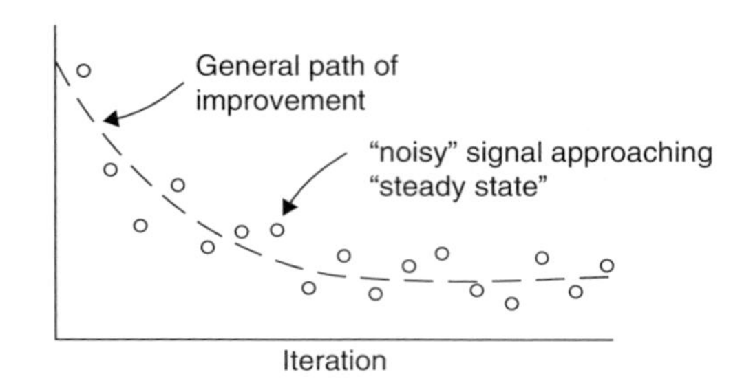

Figure 12.5 Characteristic SSD or rms from random subset approach to optimum

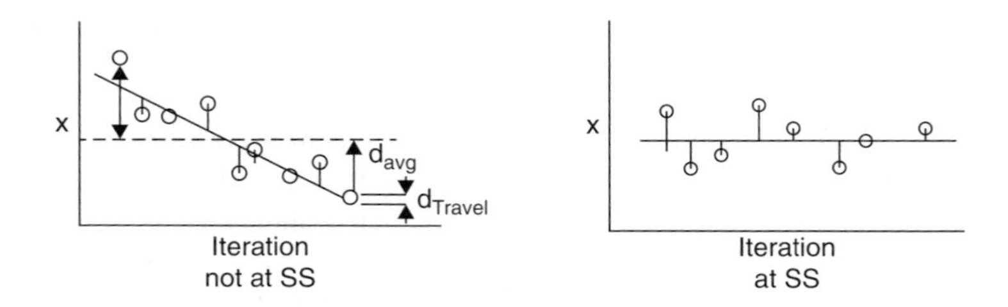

Figure 12.6 Characteristic approach to optimum

it appear as the transient of a noisy signal approaching a steady state, as illustrated in Figure 12.5.

Since the sampling is random, the "noise" on the rms subset is independent from iteration to iteration.

Note that, although this view will be observing the SSD or rms of a random selected subset of the data for determining convergence, the optimizer needs to calculate its SSD (or rms) OF from all data at each iteration to make rational trial solution sequential decisions.

At steady state (at the point you should stop optimization work) the rms_{random}, or SSD_{random}, value w.r.t. iteration shows no improvement. The plot will appear as a "flat-line," but with random, independent perturbations at each sampling. At steady state (SS) the average rms_{random} value represents the inherent data variability due to the several sources – experimental data variability and process-model mismatch. Therefore, use any technique to identify steady state to the rms_{random} w.r.t. the iteration trend.

A computationally simple approach to determine not-at-SS was developed by Cao and Rhinehart (1997b) to trigger control-related activities in process automation. It has been tested by several others on nonlinear, least-squares stopping, and is explained in Chapter 6. Here is a summary. Figure 12.6 illustrates data changing with iteration (still in a TS of improvement) on the left and that for which there is no sequential improvement on the right. The variance in the data on the left (a measure of deviation from the average of the data) is larger than the variance on the right-hand illustration because the end data (following and leading the trend) are far from the average over the window. If there was a way to know what the variance could be at steady state, then the ratio of variances could be used as a scale-free indication of being in the transient or at steady-state.

In computer code (assignment statements) there is no need to retain subscripts, because the computer executes the expression on the right side of the equal sign and then assigns that value to the variable on the left side. Here L means λ and CL means the complement to lambda, $(1 - \lambda)$:

```
D2:=L3*(X-XOLD)^2+CL3*D2
N2:=L2*(X-XF)^2+CL2*N2
XF:=L1*X+CL1*XF
XOLD:=X
R:=(2-L1)*N2/D2
```

Note the possible divide by zero in the calculation of the ratio statistic, which could happen if the denominator measure of the noise is zero, which, in regression, would indicate that the model exactly matches all data points, suggesting that there are too many adjustable coefficients in the model. This possible divide by zero will be eliminated later.

First, consider threshold values for the ratio statistic.

The variance ratio does not have a single value. At steady state its value is about unity, with a distribution as illustrated on the left side of Figure 12.7. When not at steady state the numerator (deviations from the average) are larger than the deviations between samples, giving the r-statistic higher values. The greater the departure from steady state, the larger the r-statistic value. However, at any one condition, statistical vagaries and finite sampling provide a distribution of r-statistic values.

Figure 12.8 reveals the critical values for the r-statistic of about 0.8 for the improbable low value if the process is at steady state, and of about 1.9 for the improbably upper value if at steady state. Mostly, r-statistic values are in between, when the process is at steady state.

At steady state (SS) the PDF(r) stays around unity, mostly between 0.8 and 1.9. At SS an r-value of 0.85 is reasonably expected. Such a low r-value is not reasonable from a not-at-SS period. Therefore, if $r_i \leq 0.8$ declare steady state (or 0.85 or 0.90 depending on how sure you want to be).

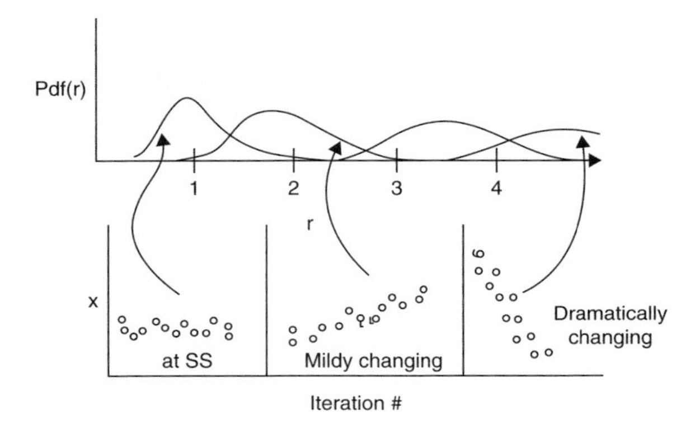

Figure 12.7 Distributions of the ratio-statistic value as process changes state

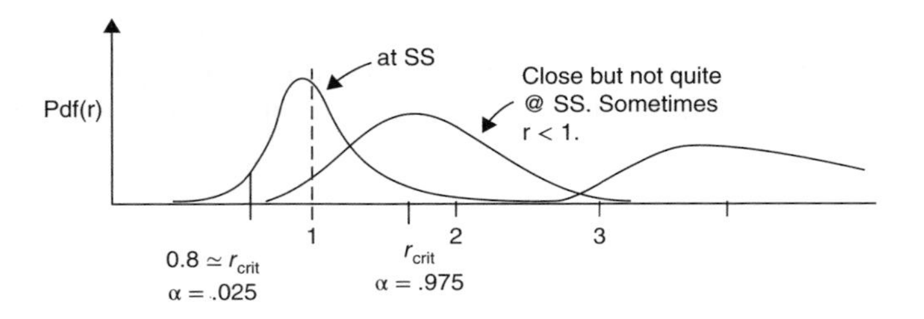

Figure 12.8 Illustration of critical values around steady state

For convenience, I initialize variables to zero. Including initialization, the critical value, and rearranging to prevent a possible divide by zero, the code for the steady-state stopping criterion is:

```
IF FIRST CALL
      XOLD := 0
      XF := 0
      D2 := 0
      N2 := 0
      L1 := 0.05
      L2 := 0.05
      L3 := 0.05
      CL1 := 1-L1
      CL2 := 1-L2
      CL3 := 1-L3
      SS := "FALSE"
ENDIF
D2 := L3*(X-XOLD)^2+CL3*D2
N2 := L2*(X-XF)^2+CL2*N2
XF := L1*X+CL1*XF
XOLD := X
IF 0.8*D2≥ (2-L1)*N2 THEN SS := "TRUE"
```

Note that this technique can also reject the SS hypothesis (accept the TS) by considering the upper value in the last line of the code, but intermediate values for the r-statistic prevent a confident determination for either SS or TS. The full featured code might be:

```
IF FIRST CALL
      XOLD := 0
      XF := 0
      D2 := 0
      N2 := 0
      L1 := 0.05
      L2 := 0.05
      L3 := 0.05
      CL1 := 1-L1
```

```
        CL2  :=  1-L2
        CL3  :=  1-L3
        SS  :=  "INDETERMINATE"
ENDIF
D2  :=  L3*(X-XOLD)^2+CL3*D2
N2  :=  L2*(X-XF)^2+CL2*N2
XF  :=  L1*X+CL1*XF
XOLD  :=  X
IF 0.8*D2 ≥ (2-L1)*N2 THEN SS  :=  "TRUE"
IF 1.9*D2 ≤ (2-L1)*N2 THEN SS  :=  "FALSE"
```

In my use of this procedure to detect both SS and TS conditions, during the indeterminate periods, I have simply retained the most recent definitive SS or TS description.

Recall the substitution of $\lambda = 1/N$ in the derivation. However, in exponentially weighted filtering there is a persistence of the influence of past data, so $1/\lambda$ is equivalent to three to four times the number of data "averaged." If you wish to "see" the steady state persist within a window of N data, use $\lambda = 3.5/N$.

Shrowti, Vilankar, and Rhinehart (2010) recommend that $\lambda = 0.05$ (effectively 20 data points in the assessment window) is a reasonable quantity in accepting SS (best balancing Type I and Type II errors). Cao and Rhinehart (1997b) recommend λ values of about 0.05–0.1 to best balance speed of identification and Type II errors in rejecting SS. Thresholds of 0.8–0.9 are fine for r_{crit}. Shrowti *et al.* recommend 0.8. Use about one-third of the data for random SS. It is not overly sensitive to any choices.

I have been pleased with initializing all computer code variables to zero. Alternately, initialize all to zero, except x_{f0}. Make x_{f0} = initial rms value. Then initial r values are large and x_{f0} tracks the total rms trend. There is no need to collect $N = 1/\lambda$ number of data points to sense the "true" average values for x_f, n_f^2, d_f^2 prior to starting to use the r-statistic stopping criterion – but you could. It also works.

Figure 12.9 illustrates a random subset rms value as it approaches steady-state convergence. In this application, the leapfrogging optimizer is seeking to determine six coefficient values for a second-order-plus-dead time transient model of a process (gain, two time constants, initial values of input and response, and time delay). The figure displays a random subset rms on a semi-log scale to provide visual identification of both end-game variability and initial model discrepancy on the same figure. The value of x_{f0} was initialized as zero, which is indicated on the figure by its initial rise up to the data trend.

How can we classify what is an iteration to trigger the stopping criterion? The norm on optimization is to declare an iteration (or epoch) when the optimizer has completed all of its operations and is ready to cycle back to the start of the next trial solution calculation. In gradient-based optimizers the number of function evaluations per iteration rises with the square of the DV dimension. However, in many multiplayer direct search optimizers an iteration or epoch is declared when all population members have had a chance to move, or evolve, or when all DVs have been adjusted. Here the number of function evaluations per iteration is linearly proportional to either the number of players or the DV dimension. In some neural network (NN) training procedures that take one data point at a time, the iteration is defined when each data set has been "presented to the NN," in which case the number of model adjustments in

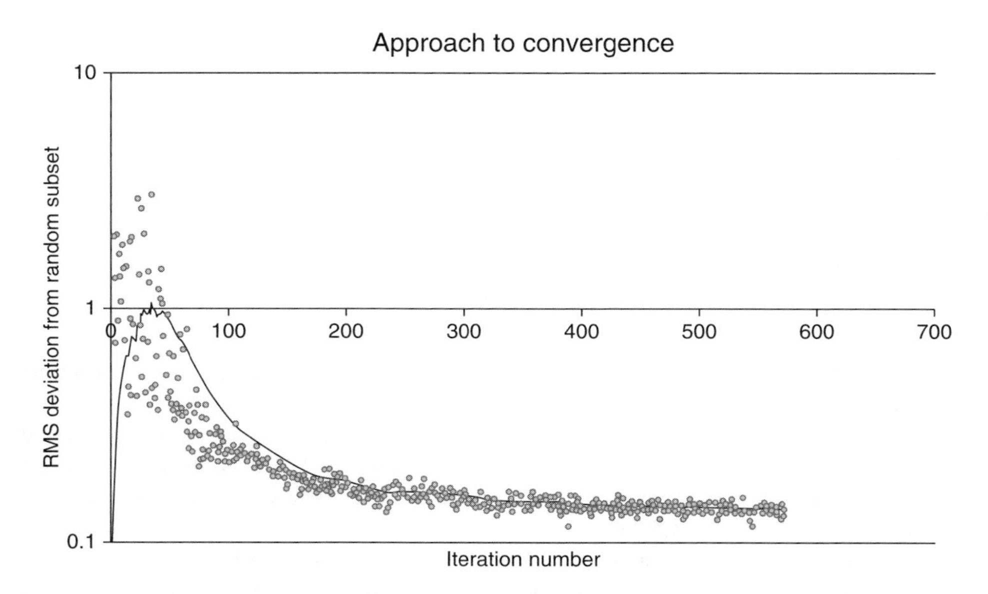

Figure 12.9 Random subset RMS deviation versus optimization iteration

an iteration is proportional to the number of data sets. In leapfrogging there are several logical options for declaring an iteration:

- *One feasible leap-over*. Each leap-over creates a trial solution with new values for each DV, so each leap-over could be considered an iteration. However, only one of *M* players would have moved, and during some stages, it may take many leap-overs to fortuitously find a better spot.
- *N feasible leap-overs*. *N* is the number of players. This permits each player to move, but not each player will be selected to leap. Also, since it is likely that several of the *N* leap-overs will have found a better spot, this infrequently assesses progress.
- *A successful leap-over*. A successful trial solution means that all DVs have changed and the new OF is the best found so far (it also must be feasible). However, the interval is not regular. There could be few or many leap-overs between successes.
- *M feasible leap-overs*. One leap-over for each of the *M* dimensions. This scales the iteration to the dimension (complexity) and provides a consistent interval.

I choose *M* leap-overs, one for each DV dimension, as an iteration. It seems to work.

A secondary benefit of this technique is that it makes some data or modeling problems visible, as revealed in Figure 12.10, from regression modeling data that contains an outlier.

Prior to about iteration number 400, all the random subset rms data seems to be clustered and progressively improving. There is an outlier in the data, but up until this point, the model was not fitting the bulk of the data very well, and the outlier was just another point not yet well modeled. After iteration 400, there is a bifurcation with one set of data at about 5 (on the log scale) and a broader set at about 2. When the outlier was included in the subset, the rms deviation is uncharacteristically high. In this case, the bimodal RS rms trace is the result of

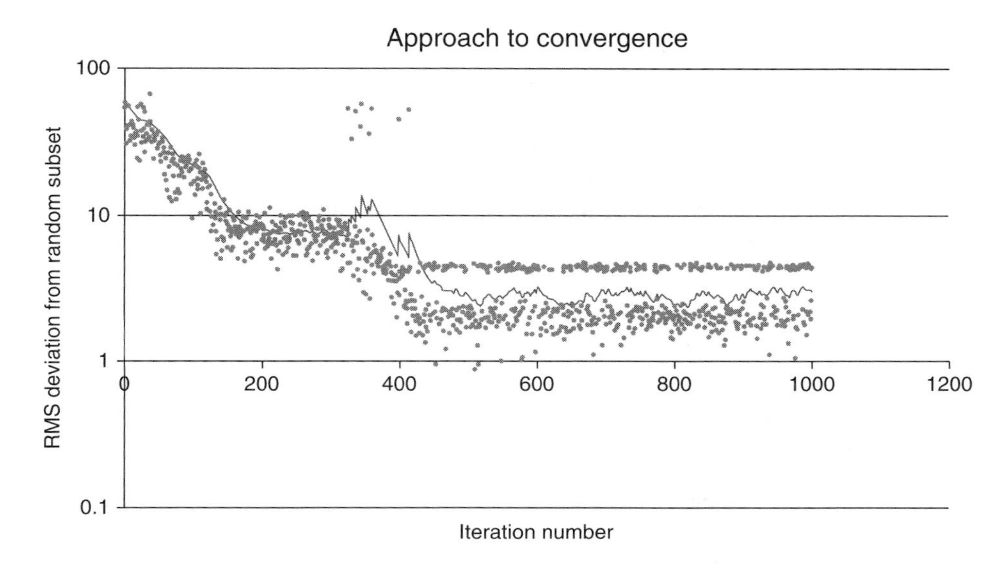

Figure 12.10 Random subset RMS deviation versus optimization iteration – bimodal trace

a single outlier data set, but it could relate to data sections that the model does not get right (either missing a mechanism that Nature has, adding a mechanism that Nature does not, or inconsistent data from different test devices). Since the problem is randomly in or out of the subset it is seen, or not.

12.7 Neural Network Validation

The tradition for determining regression convergence in training NNs is a bit different. First, take out a random subset of about 15% of the data, called the validation set. The remaining data used for regression is called the training set. Do regression on the training set – adjust model coefficient values to best fit the data in the training set. The optimizer does not attempt to fit the validation set data. With each iteration of the optimizer, the SSD on the training set will decrease. Also observe the SSD on the validation set. If the model is improving, if the optimizer is progressively adjusting coefficient values to make the model fit the overall mechanisms represented by the data, then the SSD on the validation set will also decrease. If, however, the model is being adjusted to fit noise in the training set, then this adjustment will not be fitting the mechanistic trend in the validation set, and the validation set SSD will begin to increase. Claim convergence when the validation SSD starts to increase.

As a perspective on why this approach is used, we train students on in-class examples. Then we test using new problems. If the students can get the test problems right, this validates learning. This convergence determination procedure parallels classroom education, and is accepted as the right approach within the NN community, which historically created NNs as a computational model of how the brain works.

This approach, however, only uses a subset of the data for regression and contrasts two concepts: (i) if you paid for the data, then use it for model development and (ii) the older fields of statistics and modeling use all of the data.

12.8 Takeaway

Use the steady-state identification method from Section 12.6 as your stopping criterion for regression. It is not scale-dependent. It does not require user-chosen thresholds, which require *a priori* knowledge. It is fundamentally based and designed to claim convergence when there is no improvement in the OF relative to the data variability associated with the model. It is simple to code. It is computationally simpler than the compound OF measures based on propagation of uncertainty. It does not add the computational burden of calculating derivatives.

Exercises

12.1 If the σ of replicates is 3 kg/m, what would be a reasonable stopping criterion on SSD with $N = 20$ sets of data?

12.2 For diverse applications, what might be a meaningful ΔOF value that could be used as a convergence criterion?

(a) Modeling human height w.r.t. age, with height SSD as the OF.
(b) Modeling human height at age 15 w.r.t. several diet parameters, with height rms as the OF.
(c) Modeling human adult height w.r.t. the global latitude value, with height rms as the OF.
(d) Modeling human adult height shrinkage (from age 30 to 70) w.r.t. the global latitude value, with Akaho's estimate of total least squares as the OF.
(e) Modeling drag force w.r.t. the flow rate, with force SSD as the OF.
(f) Modeling human life span w.r.t. income, with age SSD as the OF.

12.3 Implement steady-state identification (SSID) as a convergence criterion.

12.4 Use Equation 12.3 to choose a meaningful ΔOF for the convergence criteria of several cases that you have encountered.

12.5 Use Equation 12.6 or 12.9 to determine ΔDV convergence criteria of several cases that you have encountered.

13

Model Design – Desired and Undesired Model Characteristics and Effects

13.1 Introduction

The user will choose a model functional form, its architecture. Perhaps it will have been derived from first principles and left in its primitive form, or converted into dimensionless group variables, or scaled variables by range or by standard deviation, or rearranged for computational convenience. Such rearrangements may have an impact on the ability or efficiency in seeking coefficient values.

Alternately, the user might have decided to use an empirical model, such as a power series, or some set of functional relations that seem to match the data trend (power law, reciprocal, exponential), or a fuzzy logic (FL) structure, or a neural network model. In such cases the user needs guidance as to the number of terms to include in the model. This includes the number of neurons in the hidden layer of a network or the number of FL rules and membership categories.

In any case, the user will desire that the in-use form of the model be convenient to use and provide a minimum of uncertainty. This chapter reveals issues related to model structure.

13.2 Redundant Coefficients

Sometimes coefficients appear in models as a product (or ratio) or sum (or difference), and not in an independent form. Here is an example from a pseudo-component modeling of stress–strain relationship in a film. Each pseudo-component represents a particular molecular structure. The instantaneous stress, σ, is a nonlinear response to strain, ε, and is often modeled as a hyper-elastic spring:

$$\sigma_i = A_i(e^{B_i\varepsilon} - 1) \tag{13.1}$$

where the subscript, i, indicates the ith pseudo-component.

Nonlinear Regression Modeling for Engineering Applications: Modeling, Model Validation, and Enabling Design of Experiments, First Edition. R. Russell Rhinehart.

Note that the functionality imposes a constraint that the AB product must be positive. If B is negative, then the exponential decreases with strain, making the term in parenthesis become negative. However, since stress must increase with strain, then A must also be negative. A similar argument reveals that if B is positive, A must also be positive. Further, if either is zero, the model would indicate that stress is not a function of strain. Accordingly, the constraint for the optimizer seeking values for coefficients A and B would be

$$AB > 0 \tag{13.2}$$

If there are several pseudo-components operating in parallel, with each having a particular volume fraction, x, then the total stress is the sum of all individual parallel components is

$$\sigma = \sum x_i \sigma_i = \sum x_i A_i (e^{B_i \varepsilon} - 1) = \sum \varphi_i (e^{B_i \varepsilon} - 1) \tag{13.3}$$

By measuring stress and strain alone, the product $x_i A_i$ always appears as a single value, indicated by the symbol φ_i in Equation 13.3 and individual values for x_i and A_i cannot be determined.

Further, when the strain is very small, the exponential term can be represented by a linear relation ($e^{small} \cong 1 + small$) and the composite stress equation reduces to a form that prevents independent evaluation of even B_i:

$$\sigma = \sum x_i A_i (1 + B_i \varepsilon - 1) = \sum x_i A_i B_i \varepsilon = \sum \varphi_i \varepsilon \tag{13.4}$$

In this limit, only the value for φ_i is needed to relate stress to strain. It does not matter what combination of x_i, B_i, and A_i values lead to a particular value of φ_i. For instance, $1 \times 1 \times 1 = 1$ provides the same value of φ_i as $0.5 \times (-8) \times (-0.25) = 1$. If x_i, B_i, and A_i are each adjustable, then the optimizer will be able to find an infinite number of functionally identical solutions, each one misguiding the researcher.

However, you would not know that this is the situation until after determining coefficient values.

As another example consider a steam-heated heat exchanger, a condenser heating a liquid. Ideally, the heat transferred from the hot to the cold fluid across the tube material is equal to the heat picked up by the cold fluid. In a simple representation:

$$F\rho Cp(T_{out} - T_{in}) = \dot{Q}_{pick-up} = \dot{Q}_{transfer} = UA\left(T_{steam} - \frac{T_{out} + T_{in}}{2}\right) \tag{13.5}$$

If the physical system coefficients are unknown and inlet conditions are constant, then the outlet temperature of the cold fluid will depend on the fluid flow rate. Solving Equation 13.5 for T_{out} as a function of F:

$$T_{out} = \frac{aF + b}{cF + d} \tag{13.6}$$

Equation 13.6 appears to have four adjustable coefficients, a, b, c, and d. However, one can divide by "a" and re-label the coefficients that are the b/a, c/a, and d/a ratios, which reveals that there are only three independent adjustable coefficients:

$$T_{out} = \frac{F + \alpha}{\beta F + \gamma} \tag{13.7}$$

Further, if the value of βF is $\gg \gamma$ then the model is effectively

$$T_{out} = \frac{F+\alpha}{\beta F} = \delta + \varepsilon / F \tag{13.8}$$

which is a two-parameter model.

As another example, if bias is permitted to be an adjustable coefficient in a neural network, then the product of weight times bias becomes a redundant coefficient, leading to an infinite number of equivalent solutions.

As still another example, the Arrhenius relation for temperature dependence in a chemical reaction rate might be stated as

$$r = ke^{-E/RT}[c] \tag{13.9}$$

where k and E are the coefficients of unknown value, but if concentration, $[c]$, is changed with a relatively constant temperature, T, then the factor $ke^{-E/RT}$ is relatively constant, and the values for k and E cannot be independently evaluated.

As a final example, consider a first-order plus dead time (FOPDT) model often used in process control. The model nominally has three coefficients – gain, time constant, and delay. However, the first-order differential equation it represents needs an initial condition, which is a fourth model coefficient. In addition, linearized, the model represents deviations from a base or reference value in both input (forcing function) and output (state variable), adding two more coefficients and summing to six model coefficients. However, the steady-state relation defines the relation between the gain and base values for the state variable and the influence variable. Therefore, while it appears to be six coefficients in the model, only five model coefficients are independent and needed.

Redundant variables lead to an infinite number of combinations of DV* (decision variable) values with equivalent regression fit OF* (objective function) solution values.

However, sometimes this property can be exploited. In dimensionless group models, such as Equation 2.19, the value of the dimensionless Reynolds number, for example, depends on the product/ratio of experimental conditions:

$$N_{Re} = \frac{du\rho}{\mu} \tag{13.10}$$

It does not matter how N_{Re} obtains a particular value. Only its value is required to determine coefficient values in Equation 2.19. This means that an experiment conducted by adjusting d, k, and u can provide an Equation 2.19 model that also predicts how fluid density and viscosity will affect heat transfer.

13.3 Coefficient Correlation

When there are redundant coefficients, infinite combinations of values lead to exactly the same model. In this case an optimizer will find the same OF minimum, but with a variety of DV values. This situation can be illustrated in the two-dimension model represented by Equation 13.1 with A and B as the coefficient values and data from very small strain values. In Figure 13.1 the vertical axis is the least-squares OF and the two horizontal axes are the A and B values scaled to a common 0–10 range. Note that the minimum is along the curved valley with a level floor. Here a variety of A and B combinations will lead to the same OF* value.

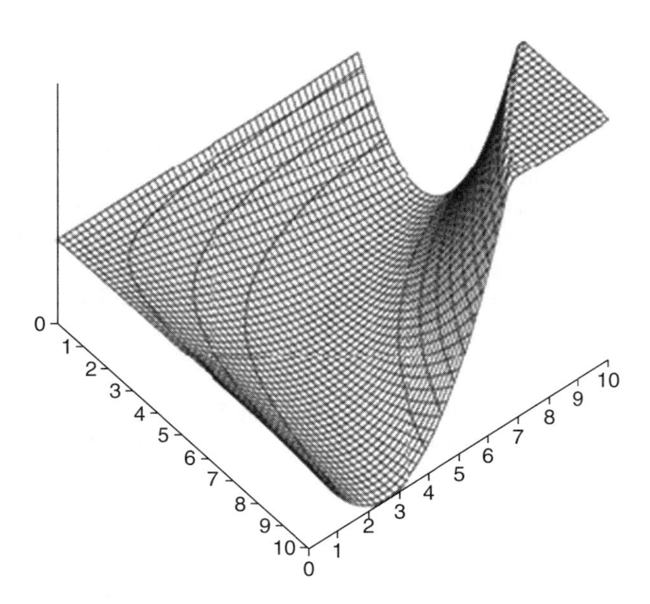

Figure 13.1 A view of parameter correlation

Figure 13.2 reveals the contour plot associated with Figure 13.1. It shows OF w.r.t. A (the horizontal DV1 axis) and B (the vertical DV2 axis) and the X markers represent converged solutions for the optimizer from 100 independent initializations. Notice the reciprocal relation of the B value to the A value in the converged coefficient values. Expectedly, if there is a single value, V, for the model represented by the AB product, then the value for B should be proportional to the reciprocal of A:

$$B = \frac{V}{A} \tag{13.11}$$

The contour plot is a convenience that helps the reader relate the locus of optimum solutions to the three-dimensioanl shape of the OF in Figure 13.1. However, to see either the three-dimensional or contour graph requires the computation of the OF at all coefficient combinations. That is a lot of computational work. Optimization seeks to find the best values by minimizing the computational burden, so the entire surface will not be visible. However, multiple optimizations initialized at random coefficient values will produce multiple coefficient sets. The X markers in Figure 13.2 reveal the A and B solutions that resulted in the same minimum OF. If a graph of one coefficient w.r.t. another shows a clear pattern, such as in Figure 13.2, this indicates correlation in the model coefficients (termed parameter correlation) and suggests that the model has redundant parameter values.

If the pattern is a reciprocal, as in Figure 13.2, then the model coefficients are effectively combined as a product. If the pattern is linear, with a zero intercept, then the coefficients are effectively combined as a ratio. If the pattern is linear with a non-zero intercept, then the coefficients are effectively combined as a sum or difference. Such patterns may be the clues to help identify the source of the parameter correlation. Many patterns are possible (semi-log, exponential, power) depending on the manner in which coefficients are combined in the model.

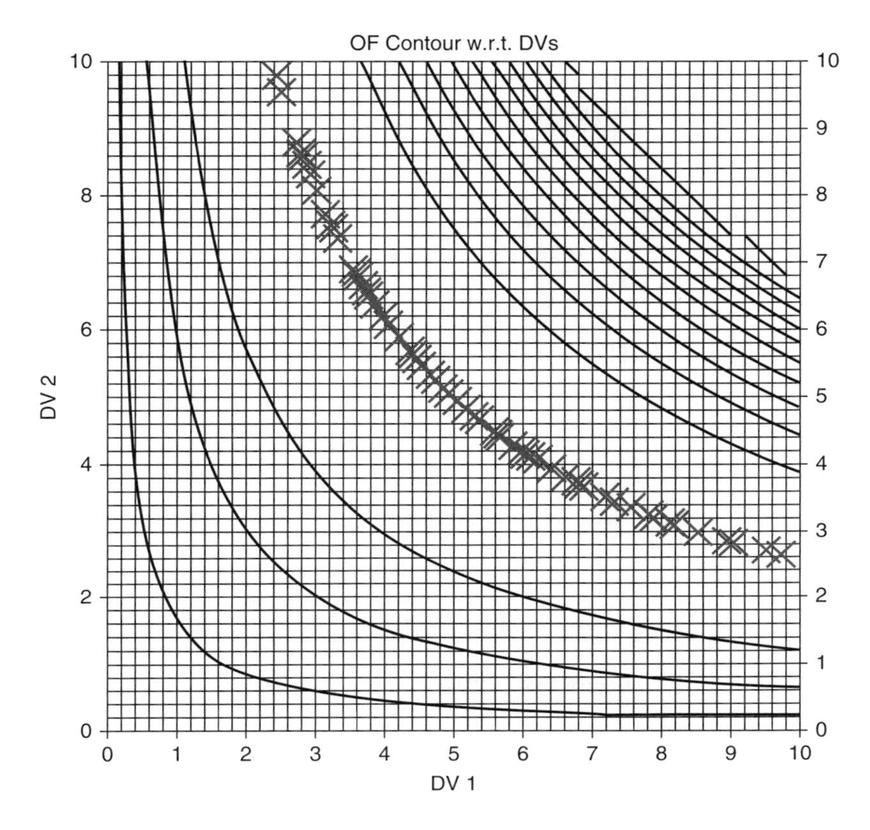

Figure 13.2 An indication of parameter correlation

To see if there are redundant coefficients, test for parameter correlation. Run multiple (perhaps 100) optimization trials from independent randomized initial coefficient values and test each parameter pair (their values from each optimization trial at the common OF*) for correlation. For an M-dimensional application, this is $M(M-1)/2$ comparisons. If seeking a visual indication, this means $M(M-1)/2$ graphs.

13.4 Asymptotic and Uncertainty Effects When Model is Inverted

Figure 13.3 illustrates the use of a quadratic-like model (illustrated with the solid line) to fit data from a process that has a monotonic asymptotic trend (illustrated with the dashed line), which generated data (markers) for regression. Seeing only the data and not having a first principles indication of the true functional relation, the user might presume that the quadratic model is appropriate. The process is invertible: given the indicated value of y there is an x value that would have the process output match the target. However, note that the quadratic model cannot provide that value of y. Inverted, solving for x given y, the model would return complex values (real plus imaginary).

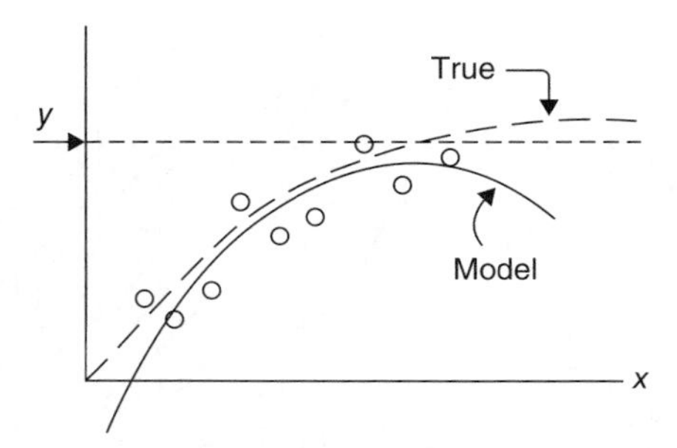

Figure 13.3 Illustration of noninvertibility and polytonic problems

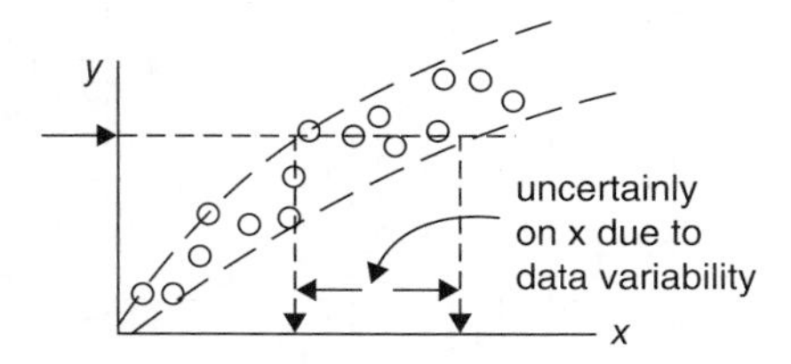

Figure 13.4 Illustration of uncertainty

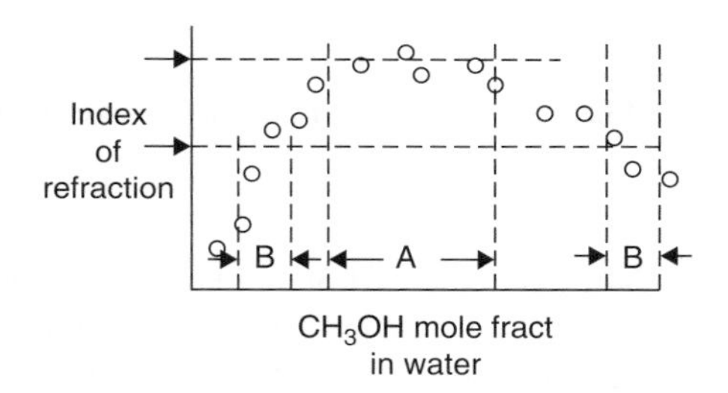

Figure 13.5 Illustration of sensitivity, uncertainty, and polytonic problems

The figure also reveals that with a lower y target, the model will return two x values, not one. Desirably, the model will provide a unique x value, matching the expected natural relation.

Figure 13.4 reveals a problem of uncertainty amplification on the inverse. Given a target for y, there is a wide uncertainty on the corresponding x value.

Figure 13.5 reveals a behavior that is polytonic. It represents the index of refraction (IR) of a methanol–water mixture. Knowing the curve, process samples can be taken, IR read, and from the curve back-out the methanol composition. The figure illustrates two problems of invertibility. First, consider what happens with the IR measurement indicated by the upper arrow. Since the sensitivity of IR to composition is essentially zero, there is a wide uncertain range on the corresponding sample composition, as indicated by the Region "A." Second, since the curve is polytonic, at the IR value indicated by the lower arrow, there are two distinct compositions, both also confounded by an uncertain region, as indicated by Region "B."

The user should consider the structure of models and choose structures to minimize the impact of the inverse use – using the model "backward" to determine the input required to generate a specified output. Often this means rearranging the first principles model to predict x (the normal independent variable) from y (the normal dependent variable).

13.5 Irrelevant Coefficients

A model might be developed with an initial concept that some variable is important, when in fact it is irrelevant. It might be hypothesized that ambient temperature has some impact on reactor yield. We know that temperature has an impact on yield and that temperature of the surroundings in an outdoors chemical storage change from day to day. However, it may be that the reaction conditions dominate the influence on yield and that the ambient conditions effectively have no impact. A simple model might be structured as

$$Y = a + bT_{amb} + cF_1 + dF_2 + eP \qquad (13.12)$$

where Y represents the reaction yield.

If the variable representing ambient temperature is irrelevant, then the coefficient, b, in the model should have a zero value. Therefore, finding a coefficient value of zero indicates an irrelevant factor. However, it is unlikely that the optimizer will exactly find that $b = 0$, or that it will find the true value for any model coefficient when fitting the model to experimental data (which is corrupted with variability). Further, being close to zero does not mean that it is zero. It could be the result of the large magnitude of the input variable and the small magnitude of the response variable. For instance, if yield is in a mass fraction and is of the order of 0.03 and pressure, an important variable, is measured by pascal, torr, or bar, the coefficient, e, might have a value of 0.000 01, seemingly small.

The generation of irrelevant model coefficients can arise from a progressive addition of model functionality, making an earlier variable irrelevant. Suppose a true y w.r.t. x phenomena is quadratic, and you start looking at model development starting with a first-order (linear) model and then by adding a higher-order quadratic variable. The linear model will provide a significant improvement over the one-term model (average) and the quadratic model will be better yet. However, when the quadratic term is added to the model, the intercept and linear terms are not needed (because the data has the quadratic dependency) and the intercept and linear term could be dropped from the model. In general, as functions are progressively added to a model old functions may become irrelevant. Irrelevant functions should be removed to enhance model simplicity and final prediction error (FPE).

Section 13.8 provides some additional situations.

If the coefficient is really zero then data sets from independent trials (independent realizations) will provide positive and negative coefficient values that include zero. By contrast, if the coefficient value is not zero, multiple data realizations will result in reproducible non-zero coefficient values. Since trials consume resources, we prefer an alternate approach to the generation of data realizations. Bootstrapping is one approach, described in Chapter 17.

13.6 Poles and Sign Flips w.r.t. the DV

The structure of some models generates poles. Here is an example considering that the coefficients are represented by a, b, and c:

$$y = a + b/(x - c) \tag{13.13}$$

The optimizer will be given (x, y) pairs and seek coefficient values to best match the data. When c has a value near x the slight change in the DV value could lead to y jumping from large negative to large positive. The sign of coefficient b can correct for the modeled c value, but a graph of the y model w.r.t. the coefficient c will have poles ($y = \infty$) at each x value.

Look at your equation and anticipate what issues it might create for an optimizer. In this case there are several options. One is to constrain the value of c smaller than the smallest x value or larger than the largest. Another is to rearrange the model and compare the predicted yx product to the actual

$$xy = ax + b + c(y - a) \tag{13.14}$$

In this case the optimization objective would be

$$\underset{\{a,b,c\}}{Min} \; J = \sum \left(x_i y_i - a x_i + b + c(y_i - a)\right)^2 \tag{13.15}$$

However, note that this mathematical functional transformation violates the guide given in Chapter 9.

13.7 Too Many Adjustable Coefficients or Too Many Regressors

Which variables should be included in a model? Consider the gas equations of state in Equations 2.1 and 2.2. If you believe n, T, and V affect P, then the models have the right input variables. If however, you did not have a fundamental model, theory, or analysis that reveals influences on P, you would have to select ones that seem appropriate to build a model. In such a case, you might think that molecular weight of the gas, the number of atoms in a molecule (He is 1, O_2 is 2, CO_2 is 3, CF_4 is 5, etc.), and the year in which the gas was discovered also affect pressure. In which case your relationships would be

$$P = f(n,\ T,\ V,\ molecular\ weight,\ N,\ Y) \tag{13.16}$$

If you chose a linear relation for each input on the response, then the model would be

$$P = a + bn + cT + dV + e\ molecular\ weight + fN + gY \tag{13.17}$$

When doing regression the influence variables are termed *regressors*. Regressors could be combinations of the primitive input variables. For instance, if $x_1 = n,\ x_2 = T,\ x_3 = V$, and so on, you could define some combination variables, such as $x_7 = n^2,\ x_8 = nT,\ x_9 = 1/T$, and so on. Replacing P with y, the model could be represented as

$$y = a_1 + a_2 x_2 + a_3 x_3 + \cdots = \sum a_i x_i \tag{13.18}$$

Even though the model appears linear in the regressors, it is nonlinear in the primitive variables.

Equation 2.15 represents a polynomial or power series model. If the function is odd (anti-symmetric about x), as opposed to even (symmetric about x), then x to even powers (x^0, x^2, x^4, x^6, etc.) will not express the y response to x. Accordingly, even powers of x should not be in the model. When choosing which regressors to use for an odd function, only use x to the odd powers. If even, choose even powers.

Equation 2.24 is an empirical dynamic model, with regressors representing past values of the process output and input. The question to a model builder is, "How many and which past u and y values are needed to have a model that adequately predicts the next process output?" If there was knowledge about the process, or a model, then the user would possibly know the process order and the delay, and appropriately choose the number of regressors. If there is a delay in the y response to u, then u_i should not be a regressor in a model for y_{i+1}. If the delay is equivalent to n samplings, then u_{i-n} should be the first in the sequence of u regressors.

In general, the question is, "Which variables need to be included in the model?" Preferentially, knowledge about the process, through its model or experience, will indicate which variables need to be included. By contrast, if you did not have confidence in such knowledge, you would have to select regressors that seem appropriate to build a model and then explore the effect of other choices for regressors.

Unfortunately, in empirical modeling, the order and functionality are not known. The best way to determine which variables and functionalities should be included in the model is grounded in trial-and-error, data-based tests.

The following application illustrates the impact of both choice and number of regressors. It comes from a study of the cycling nature of undergraduate chemical engineering enrollment. Data suggests that periods of low graduation rates lead to high salaries and multiple job offers. That information influences freshmen matriculates to choose chemical engineering. This makes class size swell. Then, when we graduate high numbers of BS ChEs, some cannot get jobs upon graduation. When this news reaches high school students, they choose alternate majors, and ChE matriculation falls. It is a dramatic effect, creating a regular cyclic pattern with about a 14-year period and a 2.2 : 1 amplitude ratio. Figure 13.6 reveals the trend. National data (solid line with square markers) is from the NCES Digest of Education Statistics. It started in the early 1970s. Data from Oklahoma State University (dashed line and triangle markers) goes back to our first graduates in 1922, but I only show data since 1940.

Forecasting the trend is important to long-range planning. Many argue that the trends are due to independent influences (baby boom, war, recession, Sputnik, energy crisis) and would seek a model based on the vagaries of external influences. By contrast, the regular, periodic cycling suggests a sustained, common feedback mechanism within a seasonal-type of model.

In seeking to identify the model, first consider which variable to use as the model input. Figure 13.7 shows a fairly clear correlation of freshman matriculates to the OSU ChE program w.r.t. the national BS ChE graduating class size. Lag 2 means a two-year time shift, for example, the number of freshman matriculates and freshman and sophomore transfers in 2015 at OSU are plotted relative to the national BS ChE graduation rate in 2013. The concept is that it would take about two years for data to become visible in the public domain and influence career advice to matriculating students.

However, when Lag 4 is used, as illustrated in Figure 13.8, the correlation is better, as indicated by the R-square value, the magnitude of the slope, and visually by looking at residual scatter from the model. The message is that a first choice of input variables to predict an outcome are not necessarily the right choice. It would be nice to have a structured approach to find

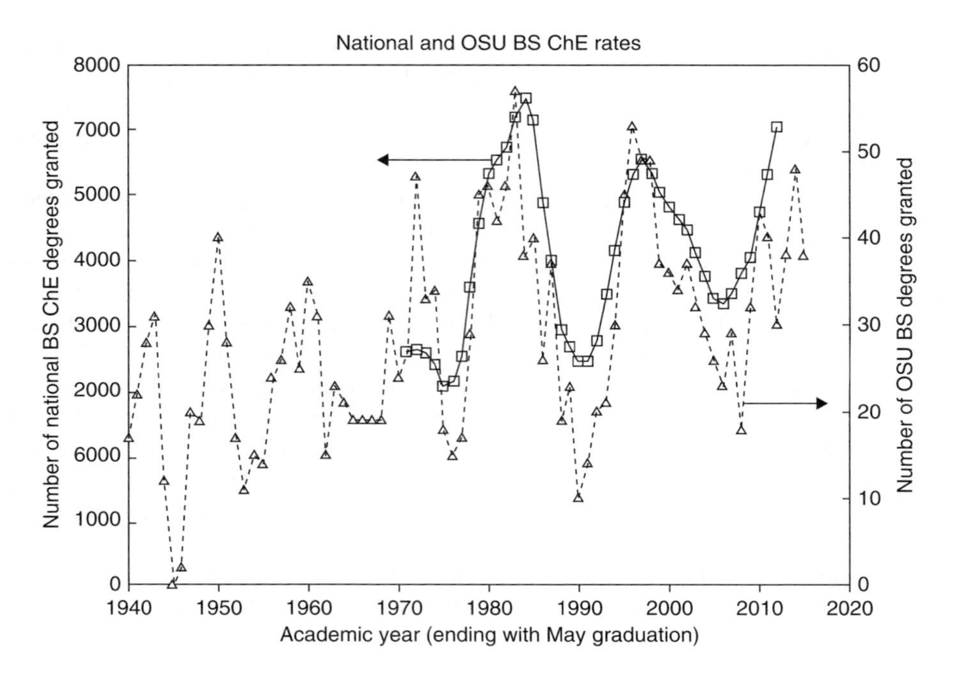

Figure 13.6 Chemical engineering enrollment cycling

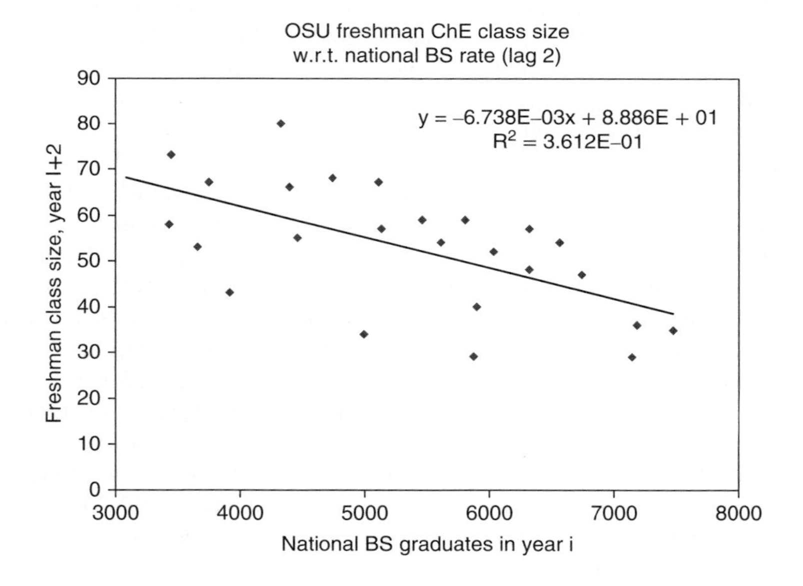

Figure 13.7 Illustration of wrong choice of input variable

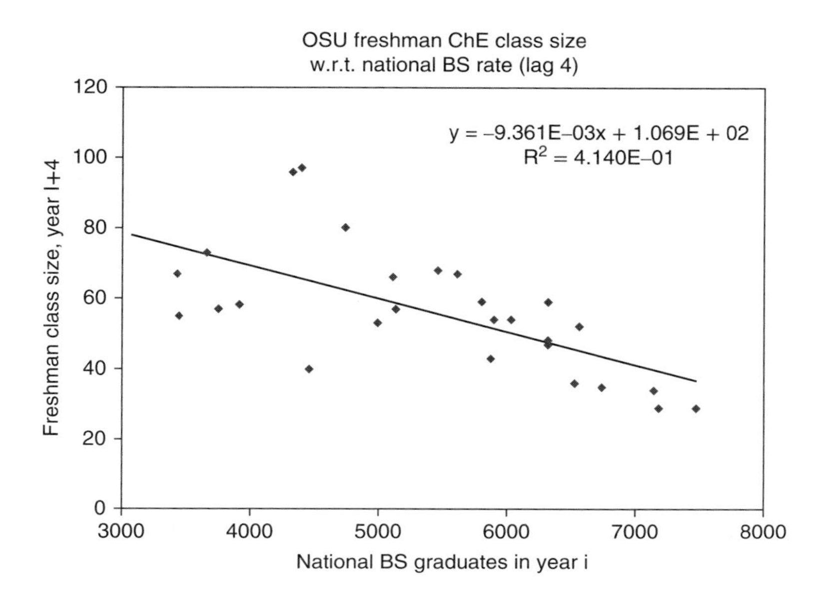

Figure 13.8 Illustration of the right regressor but too few adjustable coefficients

the right input variable. Additionally, note that identifying the right lag interval is a nonlinear optimization.

Figure 13.8 used a linear model. Does this make sense? The data seem to suggest some downward curvature, that if the graduation rate is over 7000 per year, the new student enrollment rate will drop off dramatically. A quadratic model might provide a better fit to the data.

Figure 13.9 uses a quadratic model. The R-square value improves from 0.41 to 0.54 over the linear model and visually the fit is much more satisfying. This suggests that the linear model did not have enough terms (high enough order) to properly model the data.

If a second-order model is better, what about a sixth-order model? Figure 13.10 shows the sixth-order polynomial fit. The R-square value is better yet. However, the behavior of the model is somewhat absurd, especially when considering the model-indicated increase in enrollment as graduation rates exceed about 7500 per year and the conflicting trend on either side of about 3500 per year. These features of the sixth-order model contradict a mechanistic understanding. As well, the intermediate inflections do not match our general experience with natural phenomena, which usually reveals a low-order response. Too high an order, too many input variables, creates a model that begins to fit noise, not just reflect the underlying phenomena. How can one choose the justifiable order in an empirical model?

If there is not fundamental knowledge to guide selection of input variables you might use a correlation index to choose regressors. Plot y w.r.t. each of the possible influences, or regressors (functions of the primitive variables), and see which regressors create a statistically significant correlation. Of course, the graph is not necessary to determine the correlation coefficient. This would provide a subset of regressors that are correlated to the output. However, an empirical model that is a combination of all regressors may be unnecessary. For instance, if y is proportional to x, then regressors representing x^2, $z \times x$, and $1/x$, will be likely to show

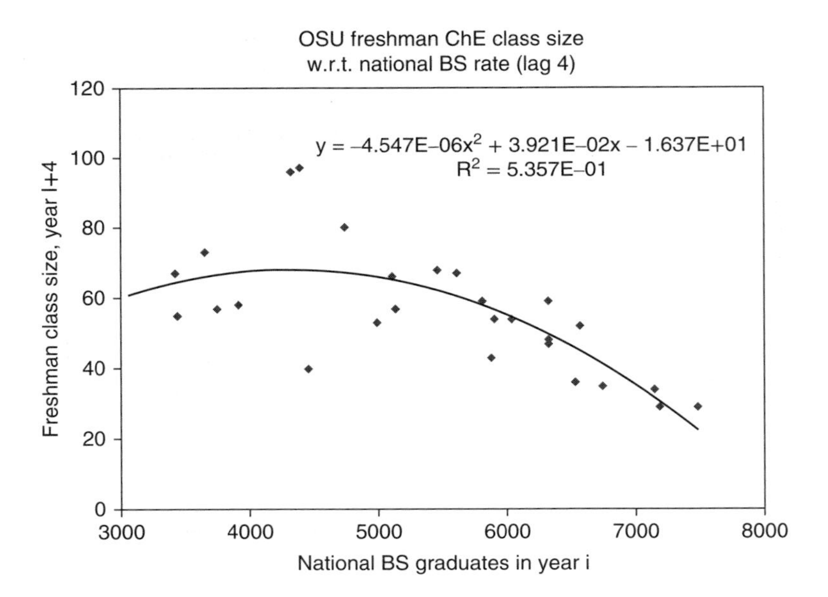

Figure 13.9 Illustration of number of adjustable coefficients

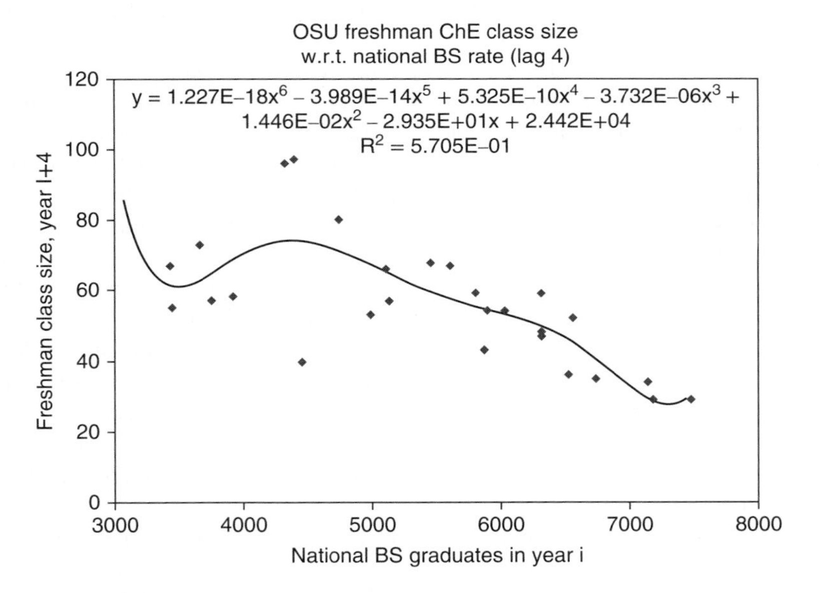

Figure 13.10 Illustration of too many adjustable coefficients

significant correlation, even if z is unrelated to y. The model should not include redundant regressors.

One approach to eliminating redundant regressors is forward selection followed by backward elimination. In *forward selection*, start with your list of possible regressors (primitive variables, and their functional forms and combinations, and a constant). Choose each, one at a time, to create a one-term model, as

$$y = a_0 + a_1 x_i \tag{13.19}$$

For each of the ith regressors there will be a sum of squared deviations (SSD_i) of the modeled y from the actual y. Select the regressor with the best (smallest, least) SSD value.

Next, add one more term. From the remaining regressors, choose each, one at a time, and add them to the model:

$$y = a_0 + a_1 x_j + a_2 x_i \tag{13.20}$$

where j is the index for the first selected regressor and i is an index for any other regressor.

For each new model, adjust a_0, a_1, and a_2 to minimize the SSD. Do not keep the prior coefficient values. Select the regressor that leads to the model with the lowest SSD.

Continue adding regressors one at a time. Expectedly, each time a regressor is added to the model, the SSD decreases, meaning that the fit to the data is improving.

Stop adding regressors when the reduction in SSD is not as good an improvement relative to any of several undesired aspects of too many terms: the increase in model complexity or the fitting of noise not underlying trends. Metrics to stop adding regressors will be provided subsequently, but first consider another problem.

It could be that the shared functionality of two or three lately added regressors provides a combination that describes the y behavior better than an early added regressor. For example, if a function is quadratic with an inverted parabola shape, such as $y = x - x^2$, but the underlying relation was not known, a view of the data for the range $0 < x < 0.4$ could lead one to suspect that $sin(x)$, $exp(x)$, or x^p functionalities should be included with the regressors. The data might be such that the $sin(x)$ functionality is first selected, then the linear x, and then the x^2 regressors. After x and x^2 are chosen, the developing model can match the data and the $sin(x)$ term is irrelevant. Therefore, after each regressor is added, starting when there are three terms in the model, test to see if one of the earlier chosen regressors has become irrelevant by performing backward elimination.

In *backward elimination* drop one term from the model. For each term (except the last added), taking one at a time, drop it from the model retaining all of the other terms and determine new values for model coefficients that minimize SSD. Consider eliminating the regressor with the least reduction in SSD. A small reduction in SSD indicates a small impact on the overall model performance.

In either forward selection or backward elimination, the evaluation of a "small" contribution to the SSD is relative to the SSD that would remain if the model were perfect, and this also is relative to the undesired aspects of a complex model. Many individuals have developed rules and criteria to determine when to stop forward selection and when to drop a regressor in backward elimination.

At what point does an increase in the number of model terms begin to add undesired aspects more than it adds improvement? Where is the point of diminishing returns with increasing complexity? I find that the FPE promoted by Ljung (1999) and built upon the work of Akaike is one of the simplest approaches and often provides a reasonable and useful assessment for determining model order for empirical models. FPE is the SSD multiplied by a complexity factor related to the relative number of coefficients to data points:

$$FPE = \frac{N+M}{N-M} SSD \tag{13.21}$$

where N is the number of data sets, M is the number of model coefficients to be evaluated by the data, and SSE is the sum-squared error between the data and model with M coefficients. Plot FPE w.r.t. M. In my uses, with low values for M, SSD reduces faster than the complexity term increases and FPE reduces with increasing M. Eventually, with additional model coefficients, the complexity term increases faster than SSD reduces and FPE increases with M. The method would be to start with a low-order model and progressively add model complexity until FPE rises; then choose the model with minimum FPE, the next to last model.

The condition, stop when $FPE_{(M+1)} > FPE_{(M)}$, can be rearranged to

$$\frac{SSD_M - SSD_{M+1}}{SSD_{M+1}/(N-M-1)} < 2\frac{N}{N+M} = \frac{2}{1+M/N} \tag{13.22}$$

The left-hand side (LHS) of Equation 13.22 represents the improvement in SSD relative to the y-variance in the data as estimated by the model with $M + 1$ coefficients. The LHS denominator is a rough approximation to inherent y-variance, but is commonly used, since there does not seem to be a better data-based measure, outside of running replicate trials.

The right-hand side (RHS) represents a model complexity factor, reducing as the M/N ratio increases

With $M \ll N$, the RHS has a value of a bit less than 2. One rule-of-thumb for empirical models is that there should be three (or more) independent experimental data sets for each regression coefficient. Accordingly, a maximum value of M should be about 1/3 of N. In this case the RHS of Equation 13.22 has a value of about 1.5, so, with the FPE criterion, the RHS of Equation 13.22 ranges between about 1.5 and 2. The criteria to stop increasing model complexity then is, if the relative improvement in SSD to y-variance is less than about 1.5 or 2, stop adding model terms.

In many applications the software reports an r-squared correlation coefficient value and not SSD. Recognizing that r-square is the relative reduction in SSD from the model compared to the variance in the original y data:

$$r^2 = \frac{SSD_1 - SSD_M}{SSD_1} \tag{13.23}$$

where SSD_M is the residual sum of squared y-deviations between the data and model and SSD_1 is the sum of squared deviations between the data y values and the average of the data y values (which is identical to $(N-1) \times$ variance). The subscripts, 1 and M, represent the number of coefficients in the model. Then substituting into the FPE in Equation 13.21:

$$FPE' = FPE/SSD_1 = \frac{N+M}{N-M}(1-r^2) \tag{13.24}$$

where FPE' is the scaled FPE. Since it is scaled by the constant SSD_1, minimizing FPE' will result in the same model choice, m, as minimizing FPE. The form of Equation 13.24 is more convenient for the user when the software returns r-square, not SSD.

Although derived from diverse reasoning, Mallows' *Cp,* Akaike information criterion (AIC), and several other indicators to stop adding regressors can each be converted to similar forms, with the RHS having values between 1.5 and 2, as Miller (1990) reveals.

For example, Mallows' *Cp* is

$$Cp = \frac{SSD_M}{\sigma^2} - (N - 2M) \tag{13.25}$$

and the rule is: find M to minimize Cp or stop at M number of coefficients when $Cp_{(M+1)} > Cp_{(M)}$. This technique uses the SSD for the full model (with all possible regressors, k number of regressors, not just the subset of M regressors) to estimate the y-variance. Mallows' criterion to stop selecting regressors is then

$$\frac{SSD_M - SSD_{M+1}}{SSD_k/(N-k)} < 2 \tag{13.26}$$

Additionally, AIC seeks to minimize the metric:

$$AIC\ Metric_M = -2(LL_M - M) \tag{13.27}$$

which is equivalent to stopping at M number of coefficients when $AIC_{(M+1)} > AIC_{(M)}$. LL_M represents the log of the likelihood function. If Gaussian, with uniform variance throughout the data range, variance allocated to y only, and estimating variance from the model with $M + 1$ coefficients, then the AIC to stop adding terms is stop at M regressors when

$$\frac{SSD_M - SSD_{M+1}}{SSD_{M+1}/(N-p-1)} < 2 - \frac{M}{N} \tag{13.28}$$

These three triggers, Equations 13.24, 13.26, and 13.28, to stop adding regressors are essentially identical, although they were independently created.

However, those criteria for determining the number of regressors are independent of the situation, and a knowledgeable user might have a different relative importance to the complexity-functionality balance. To me, sometimes those methods permit greater model complexity than I believe is justified. Observe that the RHS of Equations 13.22 and 13.28 approach the maximum value of 2, as N (the number of data sets) increases. Additionally, large N attenuates the impact of increasing M on the RHS. This means that the greater is the number of data sets, N, the higher is the value of M permitted. If the model should only reflect a quadratic response, very large N will permit cubic and quartic terms to be included in the above criteria. In my opinion, when the data set size, N, is large; FPE, AIC, and Cp permit the inclusion of too many regressors because the complexity factor is dependent on the relative impact of M to N.

There are several issues related to model complexity. A model with excessive number of terms requires additional data to determine coefficient values. Excessive number of terms makes a model that begins to fit noise, not an underlying phenomena. Excessive number of terms increases the difficulty related to the model in use (inverse, storage, display, verification, etc.).

Accordingly, I like another metric for determining model complexity, M, which is based on a situation-dependent equal concern concept. Here again, the objective is to find the model set of regressors that minimize SSD, tempered by a penalty for complexity. Equation 13.29

represents the concept, and is very similar to the recommended approach to handling soft constraints in Equation 8.8.

$$\min_{\{M\}} J = \frac{SSD_M/N}{(EC_\sigma)^2} + \frac{M^2}{(EC_M)^2} \tag{13.29}$$

here SSD_M is normalized by the number of data, recognizing that SSD doubles if N doubles, SSD/N is not dependent on data set size. SSD/N represents the average residual variability left when the model is attempting to fit the data, and our primary feature of concern is this measure of goodness of fit. The second feature of concern is the number of terms in the model. Accepting that the penalty for badness should scale with the square of the deviation from zero badness, the second term in the OF represents the penalty for complexity. The EC factors represent the values of M and model standard error (standard deviation of the residuals) which create equal concern. The user chooses EC values for M and σ that create equal concern for model complexity, M, and model residual standard error. These values will be situation specific.

For example, a model outcome with an SSD/N which represents two times the inherent variance in the data (or a residual standard deviation which is 1.414 times the σ_y or about 40% un-accounted error) may have the same concern as two excessive model terms over the number anticipated. For instance if a quadratic model is expected, three terms, would be acceptable with no concern, a model of $3 + 2 = 5$ terms may have the same concern as a model that leaves 40% error (100% of the inherent variance) unaccounted for. In this case $EC_\sigma = 1.414\sigma$ is paired with $EC_m = 5$.

For the rule, "Stop adding variables when J of Equation 13.29 increases" is equivalent to stop at M regressors when $J_M < J_{M+1}$, which reduces to stop with M when:

$$SSD_M - SSD_{M+1} < \frac{N(EC_\sigma)^2}{{EC_M}^2}(2M + 1) \tag{13.30}$$

If the model is close to being correct (representing the truth about Nature), then the y-measurement variance can be estimated as $SSD_{M+1}/(N - (M + 1))$ as is indicated in Equations 13.22, 13.26, and 13.28, then the equal concern rule is stop when:

$$\frac{SSD_M - SSD_{M+1}}{SSD_{M+1}/(N - M - 1)} < \frac{N(EC_\sigma)^2}{SSD_{M+1}{EC_M}^2}(2M + 1)(N - M - 1) \tag{13.31}$$

If the average variance in the measured y value can be known from other approaches, such as replicates, it could be used for the $SSD_{M+1}/(N - M - 1)$ term. Additionally, if the equal concerns are chosen as: "When the number of model coefficients is as great as one-fourth of the number of data sets, this creates the same high level of concern as if the model residual variance is as great as 1.25 times the variance of replicates" (which means $EC\sigma^2 = 1.25{\sigma^2}_{replicates}$ and $EC_M = N/4$), then, when the model is good and the standard error of the residuals is matching that of the replicates, Equation 13.30 becomes

$$\frac{SSD_M - SSD_{M+1}}{SSD_{M+1}/(N - M - 1)} \cong \frac{SSD_M - SSD_{M+1}}{{\sigma^2}_{replicates}} < \frac{20}{N}(2M + 1) \tag{13.32}$$

When selecting EC values that represent my concern about the goodness-of-fit versus complexity issues, in my opinion Equation 13.32 stops the addition of new terms (model complexity) at reasonable places. By contrast, when I have an opinion about the competing concerns, the FPE, Cp, and AIC criterion usually permit the inclusion of too many terms. My personal preference seems to have a stronger penalty for excessive complexity. The equal concern balance of Equation 13.29 permits you to impose your own balance.

In the data presented in Figures 13.8 and 13.9, the equal concern approach says to stop with a linear model (Figure 13.8), while the FPE, Cp, and AIC indicate to stop with a quadratic model (Figure 13.9). With this $N = 25$ case, either model is fine, but for large data sets, the $(N+M)/(N-M)$ term is insensitive to M and provides no guidance as to model order. I prefer the equal concern approach.

If you have a feel for what is an appropriate number of terms and desired reduction in variance, then use the equal concern stopping criterion. Alternately, if you have no feel for what is the right balance for your application, use FPE, the simpler equivalent of the three criteria (FPE, Mallows Cp, and AIC).

Another approach to trigger the stop of adding regressors is to include a regressor that uses a random number for each of its values. The random number will have no relation to y and, once it is selected, you have an indication that all other regressors have even less relation to y. However, I think this leads to excessive regressor additions. For instance, it will permit a regressor to be included that is barely better than random.

Miller (1990) calls the value on the LHS of Equations 13.22, 13.26, 13.28, and 13.31 F-to-enter. The LHS statistic is a ratio of variances, as is an F-statistic. However, also as Miller indicates, it is not F-distributed. F-to-enter (the threshold to add a regressor) is in contrast to F-to-eliminate, the RHS ratio that would indicate justification to drop a regressor (backward elimination). I would suggest using the same value for F-to-eliminate as F-to-enter.

Such a procedure does not identify the model that represents the $y = f(x)$ mechanistic truth about Nature. Such a procedure does not identify the best model for utility of use. Such a procedure cannot reveal when model complexity exceeds that of Nature and begins to fit noise. Such a procedure does not provide a model that passes logic-based validation criteria. FPE may permit the acceptance of high-order models with inflections that would not represent our understanding of Nature. FPE is just one of many approaches for balancing complexity and model closeness to data. For empirical model development (polynomial type models, power series models, orthogonal functions, wavelets, fuzzy logic, neural networks, and such) I believe these are reasonable guides.

In a personal communication, Dennis Williams (Consulting Engineer, Lyondellbasell, 2013) comments, "*This section discusses primarily the question of whether additional nonlinear model terms that are functions of a single independent variable should be added. In my experience, unless physics or chemistry supports some particular nonlinear form, anything beyond* x^2, $\sqrt{x}$, *ln(x) tends to lead to unusual (and clearly wrong) extrapolation and sometimes interpolation. In multivariable problems, detection of being outside of the independent variable region of the data supporting a model can be very difficult. Also, my experience is that the usual question of interest is whether additional independent variables should be included.*" I agree.

13.8 Irrelevant Model Coefficients

The following subsections present several approaches to detect this situation and trim no-longer-needed functions. Note that these techniques work only if there is one unique global optimum with one unique set of coefficient values.

13.8.1 Standard Error of the Estimate

In one traditional approach for linear regression, use the standard error of the estimate of each coefficient to determine if its value is significantly different from zero. If not significantly different, drop that functionality.

13.8.2 Backward Elimination

In another traditional approach, after each new term is added, remove earlier terms, one at a time, and redevelop the model coefficients with the one-term-less model. If an F-test or FPE-like measure indicates that the dropped function should not have been added then keep it out.

13.8.3 Logical Tests

I come from a process and product engineering background. It seems that for a model to be useful, to provide a practicable basis for making decisions, then the model must represent more than just a statistically defensible form (architecture, mathematical functionalities) and have coefficients that are confidently not zero. The model needs to reflect the underlying cause and effect mechanisms. Desirably the model has the functionality that mechanistically matches the phenomena, but this true-to-the-mechanism mathematical functionality is not needed if power series, wavelet, neural network, fuzzy, or any number of empirical "black box" or "gray box" approaches capture the data trends, the human understanding of the mechanism, and the in-use expectations of the model as a prediction, design, or control tool. You could use a test of trends or sensitivities to determine if the model is a rational representation of the process. At any set of conditions, changing the model input value (regressor, independent variable) should make the output change in an expected way. This approach will be detailed in Chapter 16.

13.8.4 Propagation of Uncertainty

You can also check for the sensitivity of the model output to each coefficient. If a coefficient is irrelevant (if the contribution of its functionality to the output is insignificant) that functionality could be trimmed or eliminated. Some people prefer to keep all functionalities. However, taking an application view, simplicity of the model is a strong value. Therefore, any nonessential term should be trimmed. Also, if an empirical flexible model is used, then there is no basis to think its mathematical forms represents the true mechanism. Accordingly, since terms in empirical models cannot be defended as mechanistically valid, and since simplicity is a value, then inconsequential terms should be trimmed or deleted.

For any chosen functionality within the model, investigate propagation of uncertainty about the influence variable on the prediction variable (dependent variable) throughout the entire independent and dependent variable ranges. If the propagation of uncertainty of the influence is inconsequential everywhere, then there is cause to question the model functionality (the input/output relation) or whether the particular input variable should be included in the model.

13.8.5 Bootstrapping

If a model coefficient value is zero, or nearly so relative to its uncertainty, that indicates that the functionality that the coefficient represents is irrelevant and should be trimmed from the model. You can test the confidence limits on a model coefficient to see if it is not zero in a technique called bootstrapping. Sample a subset of the data (with replacement – retaining the data set in the draw-from original set) to create a new set of data. The new set should have the same number of items in the original, but items in the new set will likely be duplicates and some of the original data will be missing. Determine the model coefficient values on the new set. Repeat this process many (perhaps over 1000) times (generate a new subset with replacement, determine model coefficient values) and create a histogram to reflect the distribution of values for each model coefficient. The variability of the coefficients will indicate the vagaries within the data. If the distribution for any coefficient does not include zero (perhaps within the 95% confidence interval), then it has a value that is significant, implying that its functional contribution is significant.

This bootstrapping approach presumes that the original data has enough samples covering all situations so that it represents the entire possible population of data. Then the new sets (sampled with replacement) represent legitimate realizations of sample populations. Accordingly, the distribution of coefficient values from each re-sampled set represents the distribution that would arise if the true population were independently sampled.

From either bootstrapping or from truly independent sets of experimental samplings from the true population, the coefficient values are probably correlated, not independently distributed. If one sample set makes one coefficient value change for a best fit, then other coefficient values are likely to make compensatory changes. A graph of coefficient values will probably reveal correlation.

Although bootstrapping provides a measure of variability of the coefficient, the coefficient uncertainty, do not use that value in an analytical approach to propagation of uncertainty to estimate the collective uncertainty of the model coefficient values on the model prediction. The coefficient correlation in bootstrapping negates the assumption of independence needed in propagation of uncertainty.

13.9 Scale-Up or Scale-Down Transition to New Phenomena

We often generate models from data that have been generated on lab- or pilot-scale units. It is much cheaper to run small-scale experiments than to generate data from full- or commercial-scale units. The small-scale data can be very revealing, and the models very useful, but as units are scaled up in size, new mechanisms evolve that affect the internal process mechanisms. The extent of the change may wholly invalidate the model. Here are some examples:

- Processes are usually rate-limited by something, such as heat transfer, and scale-up can cause the rate-limiting step to change. For example, much higher flow rates can eliminate diffusive transport, such as the rate-limiting step in a catalytic reactor, making surface reaction become rate-limiting. This would change the reaction kinetic model.

- Vessel or particle size can shift the volume-to-surface-area ratio such that assumptions related to internal uniformity are no longer valid and must be replaced with a spatial distribution of internal temperature or concentration conditions.
- Size scale-up can shift the fluid flow distribution significantly, for example, laminar to turbulent within a pipe or to multiple recycling cells in a fluidized bed, and these new mechanisms substantially change the process phenomena.
- Dust comprised of large particles may pose no problem, but the same mass dispersed as fine particles can create an explosive situation. The model would need to include new mechanisms.

13.10 Takeaway

Regression model design is iterative. You choose the model architecture and choose the regressors (which and what lag). For each choice, use regression to determine the model coefficient values. Evaluate the result in the light of the multiple desired aspects related to model functionality, validity of terms, and fit to data. Adjust the architecture to balance simplicity, utility, and representation of data.

Plot one parameter value w.r.t. another after optimizer convergence. Does there seem to be a clear trend? If yes, an issue you may have is that the model structure is not quite right for the situation. Often the trend in parameter values provides a clue as to the issue. It could be that the DVs are essentially additive, or multiplicative, or ratioed. In such cases the OF has a nearly flat optimum with a very gentle slope toward extreme coefficient values.

Inspect your models and avoid redundant coefficients and noninvertible effects. Be aware of scale-up limitations. Carefully choose input variables and model order. If you have a feel for what is an appropriate (or desired) number of terms and the ultimate variance, then use the equal concern stopping criterion of Equation 13.29 to determine model complexity. If you have a feel for the appropriate number of terms and a variance reduction, but do not know the value of the minimum variance, use Equation 13.30 or 13.31. Alternately, if you have no feel for what is the right balance for your application, use FPE of Equation 13.21, the simpler equivalent of FPE, Mallows Cp, and AIC.

The model coefficients are usually continuous-valued, but the delay on a variable or the order (architecture) of a model would have integer values. This makes the optimization a mixed real-integer application. However, the model architecture may also be comprised of class variables. If the optimizer is seeking a choice of which input variables to use (temperature, pressure, etc.) or of what functionalities to use (radial basis, sigmoidal, linear, etc.), the DVs would be classifications. Therefore, the regression would be mixed real-integer-class optimization. Choose an optimizer that can cope with the DV types.

Exercises

13.1 Derive Equation 13.22 from Equation 13.21.

13.2 Derive Equation 13.24.

13.3 Derive Equation 13.30 from Equation 13.29.

13.4 (A) Choose a common nonlinear constitutive relation from your discipline. For example, the ideal gas law, $PV = nRT$, would indicate that pressure is inversely related to volume, $P = nRT/V$. Such models might be the temperature dependence on resistivity, the displacement response to load, the viscosity response to shear rate, and so on. Nature is not as ideal as the simple model and each of such relations has a progression of one-better models. The van der Waals relation is one-better than the ideal gas law, $P = nRT/(V - nb) - a$. (B) Use your one-better relation to generate a nonlinear response (such as P versus V) and simulate an experiment in which you would change the input variable and measure the noisy response (Nature masks the true response with NID(0, σ) noise). Simulate taking 10 measurements over a range of conditions that might be practicable. For example, a volume range of 0 to infinity is possible but not practical. (C) Use FPE to determine the best order of a power series model that represents your output (response, result, dependent variable, y) to input (influence, cause, independent variable, x) response. A power series model is $y = a + bx + cx^2 + dx^3 + ex^4 + \cdots$.

13.5 I claim that Ljung's FPE is a good method for selecting model order in empirical models. What do you think? Provide evidence to support your conclusion. Here is one approach. Use a linear y w.r.t. x relation to generate data and add noise to the y data. Then let Excel (or any software) find the best linear model ($M = 2$), then the best quadratic model ($M = 3$), then the best cubic model ($M = 4$), and so on. At each M, use the r-square as a measure of SSD. You can show that $SSD_M = SSD_1 \times (1 - r\text{-square}_M)$. Since FPE′, FPE scaled by a positive multiplier, does not change the DV optimum, you can use $FPE' = [(N + M)/(N - M)] \times (1 - r\text{-square}_M)$ w.r.t. M to determine the best value of M. Because you know that the generating model was linear, if the method is good, then the results should show that the minimum FPE′ occurs at $M = 2$. If the FPE method is good, then the value of M should be consistent with the generating model (for example, $M = 4$ if the generating model was cubic). The approach should be independent of either the number of data, or the amplitude of the noise, or the slope of the generating line, or values of the other model coefficients. Note that you can also show that $r\text{-square}_1 = 0$. I suspect that the results will not be deterministic. You will not get a clear "Yes, it works perfectly, there was no error in the optimum M" result. The statistical vagaries of noise perturbations on the true y values in one particular test may make it appear that a cubic model is best when the truth is quadratic. Therefore, your assessment and conclusion as to whether FPE works or not cannot be based on an expectation that the method is universally perfect.

14

Data Pre- and Post-processing

14.1 Introduction

Data processing means that you, as the user, decide to discard data, adjust data, or create data. In general, these sorts of manipulation are counter to the scientific method and honesty in investigation, but often data processing can be justified. However, one must be careful not to transgress propriety.

Data from a continuous running process is often noisy. It would seem that filtering the signal to temper the noise, to better see the true process value, is a good thing. I think this sort of pre-processing is widely accepted. It is a good thing at steady conditions, but filtering corrupts trends with a lag during transient conditions. However, if data are noisy, judgment of when it is at steady state will be contaminated with human choices.

When experimental conditions are changed, continuous processes move through a transient on their way to a steady state. If the model is a steady-state model, then the user should select data from the steady-state period and discard data from the transient period. This sort of pre-processing is appropriate.

Sensors are usually not perfectly calibrated and a sensor might be in the early stages of failing, reporting an erroneous value. Often, we use collections of on-line measurements and models to correct sensors (data reconciliation) or to impute data (fill in missing data with soft sensors or override measurement when there is greater confidence in the model). This sort of data override is appropriate, I think.

Those are pre-processing activities. However, data can also be post-processed.

Some data sets are erroneous, but there is no indication of data being an outlier until it is compared to the model. For instance, when 19 out of 20 data sets are close to the model, but one is far away, you might have adequate evidence to reject that one data set and regenerate the model on the remaining 19. I think that rejection of outliers is appropriate, but this also requires caution. A human might have a tendency to claim that data are a discardable outlier when this contradicts a preconceived (and erroneous) model.

Nonlinear Regression Modeling for Engineering Applications: Modeling, Model Validation, and Enabling Design of Experiments, First Edition. R. Russell Rhinehart.

There are diverse opinions related to pre- and post-processing of data. Some people are of the opinion that it should not be done, that the influence of outliers, noise, sensor error, and so on, should be kept. They might say, "It all represents the truth about Nature, and human bias should not infect reality." On the other hand, some fully accept filtering, data reconciliation, imputation, and data rejection as practicable and valid when acknowledged, grounded in legitimate analysis, and not contrived to spin the truth.

This chapter will present data pre- and post-processing techniques that I feel are appropriate.

14.2 Pre-processing Techniques

Pre-processing refers to data adjustment prior to using it in regression modeling.

14.2.1 Steady- and Transient-State Selection

Steady-state models are frequently used to analyze and design processes and products. In steady state, there is no accumulation of inventory (material, energy, momentum) and all measurable variables (both input and response) remain constant in time. In batch processes this would mean equilibrium, but in a continuous flow process materials might not have come to thermodynamic, phase, or chemical equilibrium when they leave the steady-state process.

However, during transient periods, when a process or device is in the process of responding or adjusting to a new influence, inventory changes and response variables (volume, temperature, speed) also change in time. For example, if you move the car accelerator pedal from 25% (with a speed of 45 mph) to 35% the car will begin to accelerate – first 46 mph, then 48, then 50, then 53, …, and eventually leveling at 60 mph. If you are developing a steady-state model of how accelerator position affects speed, the 25% pedal position goes with 45 mph and the 35% goes with 60 mph. The 46 or 48 or 50 mph do not go with the 35%. Steady-state models are usually easier to derive and understand. For nonlinear regression, data from transient periods should not be used to attempt to develop steady-state models.

Often the human selects the steady-state periods that generate data for regression. Unfortunately, steady-state data are usually noisy, and the human must interpret when a transient has ended. This choice can be biased by the human's desire to get through an experimental plan quickly and to accept steady state when things are still changing in time.

By contrast, coefficient values in transient models should not be developed with extended periods of steady-state data. At steady state the data are noisily representing the same value and excessive samples at steady state can dominate the regression objective function and weaken the ability to properly quantify the coefficients in the model that represents the transient behavior.

Chapter 6 presents techniques for unbiased selection for steady and transient periods.

14.2.2 Internal Consistency

A technician doing an analysis might be self-consistent, but when different technicians use different analysis devices in different labs they might not provide consistent results. When a new analysis device or procedure replaces an old one, the new data might not be consistent with the prior measurements. If one is attempting to use data to quantify trends, then the data must be internally consistent.

Evidence of consistency, or inconsistency, could be based on duplicate analysis of the same material. If there is a consistent bias, perhaps as indicated by a t-test of values from the different treatments, then the labs/treatments/procedures are inconsistent.

Inconsistency could also be revealed by the residuals from a model, the difference between the model and data. For instance, if all the data from one time period are above the model and from another time period are below, then there is strong evidence that the data are not internally consistent. Alternately, if the residuals from one operator are generally negative and those from another are generally positive, it suggests inconsistency.

Accordingly, to see the individual trends, segregate data into controls of common labs, operators, raw material batches, historical period, procedures, plants, and so on and look for patterns in the residuals that indicate different treatments. Alternately, if you are interested in an overall behavior that blends the vagaries of life, keep the data in a common set. Chapter 16 will discuss observing model residuals w.r.t. each possible treatment.

Either way, report your decision.

14.2.3 Truncation

The molecular weight of water is 18.011, but we often use 18. The value of pi is 3.14159265358979 … , but we often use 22/7. When converting temperature from Fahrenheit to absolute, it is convenient to use 460 instead of 459.67.

However, truncating coefficients changes their values and regression will have to adjust values of model coefficients to counter the truncated (wrong) values. Truncating to convenient values is a form of data pre-processing. Truncate to a minimum number of digits for convenience and utility, but not so much as to create a significant compensating bias on regression coefficient values. Truncate so that the error introduced by truncation is at least an order of magnitude lower than the uncertainty in the equation application. A rule of thumb is to keep two digits more than you think is significant.

14.2.4 Averaging and Voting

If there is only one measurement device, then you have no evidence that it is providing a wrong value; there is no evidence that there is a reproducible bias, even though you know that calibration is imperfect and that the sensor is subject to diverse perturbations. If two devices are measuring the same property, it is likely that they will not report identical results. If you believe that both devices are valid, you might average the two readings to generate a value that is hopefully more representative of the truth than either individual measurement.

If a signal is noisy, you might average a series of values from the most recent time-window to report the average (see Section 14.2.6).

If there are three devices measuring the same property, you might average all three measurements. However, it is likely that three redundant devices are installed because they have a history of failing. Therefore, rather than averaging all three, which risks including a wrong value from a failing instrument, take the middle of the three readings as the measurement. This middle-of-three technique is called "voting" and is a common practice with on-line process pH measurements.

In a time series, taking the middle-of-three sequential data is a median filter (see Section 14.2.7).

14.2.5 Data Reconciliation

Those techniques (averaging and voting) override a single measurement with a correction based on replicate sensors. In data reconciliation the correction is based on redundant measurements (but not at the same location) and a model. For instance, if fluid is flowing into a tank, what flows in minus what flows out must represent what accumulates. If the tank has a uniform cross-sectional area and the fluid is incompressible, then a dynamic model is

$$F_{in} - F_{out} = A\frac{dh}{dt} \tag{14.1}$$

If it is expected that the value of h comes from a reliable sensor, but the flow sensors provide unreliable data, then their values can be corrected by a deviation:

$$(F_{in} - \varepsilon_{in}) - (F_{out} - \varepsilon_{out}) = A\frac{dh}{dt} \tag{14.2}$$

and the error in that material balance is calculated as

$$\varepsilon_1 = (F_{in} - \varepsilon_{in}) - (F_{out} - \varepsilon_{out}) - A\frac{dh}{dt} \tag{14.3}$$

Similar models can be constructed using energy balances or other mass balances, and the balance deviation can be calculated for multiple time intervals in any one day. In data reconciliation there are several sets of redundant sensors and more balances than there are individual sensors. In addition, the objective is to find the deviations on each that minimize the sum of squared deviations (SSD) of the various balances. Although this seems to only represent fundamental first principles phenomena, data reconciliation requires user choices. The units on an energy balance need to be weighted to be combined with the units on a mass balance. The weighting of various model equations needs to reflect reliability of the devices and the model. The permissible range of the device errors also needs to be tempered by expectations of calibration drift.

After the ε_{device} (ε_{in} and ε_{out} in Equation 14.3) values are found to minimize the sum of squared ε_{model} deviations (ε_1 of Equation 14.3), the pre-processed measurements are those corrected by the ε_{device} values.

In general, with x representing process inputs and y representing process outcomes, the process models (representing material and energy balances, equilibrium relations, kinetic relations, etc.) are

$$\begin{aligned} \tilde{y}_{1,t} &= f_1(x_1, x_2, \ldots, x_i, \ldots, x_n)_t \\ \tilde{y}_{2,t} &= f_2(x_1, x_2, \ldots, x_i, \ldots, x_n)_t \end{aligned} \tag{14.4}$$

where the subscript t represents a particular sampling, i an input. With j representing an output, $\tilde{y}_{j,t}$ represents the numerical solution to a modeled output. With deviations from the measurement indicated as ε, the corrected y and x values are

$$x'_i = x_i + \varepsilon_{xi}$$

$$y'_j = y_j + \varepsilon_{yj} \tag{14.5}$$

Then data reconciliation will seek the ε values to minimize deviations from the measurements, tempered by the expected standard deviation (an equal concern factor):

$$\underset{\{\underline{\varepsilon}\}}{Min} \; J = \frac{1}{N_t}\frac{1}{N_y}\sum_t\left[\sum_j\left(\frac{\tilde{y}_j(\underline{x}') - y'_j}{\sigma_{yj}}\right)^2\right] + \frac{1}{N_x}\sum_i\left(\frac{\varepsilon_{xi}}{\sigma_{xi}}\right)^2 \tag{14.6}$$

The first term on the right-hand side (RHS) represents the "closure" on the data, the deviation between the data and model. The objective is to reduce this. It is scaled by the expected uncertainty in the y-measurement data. It is summed over each y-variable and summed over N time intervals. The reciprocal N scaling accounts for the number of y-variables and time samplings. The second term in the RHS of Equation 14.6 represents the scaled value of the correction, scaled by the expected random uncertainty of the measurement. It is a penalty for large corrections to the measurement. It is summed over all x values and each time increment. The reciprocal N scaling accounts for the number of x-variables. The ε_{xi} correction for the x-measurement and ε_{yj} for the y-measurement are common values over all the time intervals in the data reconciliation period.

Data reconciliation is commonly used as a pre-processing correction to measurements. It is justified when data could be unreliable, and procedures are grounded in legitimate science and engineering. In spite of the fact that human preferences are involved (Average of how many? Voting from how many? What weighting in data reconciliation?), these pre-processing techniques help make sense of unreliable data.

14.2.6 Real-Time Noise Filtering for Noise Reduction (MA, FoF, STF)

Measurements are often noisy. Even at steady state the measurement does not remain at one fixed value, but is perturbed by independent fluctuations at each sampling. Figure 14.1 represents the concept. The process is at steady state and the output value, y_{true}, is constant in time. The measured value, however, is corrupted with additive noise as random independent perturbations. Although the concept is as illustrated, the process true value might be providing the noise as a result of imperfect mixing, turbulence, and so on. Regardless of the mechanism for noise at steady state, the desire is to temper the noise level in order to see the true value better.

A moving average (MA) is a common approach to temper noise. It is a conventional average, applied to the most recent N data. Since the window of the most recent N data moves in time, it is termed a moving average:

$$X_{MA,j} = \frac{1}{N}\sum_{i=1}^{N} X_{j-i+1} \tag{14.7}$$

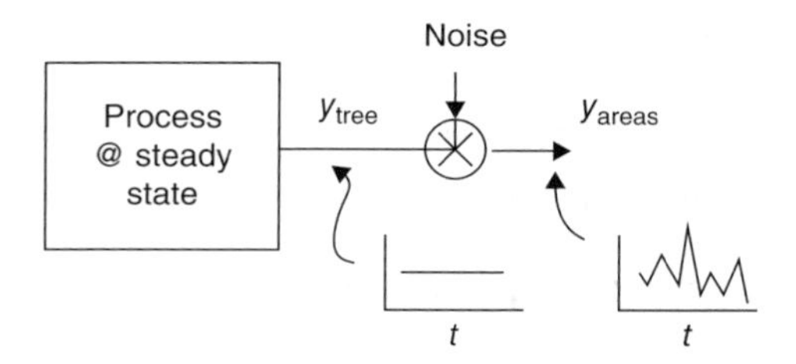

Figure 14.1 Concept for a noisy steady state

The subscript j represents the current sampling count and i the number of samples back.

Propagation of variance on Equation 14.7, assuming that the noise perturbations are independent and retain the same variance, reveals that the MA reduces the noise on the average, but cannot eliminate it:

$$\sigma_{MA} = \frac{\sigma_X}{\sqrt{N}} \tag{14.8}$$

The user needs to choose a value of N that tempers noise to a desired value. It would seem from Equation 14.8 that larger is better. However, large N means that it takes the MA signal a longer time to respond during transients. This is illustrated in Figure 14.2 for a process that makes an abrupt change. Since a MA averages the past N values, it does not reflect the steady-state process value until N samples after steady state is attained. Therefore, the user must choose a value of N that balances the benefit of tempering noise to the penalty of responsiveness of tracking the process during transients.

Also note in Figure 14.2 that the MA value during steady state wiggles a bit. Variability is not removed; it is reduced by filtering.

The MA is a computational burden, since in its primitive implementation it must retain N values and at each sampling update the array values and sum each element. Accordingly, the improvement that is commonly implemented is a first-order filter (FoF).

There are a variety of ways to derive the filter equation. I will expand the familiar MA of Equation 14.7 by moving the most recent data outside of the sum

$$X_{MA,j} = \frac{1}{N}X_j + \frac{1}{N}\sum_{i=2}^{N} X_{j-i+1} \tag{14.9}$$

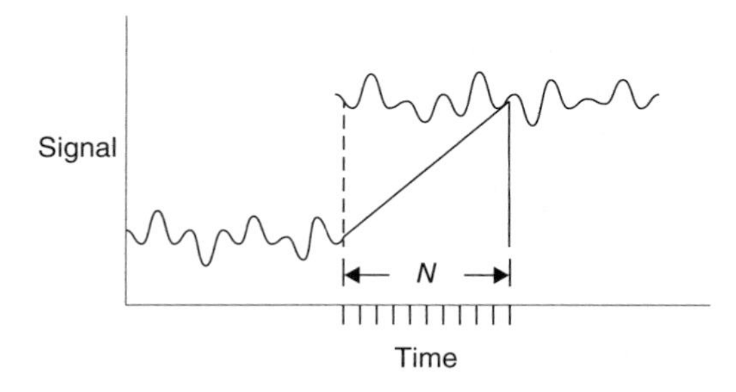

Figure 14.2 The delay of N samples with a moving average

Then recognize that at steady state the average is not changing:

$$X_{MA,j} \cong X_{MA,j-1} \tag{14.10}$$

where the approximately equal sign indicates that the vagaries of noise do make the average change in time.

If the process is at steady state, then the sum of the prior $N - 1$ samples is $(N - 1)$ times the prior average. Therefore,

$$X_{MA,j} \cong \frac{1}{N}X_j + \frac{N-1}{N}X_{MA,j-1} \tag{14.11}$$

This is also the formula for a FoF, which represents a classic resistor–capacitor circuit for electronic filtering of noise. It is also the explicit numerical solution to a first-order differential equation (Euler's method). It is variously represented. The $1/N$ term is often replaced with λ and the filtered value is defined as

$$X_{f,j} = \lambda X_j + (1 - \lambda)X_{f,j-1} \tag{14.12}$$

where

$$\lambda = \frac{1}{N} = 1 - e^{-\Delta t/\tau} \tag{14.13}$$

where λ is the filter factor, a measure of the reciprocal number of data averaged, Δt is the sampling interval, and τ is the time constant. This is also termed an exponentially weighted moving average (EWMA) because expanding the RHS sequentially to past measured values reveals an MA of all past terms that are exponentially weighted with the $(1 - \lambda)$ factor. For a long data collection period of j samples, when the memory of the initial state has faded, in which the term $(1 - \lambda)^j$ is negligible, the expression for $X_{f,j}$ is that of a conventional average, but with each term scaled by a weighting factor that is an exponential of the term $(1 - \lambda)$:

$$X_{f,j} = \frac{1}{N}\sum_{i=0}^{\infty}(1 - \lambda)^i X_{j-i} \tag{14.14}$$

In a similar manner to the MA, the EWMA does not remove noise, but tempers it. Propagation of variance on Equation 14.12 with the assumption that the process is at steady state and that sample-to-sample perturbations are independent with a uniform σ provides the tempering relation to λ:

$$\sigma_{EWMA} = \sigma_x\sqrt{\frac{\lambda}{2 - \lambda}} \tag{14.15}$$

Also, as with averaging, the FoF (or EWMA) does not immediately respond to a change in the signal. As revealed in Figure 14.3, the EWMA path to a step in the signal is not a linear rise that gets there in N samples, but an asymptotic approach with a time constant of τ and a settling time ($\approx$95% of the move complete) after a time period of three time constants. Given a choice of λ, Equation 14.13 can be solved for τ and, using the 95% settling time,

$$ST_{0.95} \cong 3\ \Delta t\ \ln(1 - \lambda) \tag{14.16}$$

Again, the user must choose the value of λ (or τ) to provide the desired speed of response and noise reduction. Depending on the product provider's choices, the definition of λ and $(1 - \lambda)$ terms might be switched. Check the manual.

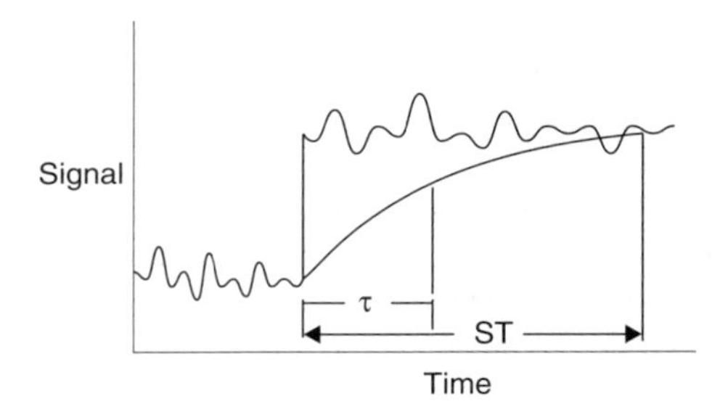

Figure 14.3 Illustration of the lag of a first-order filter

There are many examples in which the process noise level changes in time. The noise on an orifice flowmeter is due to turbulence in the fluid, which increases the amplitude with flow rate. Simultaneously, the square-root functionality in the flow device, which converts the sensor differential pressure signal into a "measured" flow rate, tempers more at higher flow rates and the noise level can increase or decrease depending on the flow rate. pH noise can be produced by mixing irregularities that let acid-rich or base-rich packets of fluid pass the sensor. The amplitude of the pH noise response to mixing vagaries depends on the titration curve, which changes in time.

If the process noise level decreases, then a user would reduce N in an MA calculation of Equation 14.7 or increase λ in an EWMA calculation of Equation 14.12 to keep the noise on the reported value within the desirable bounds, but to minimize the delay (of N samples) or lag (with a settling time of about of 3τ).

A self-tuning filter (STF) (Cao and Rhinehart, 1997) automatically adjusts the time constant of a FoF to preserve the desired variance on the filtered variable with minimum lag. When the noise is independent, at SS the variance of the noise can be calculated as an exponentially weighted moving variance (EWMV) based on differences in successive data:

$$\delta^2{}_{f,i} = 0.1(X_i - X_{i-1})^2 + 0.9\delta^2{}_{f,i-1} \tag{14.17}$$

where $N = 10$ $(\lambda = 0.1 = 1/N)$ and $\delta^2{}_{f,i}$ is a measure of the variance. Then λ for Equation 14.12 is calculated as

$$\lambda = \min\left\{\left(\frac{1}{2} + \frac{1.1668\delta_f^2}{E^2}\right)^{-1}, 1\right\} \tag{14.18}$$

where E is the user-defined desired 95% confidence half-interval on the filtered value. At steady state, filtering does not remove noise; it tempers noise. Choose a range for the filtered value at steady state that bounds the desired + or − limits of the filtered value from the true value. E in Equation 14.18 is the user-desired 95% + or − deviation limit.

14.2.7 Real-Time Noise filtering for Outlier Removal (Median Filter)

Like "voting" in redundant measurements, a median filter accepts the middle of several values. However, unlike voting of independent sensors, this looks at the time series of data, the recent measurements from one sensor. If the window is three samples, a median filter takes the data with the middle value (not the middle in the sequence). In this manner, if there is one spurious data value, it is rejected. By contrast, in averaging or FoF, the spurious value is included in the average, and corrupts the reported process value.

If the user anticipates that the spurious event might last for several samples, a median filter can be set to report the median of 5, or of 7, and so on, samples. This makes it more robust to the persistence of a spurious event. The median filter adds a delay of approximately half of its window. The median filter does reduce noise at steady state, but its primary role is to reject spurious events, not to temper noise.

14.2.8 Real-Time Noise Filtering, Statistical Process Control

In statistical process control (SPC) the decision to change something is tempered by waiting until there is 99% confidence that change is needed. This philosophy can be applied to time series of data. Hold the prior reported value until there is adequate evidence to report that it has been changed.

If the process variable (PV) is "stationary" its mean is constant (the process is at steady conditions) and "noise" can be considered as independent, random influences added to the mean at each sampling. If the noise is normally (Gaussian) distributed, there is only a 0.27% chance that data will appear in the tails of the histogram (more than 3 $\hat{\sigma}_x$ from the average). Only 1 out of 370 points will fall in the extreme tails.

If the PV "really" changes, then the average will shift, and there is a good chance that the new process will place points outside of the old 3 $\hat{\sigma}_x$ limits. The SPC philosophy is that if x_i is more than 3 $\hat{\sigma}_x$ from the old average, then one will accept that the process really has changed. Although one would be wrong 0.27% of the time, the basic SPC philosophy is:

$$\text{If} |x - \bar{x}_{old}| > 3\hat{\sigma}_x, \text{ then respond.} \tag{14.19}$$

However, the action from Equation 14.19 waits until one point is in the improbable tail region. If the process shift is small compared to σ_x, then it may take many points before the first measurement is more than 3 $\hat{\sigma}_x$ from the average. However, if there is a shift in the mean, then the measurements will show a systematic bias from the old average.

Define the variable "CUSUM" as the cumulative sum of deviations from the reported average scaled by data variability:

$$CUSUM = \sum \frac{(x - \bar{x}_{old})}{\hat{\sigma}_x} \tag{14.20}$$

If the process has not changed and noise is independent at each sampling, then we expect CUSUM to be a random walk variable starting from an initial value of zero. However, if the process has shifted, then $|CUSUM|$ will steadily, systematically grow with each sampling. CUSUM is the cumulative number of $\hat{\sigma}_x$'s that the process has deviated from $\bar{x}_{old}$. CUSUM could have a value of 3 if there is a PV violation of the 3 $\hat{\sigma}_x$ limit. Regardless as to whether it is a single or accumulated value of 3, for this procedure,

$$\text{If } |CUSUM| > 3\sqrt{N} \text{ take action} \tag{14.21}$$

The variable N is the number of samples for which the variable CUSUM is calculated and the square root functionality arises from the statistical characteristics of a random walk variable. The trigger value, 3, representing the traditional SPC 99.73% confidence level in a decision, is not a magic number. If the trigger value was 2, approximately representing the traditional 95% confidence level of economic decisions, then action will occur sooner, but it will be more influenced by noise: 5% of data in a normal steady process will have a $\pm 2\sigma_x$ violation. If the trigger value was 4, representing the 99.99% confidence level, then the reported average will be less influenced by noise, but it will wait longer to take action. Values from 2 to 4 are generally chosen to balance responsiveness and false alarms for particular SPC applications. For data filtering, I find that the traditional economic decision trigger value of about $\pm 2\sigma_x$, representing a 95% confidence in a decision, gives the best value for on-line filtering. Choose TRIGGER = 2 and

$$\text{If } |CUSUM| > \text{TRIGGER} * \sqrt{N}, \text{ then take control action} \tag{14.22}$$

The action will be to report a change in the mean of the noisy PV. How much change should one take? Short of detailed process model inference procedures, primitively credit that $\bar{x}$ had sustained an average offset since the last change in $\bar{x}$. Since the previous value of CUSUM at the true $\bar{x}_{old}$ was about zero, with a shift to a value of $\bar{x}_{new}$, the CUSUM is now $N(\bar{x}_{new} - \bar{x}_{old})/\sigma_x$. Accordingly,

$$\begin{gathered} \text{If } |CUSUM| > \text{TRIGGER} * \sqrt{N}, \text{ then} \\ \bar{x}_{new} = \bar{x}_{old} + \hat{\sigma}_x \cdot CUSUM/N \\ \text{and reset } N \text{ and } CUSUM \text{ to zero.} \end{gathered} \tag{14.23}$$

Since my preference is to eliminate possible instances of a divide-by-zero (which may occur if the PV value is "frozen" and $\hat{\sigma}_x$ becomes zero), re-define

$$CUSUM_{new} = CUSUM_{old} + (x - \bar{x}_{old}) \tag{14.24}$$

Then the trigger is on TRIGGER * $\hat{\sigma}_x$

The method requires a value for $\hat{\sigma}_x$ and the value of $\hat{\sigma}_x$ should automatically adjust when the process variability changes. Based on the EWMV technique, define

$$\hat{\rho}^2_{f new} = \left(\frac{M-2}{M-1}\right) \hat{\rho}^2_{f old} + \left(\frac{1}{M-1}\right) (x_{new} - x_{old})^2 \tag{14.25}$$

where M is the number of samples to be averaged to determine the variance and $\lambda = 1/(M-1)$ is the FoF factor.Then estimate the data standard deviation from Equation 14.25:

$$\hat{\sigma}_x = \sqrt{\hat{\rho}^2_{f new}/2} \tag{14.26}$$

I find that $\lambda \cong 0.1$ (or $M \cong 11$) produces an adequate balance of removing variability from the $\hat{\sigma}_x$ estimate, yet remaining responsive and having a convenient numerical value.

Including in initialization of variables, the algorithm is:

```
          IF (first call) THEN
                N := 0
                XOLD := 0.0
                XSPC := 0.0
                V := 0.0
                CUSUM := 0.0
                M := 11
                FF2 := 1.0/(M-1)/2.0
                FF1 := REAL((M-2)/(M-1))
          END IF
Obtain X
          N := N + 1
          V := FF1* V + FF2*(X - XOLD)^2
          XOLD := X
          CUSUM := CUSUM + X - XSPC
          IF (ABS (CUSUM).GT.TRIGGER*SQR (V*N)) THEN
                XSPC := XSPC + CUSUM/N
                N := 0
                CUSUM := 0.0
          END IF
```

14.2.9 Imputation of Input Data

Of the several techniques discussed in this chapter, imputation is likely to be the most controversial. Imputation is the act of creating missing data.

Calibration of measurements is well accepted. The orifice sensor–transducer sends a 4–20 mA electrical signal into the control computer, which uses a model to convert the milliampere reading to a process flow rate. In a sense this is imputation, using something measurable and a model to predict something not so easily measurable.

We have become comfortable with soft sensors (inferential measurements) in which a model is developed from data and then used as a surrogate for a measurement that might otherwise be periodic, expensive, or unreliable. In Kalman filtering, we statistically analyze noisy data and compose an estimate of the process state that is weighted by either the data or the model, depending on the certainty of the data. Both of these practices could be called imputation. They both use data and a model to report a "right" value of something for which the true value is not easily obtained.

Let us consider a more controversial application of imputation to create missing data. Often in experimental data, especially from social science applications, there are missing elements in a nearly complete data set of influences. As an example, I have developed a model aimed at predicting student performance in the upper-level chemical engineering classes based on the student's performance in the lower-level classes. We use the model predictions to help with advising students on individual preparation needed for success in the upper-level work. Although the model provides strong correlation, the behavior and commitment of a 20-year old within their chosen major and within the upper-level complexity is different from that of the 18-year old, two years ago, in general STEM classes. As a result there is considerable prediction uncertainty, and the model GPA prediction has a 95% confidence interval of about ± half a letter grade.

I found that a key indicator of upper-level success is the student's grade in Calculus I. Unfortunately, out of 150 students in the data base, about 30 students did not have a grade for Calculus I because they received advanced placement credit for any number of reasons and did not have to take the course. However, grades in many other classes are also relevant predictors (Calculus II, III, Physics II, Chemistry II, fluid dynamics, thermodynamics, material, and energy balances), and those 30 students missing a Calculus I grade have the other grades for the other relevant courses.

Should I reject the data from the 30 students because it is incomplete? This might be culling out the top performers who were able to skip the college Calculus I class and bias the model.

Should I include the 30 students, but not use Calculus I as a regressor in the prediction model? This might miss some relevant information and reduce the prediction precision of the model.

There is a third option: There is a strong correlation between the grades that any one person gets in the STEM courses. If one believes that the grades are based on the individual (motivation, ability, dedication) and consistently reflect that individual (this is the belief underlying the study to use lower-level grades to predict upper-level performance), then one could use the data from the 120 students with a complete set, model their Calculus I grade based on the other STEM grades, and then use that model to create a representative grade for the remaining 30 students. This is imputation. It has the advantage of permitting the 30 sets of data with the missing Calculus I grade to be part of the data set. It has the disadvantage of inferring the 30 missing Calculus I grades from the performance of the others.

Should I?

My choice was to exclude Calculus I from the analysis and use the 150 sets of data, without imputation. However, I will not know whether it was the right choice until after I explore imputation of the Calculus I grade.

Note also that my choice to not use imputation and to exclude Calculus I as a possible regressor is a human choice, is data selection, is pre-processing.

14.3 Post-processing

Once the data has been selected and/or adjusted, develop the model. Now view patterns in the residuals. Diagnostic indications may reveal that there may be errors in the data. This is post-processing, analysis of data in the light of the model. Post-processing may lead you to a decision that a data point is bad, should be discarded, and the model redeveloped with the remaining $N - 1$ data. When data are bad (erroneous, mistaken, corrupted, spoiled) they should not be used to influence the model.

As a caution, a bad model might make a good data point seem erroneous. Accordingly, some people have strong objections to post-processing. If there are enough data points, one slightly erroneous one will not wholly corrupt the model. Although I think post-processing is legitimate and justifiable, if your customers do not, then do not.

14.3.1 Outliers and Rejection Criterion

An outlier is a data point that is unusually far from the place it should be – unusually far from the cluster, average, or from the model – with unusually far meaning relative to the variation exhibited by the other points. If the data point is representative of an error, not the normal outcomes of an experiment, then you should remove it from the data set so that its error does not influence the model. You do not want the model to be based on the erroneous points. Such errors could be transliteration of numbers, transposing the influence conditions and resulting outcome, a spurious influence, a contaminated sample, and so on. These would be rare events in a carefully performed experiment.

However, if the data provide an unexpected, but true and consistent representation of the process, you should keep it. It may be incompatible with the model or the human concept behind the model, and this may provide the clue to improve the model. The action to either keep or discard the data point depends on the strength of your belief that it is truly an outlier, a misrepresentation of the behavior you are seeking to model.

If you spot a point that you feel is obviously incorrect, first review all your calculations and data transcriptions to be sure that the suspected point is not the result of a mistake in arithmetic, number transposition, and so on. If you find no obvious error, your second review should include all the data of the original process, production, research, and so on. Look for indications of an incorrect procedure, an upset in performance due to some uncontrolled event, and so on. The results of these reviews may allow you to either discard the outlier with valid mechanistic justification or correct its value to confidently repeat the regression analysis with either the remaining or the corrected data.

If you still believe that the data point is incorrect (believe without mechanistic evidence, believe because it seems an outlier), you can construct a $(1-\alpha)100\%$ confidence interval parallel to the y-axis around the y-model value and discard the data point in question if it is not within the confidence interval. For normal, economic-based statistical decisions, 95% confidence is usual. For this two-sided test, this would make $\alpha = 0.025 = (1-0.95)/2$. Where a more conservative approach is justified, for example, you want to be 99.9% confident in rejecting a point, $\alpha = 0.0005 = (1-0.999)/2$.

In the absence of a justifiable level of confidence in rejecting the data point, use Chauvenet's criterion and use $\alpha = 1/(2N)$, where N is the number of data points.

Note that for the $N = 10$-point data set, Chauvenet's criterion also specifies a $(1-\alpha)100\% = [1-1/(2N)] \times 100\% = [1-1/20] \times 100\% = 95\%$ confidence interval for data rejection.

The procedure just described follows that of ASTM Standard EI78-80. That standard also contains a method for examining multiple outliers. Unfortunately, no guidance is given there for the selection of the significance level, α. I recommend Chauvenet's method as previously described.

What distribution of residuals (deviations from the model) should you assume? The Gaussian (bell-shaped curve, normal) distribution is common, but this may not represent the log-normal, Cauchy, or other distribution of residuals in your data. Rather than presume that the residuals are normally distributed, you might first determine the distribution that best represents them and then use the $(1-\alpha)100\%$ CL of that distribution to determine if the data point is an outlier.

If one data point out of 10 or 20 is rejected, that seems reasonable for a new experimental procedure. If two are rejected, this should raise a warning flag related to the validity of the experimental procedure.

Your model might be fundamentally wrong and the rejection of a data point might be because the model is wrong – the model does not fit the legitimate data. If the x value of the rejected data is in close proximity to not-rejected data, then the other data can be used to increase the belief in the propriety of the model. However, if the rejected data point is isolated in an x-region without other data to corroborate the model (especially an extreme) then reconsider the model derivation and consider taking more data in that region.

Keep a record. You may learn something that eventually requires that data point to be included. You might encounter another such data point that would weaken the belief that it is an outlier.

Reveal the outlier in your reports. Let others understand why it was rejected, that it was not used in generating the model, but that it exists. As data and models evolve, the outlier might become viewed as true, and subsequently included.

14.3.2 Bimodal Residual Distributions

Figure 14.4 illustrates data, a model, a plot of residuals w.r.t. the independent variable and the distribution of the residuals. If the data is "normal," if the variation is due to many, independent, small perturbations, then the distribution of residuals will have a Gaussian (bell-shaped, normal) distribution.

Alternately, as Figure 14.5 illustrates, the distribution of residuals might be bimodal as there are two distinct humps in the distribution. This indicates two distinct treatments, processes, analysis, operators, time periods, and so on. The data are not internally consistent.

The distribution of the residuals might not have a visually distinct bimodal character as illustrated in Figure 14.5. If you suspect that there are two distinct treatments, then use any number of statistical tests to confidently claim it. Kurtosis is the flatness property of a distribution, the

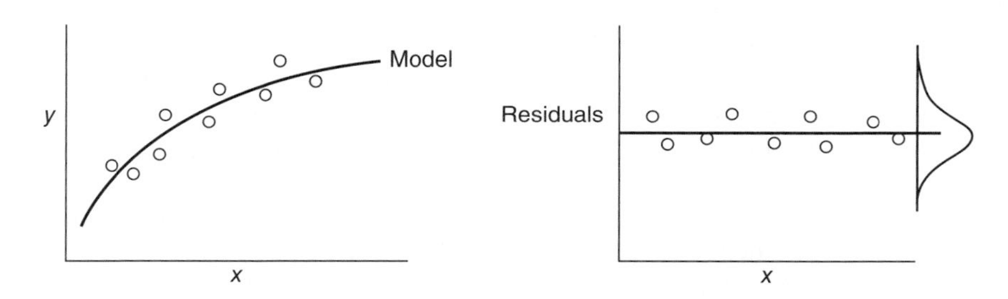

Figure 14.4 An illustration of normal residuals

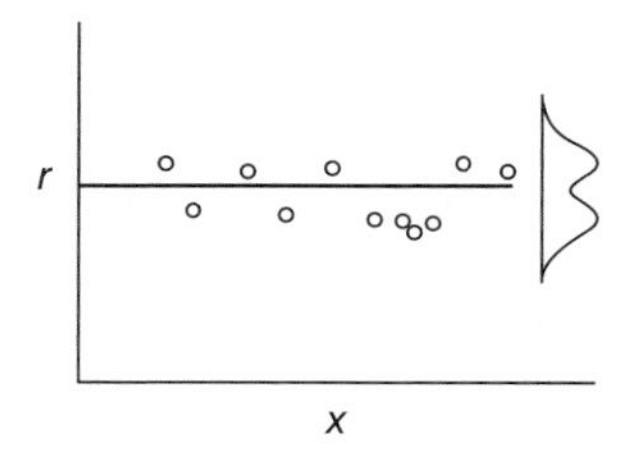

Figure 14.5 An illustration of bimodal residuals

fourth moment about the mean:

$$k = \frac{\frac{1}{n}\sum_{i=1}^{n}(x_i - \overline{x})^4}{\left(\frac{1}{n}\sum_{i=1}^{n}(x_i - \overline{x})^2\right)^2} - 3 \tag{14.27}$$

For a normal distribution, the kurtosis value is zero.

As an alternate approach to detect bimodality, form a chi-square contingency test. Generate a table of the number of residuals of either treatment with positive and negative deviations. The null hypothesis is that the data from each treatment are equally distributed above and below the model. The expectation is that the number of data in any category is half the total. In Table 14.1, there are 20 data points from Treatment 1 and 11 from Treatment 2. We expect that 10 from Treatment 1 will be in the + and − categories and 5.5 from Treatment 2. The experimental results have 14 and 6 and 4 and 7. The chi-square contributions are deviations from expected squared and divided by the expected, and the chi-square statistic is the sum of all contributions:

$$\sum_i \sum_j \frac{(O_{ij} - E_{ij})^2}{E_{ij}} \tag{14.28}$$

The degrees of freedom are based on the number of categories:

$$\nu = (r-1)(c-1) \tag{14.29}$$

where r and c represent the number of rows and number of columns, respectively.

The chi-square contingency test is appropriated from a continuous analysis, but this use has a finite number of samples and categories. Customary rules associated with the chi-square analysis to ensure a sufficient number of data and categories to adjust the experimental value to make it consistent with the chi-square distribution are: (i) if the expectation in any category is fewer than 1, combine categories; (ii) if fewer than 20% of the categories have an expectation with 5 of fewer samples, combine categories; (iii) if the degrees of freedom is unity, then decrease the deviation in Equation 14.27 by 0.5 before squaring. The data in this example violate the third rule, so the deviations need to be reduced; the revised analysis is given in Table 14.2.

The null hypothesis will only be rejected if the distributions are too far from expected. Accordingly use a one-sided test. The critical value for the chi-square statistic for 95% confidence and DoF = 1 is 3.84. Since the experimental statistic of 2.81 is not larger than the 3.84

Table 14.1 First consideration of a chi-square contingency analysis

	Treatment 1	Treatment 2
Number of + residuals	14	4
Number of − residuals	6	7
Expected number of + residuals	10	5.5
Expected number of − residuals	10	5.5
Contribution of + residuals	1.6	0.2727
Contribution of − residuals	1.6	0.2727
Sum of contributions		3.745
DoF		1

Table 14.2 A chi-square contingency analysis

	Treatment 1	Treatment 2
Number of + residuals	14	4
Number of − residuals	6	7
Expected number of + residuals	10	5.5
Expected number of − residuals	10	5.5
Contribution of + residuals	1.225	0.1818
Contribution of − residuals	1.225	0.1818
Sum of contributions		2.81
DoF		1

critical value, the null hypothesis is not rejected. There is inadequate experimental evidence to confidently claim that the data reflects two distinct treatments.

14.3.3 Imputation of Response Data

It is not unlikely that some response data may be missing. This could be the result of a measurement fault, lost data, measurement oversight, and so on. If the regression model is used to impute the missing response data values, then there is zero deviation between the imputed data values and the modeled values, which means that imputation has no impact on the SSD. This might seem good, but with no impact on the SSD, that data might as well be nonexistent. Using the regression model to impute values for missing data is equivalent to excluding those values from the regression. An advantage is that the data set will seem contiguous. A disadvantage is that the implied data might misrepresent the experimental validity.

14.4 Takeaway

You want only legitimate data for the basis of model development. This includes data obtained during steady-state periods for steady-state models, internally consistent data, and data that reveals the process without corruption from the many forms of faults that could infect experimental procedures. Pre- and post-processing of data is essential to correct and eliminate errors. However, choices of what correction is needed and what data should be discarded could be driven by a human desire for a particular outcome (purposeful) or a bias that has origins in naivety (not purposeful). Therefore, ground such choices in quantitative analysis, usually statistical tests. Preserve the uncorrected or rejected data and report the choices, so that a reader can understand and either accept or re-correct the action.

Exercises

14.1 Derive Equation 14.8 for variance reduction.

14.2 Derive Equation 14.15 for variance reduction.

14.3 Derive Equations 14.12 and 14.13 for a FoF. Start with a first-order differential equation.

14.4 Apply Equation 14.6 to a case study.

14.5 Derive Equation 14.14 for the EWMA.

14.6 Show that the 95% settling time for a FoF response to a step-and-hold input is $\approx 3\tau$.

14.7 Derive Equation 14.26. Start with the traditional definition of a variance. Add and subtract the prior measurement from the squared term. Separate the binomial terms. Argue that if the noise is independent that in the limit of a large number of terms (or alternately the expectation of) the sum of mixed residuals tends to be zero. Argue that if at steady state the expectation of the sum of squared deviations should remain constant in time.

15

Incremental Model Adjustment

15.1 Introduction

Customarily in regression, the user collects many data sets and then uses optimization to adjust the model coefficients to make the model best fit the entire batch of data. It is a batch operation. However, it is not uncommon to update a model as new data are acquired, to incrementally adjust the model coefficient values. This practice is common in continuously operating processes in which attributes progressively change in time. Examples of some time-dependent attributes include:

- catalyst reactivity,
- heat exchanger fouling,
- feed raw material composition,
- air density and humidity,
- accumulation of poisons or pathogens in a batch reaction,
- viscosity impacted mixing in a batch polymerization,
- human attitude,
- group morale or preferences,
- process gain change with operating conditions,
- viable cell growth factor,
- reaction yield,
- average distillation tray efficiency,
- insulation effectiveness, and
- piping assembly friction factor response to screen blockage or piping rearrangements.

The processes might be classified as either continuous or batch, but as it operates in time, attributes change in time and values for the attributes are needed for analysis or control.

In such cases, the coefficient of interest changes in time. If one were to collect all historical data for regression, earlier data would express one value of the time-changing coefficient, and recent data will express a different value. Rather than adjusting a model on old data and reflect an outdated property value, models are often incrementally adjusted to match the most recent data.

Nonlinear Regression Modeling for Engineering Applications: Modeling, Model Validation, and Enabling Design of Experiments, First Edition. R. Russell Rhinehart.

Process features or attributes that change with time require the model coefficient values to change in time. Such models are termed nonstationary. Process owners observe the model coefficient values that reflect such factors to inferentially monitor process condition, performance, or health, and use that information to trigger events, schedule maintenance, predict operating constraints and bottleneck capacity, and so on.

15.2 Choosing the Adjustable Coefficient in Phenomenological Models

The choice of which model coefficient is to be adjusted should be compatible with the following five principles:

1. The model coefficient must represent a process feature of uncertain value. (If the value could be known or otherwise calculated, incremental adjustment would not be needed.)
2. The process feature must change in time. (If it did not change in time, adaptation would be unnecessary.)
3. The model coefficient value must have a significant impact on the modeled input–output relation. (If it has an inconsequential impact, there is no justification to adjust it.)
4. The model coefficient should have a steady-state impact if data are to come from steady-state periods. (If it does not, if, for instance, the coefficient is a time constant, then when the process is at, or nearly at, a steady state, the model coefficient value will be irrationally adjusted in response to process noise. Since continuous processes mostly operate at steady conditions, this is important.)
5. Each coefficient to be adjusted needs an independent response variable.

If one or more of the five principles are not true, there is no sense in adapting that model coefficient on-line.

15.3 Simple Approach

There are many ways to update model coefficient values, to adapt models in real-time, which have led to a variety of adaptive control methods. A simple and effective approach to adjustment of a phenomenological model is to desire that the model coefficient be adjusted so that the process-to-model mismatch (pmm) approaches zero in a first-order manner when the process is near steady conditions. The mathematical statement of that desire is:

$$\tau_{pmm}\frac{dpmm_{SS}}{dt} + pmm_{SS} = 0 \tag{15.1}$$

where

$$pmm = y - \widetilde{y} \tag{15.2}$$

Representing the transient process,

$$\frac{d\widetilde{y}}{dt} = f(\widetilde{y}, u, d, p) \tag{15.3}$$

in which $\widetilde{y}$ is the modeled response, u is the controller influence on the process, d represents measurable disturbances, and p represents the coefficient that will be incrementally adjusted.

At steady state

$$0 = f(\tilde{y}_{SS}, u, d, p) \tag{15.4}$$

which can be rearranged to solve for $\tilde{y}_{SS}$:

$$\tilde{y}_{SS} = g(\tilde{y}_{SS}, u, d, p) \tag{15.5}$$

Equation 15.5 might be an explicit relation for $\tilde{y}_{SS}$, or possibly an implicit solution as indicated.

If p, the coefficient value, changes, then $\tilde{y}_{SS}$ will change. If near to a steady-state (SS) condition, the d and u values are not changing. Then the change in $\tilde{y}_{SS}$ w.r.t. time is

$$\frac{d\tilde{y}_{SS}}{dt} = \frac{\partial g}{\partial p}\frac{dp}{dt} + \frac{\partial g}{\partial \tilde{y}_{SS}}\frac{d\tilde{y}_{SS}}{dt} \tag{15.6}$$

Rearranged this is

$$\frac{d\tilde{y}_{SS}}{dt} = \left(\frac{\partial g}{\partial p}\frac{dp}{dt}\right) / \left(1 - \frac{\partial g}{\partial \tilde{y}_{SS}}\right) \tag{15.7}$$

Substituting Equation 15.7 into Equation 15.1 and rearranging gives

$$\left(\frac{\partial g}{\partial p}\frac{dp}{dt}\right) / \left(1 - \frac{\partial g}{\partial \tilde{y}_{SS}}\right) = \frac{pmm}{\tau_{pmm}} \tag{15.8}$$

Rearranging for incremental model coefficient adjustment using a simple numerical solution (Euler's explicit finite difference) gives

$$p_i = p_{i-1} + \left[\frac{\Delta t}{\tau_{pmm}}\left(1 - \frac{\partial g}{\partial \tilde{y}_{SS}}\right)\right]\frac{pmm_i}{\partial g/\partial p} \tag{15.9}$$

Equation 15.9 is recognizable as Newton's method of incremental, recursive coefficient updating (root finding), with the bracketed term as a tempering factor. The user's choice of the value of τ_{pmm} prevents excessive changes in the model coefficient value and tempers measurement noise effects. The sampling interval, Δt, is in the same time units as the time constant and provides a normalizing factor for the frequency that the model is updated. If the model of Equation 15.5 is explicit in $\tilde{y}_{SS}$ then the $\partial g/\partial \tilde{y}_{SS}$ term in Equation 15.9 has a value of zero.

In computer assignment statements the subscripts are unnecessary, because the past values are used to assign the new value.

The value of τ_{pmm} should be large compared to model dynamics (so that model adjustment does not interact with control calculations), but small compared to the time period over which the process attribute changes (for rapid tracking of the process).

Example 15.1 Mixing in a Continuous Flow-Through Tank As a specific example, consider a mixing tank model in which two streams of concentrations c_1 and c_2 enter a tank at a stationary liquid level. Perhaps F_1 and c_1 represent wild flow properties and F_2 is the controlled flow. The net inflow concentration is

$$c_{in} = \frac{c_1 F_1 + c_2 F_2}{F_1 + F_2} \tag{15.10}$$

Note that c_{in}, the net inflow concentration, is a nonlinear response to either F_1 or F_2. Usually, the mixing dynamics in the tank defines the outflow concentration as

$$\frac{V}{F}\frac{dc}{dt} + c = c_{in} \tag{15.11}$$

where V is the tank volume and F represents the total flow rate ($= F_1 + F_2$), and the ratio V/F has the units of a time constant.

This represents a static nonlinearity of Equation 15.10 followed by linear dynamics of Equation 15.11, which is often termed a Hammerstein model. The two equations can be combined to represent one nonlinear model:

$$\tau\frac{dc}{dt} + c = \frac{c_1F_1 + c_2F_2}{F_1 + F_2} \tag{15.12}$$

While developed here as separate equations, then combined, the same result, Equation 15.12, could be derived from a transient material component balance on the tank.

Consider that $c1$ is not measurable, that it changes in time, and is chosen as the model coefficient to be incrementally updated. Then applying Equation 15.9, the adjustment rule becomes.

$$c1 := c1 + \Delta t * pmm * (F1 + F2)/(tauwpmm * F1) \tag{15.13}$$

The reader should recognize that if $F1$ is zero the calculation cannot be executed. Should $F1$ be zero, there is no information that can be used to adjust $c1$ by the process output. Accordingly, there is no justification to attempt to use the method to update the model coefficient $c1$. The code should be:

```
If F1>threshold THEN
```

$$c1 := c1 + \Delta t * pmm * (F1 + F2)/(tauwpmm * F1) \tag{15.14}$$

```
END IF
```

15.4 An Alternate Approach

The model may be a dynamic model represented as

$$\frac{d\widetilde{y}}{dt} = f(\widetilde{y}, u, d, p) \tag{15.15}$$

At steady state the derivative is zero, leaving

$$f(\widetilde{y}, u, d, p)|_{SS} = 0 \tag{15.16}$$

This might be a functional relation that can be solved explicitly for the steady-state model-y, leading to an Equation 15.5 representation. If so, follow the procedure of Section 15.3. However, if Equation 15.16 does not admit an explicit relation, it can be solved by an iterative root-finding algorithm such as Newton's method. If the guess for $\widetilde{y}_{SS}$, $\widetilde{y}_g$ is close to the true value, then one iteration will lead to a very close representation of the $\widetilde{y}_{SS}$ value:

$$\widetilde{y}_{SS} = \widetilde{y}_g - \frac{f(\widetilde{y}_g, u, d, p)}{\left.\frac{\partial f}{\partial \widetilde{y}}\right|_g} \tag{15.17}$$

Again, using Equation 15.1 to express the desire that the model coefficient be adjusted so that the pmm approaches zero in a first-order manner when the process is near-steady conditions, and using Equations 15.17 and 15.2, the chain rule of differentiation and the finite difference approximation to the time derivative is

$$p_i = p_{i-1} + \left[\frac{\Delta t}{\tau_{pmm}} \frac{\partial f}{\partial \widetilde{y}_g} \right] \frac{pmm_i}{\left(\partial f / \partial p - f \left(\frac{\partial^2 f}{\partial p \partial y} \right) / (\partial f / \partial y) \right) \Big|_g} \tag{15.18}$$

A good value for $\widetilde{y}_g$ would be either the prior $\widetilde{y}_{SS}$ or the current process steady-state value.

15.5 Other Approaches

Classic applications of incremental model adjustment include the back propagation approach to the adjustment of weights in training a neural network, the use of statistical confidence in Kalman filtering, recursive system identification with linear empirical models, and simply waiting for steady state to collect data to adjust the model. The applications are diverse. This chapter is intended as an introduction to the concept and to reveal the issues in selecting the model coefficient to be adjusted and in overriding the adjustment with illogical situations.

15.6 Takeaway

Although it is desirable that data processing is devoid of human choices, this example reveals that humans must make many choices – choosing the coefficient to be adjusted, choosing the model, choosing the method of solving the model, choosing the time constant for the rate of model adjustment, and recognizing and deciding action to overcome potential execution errors. Regardless of the approach taken, the human needs to understand the method in order to make the right choices. This is as important in the use of software that magically does it for you as it is when you are writing your own procedure.

Exercises

15.1 Derive Equation 15.9.

15.2 Derive Equation 15.18.

15.3 Rework the example to adjust c2.

15.4 A simple speed model for a car on a level road is

$$\frac{m}{g_c} \frac{dv}{dt} + \alpha v^2 = ku$$

where m is the mass of the car, v is the velocity, u is the accelerator pedal position, and the combined friction and drag is modeled as αv^2. The coefficient, α, will change with wind direction and road surface roughness. Derive the rule to incrementally adjust α.

16

Model and Experimental Validation

16.1 Introduction

There are four main expectations of a model that claim to properly represent some physical system.

1. There should not be any trend in residuals w.r.t. any variable (input, output, chronological order, or treatment). The model should go through the center of the data, throughout the entire range.
2. The variability expectation from propagation of uncertainty in the data model should match the magnitude of the residuals.
3. The model should pass logical tests. Both asymptotic limits and local trends projected from the model should be consistent with phenomenological expectations. This is for both coefficient and input variable values.
4. Coefficient values should match expectations from literature – homolog interpolation or extrapolation, similar conditions, and so on.

This chapter looks at the model to see if it captures the truth about Nature and provides procedures to reject or accept the model. However, a model may be true, but dysfunctional in use, for example, because the inverse does not return a unique value. Issues of model utility are discussed in Chapter 19.

16.1.1 Concepts

The mathematical model has a *structure*, which is often referred to as design, order, functional form, or *architecture.* A linear model has a different structure from a quadratic model. The quadratic model of Equation 1.1 has the squared x-functionality as well as the constant and x-functionality of a linear model. In empirical models, if you add more states to a state-space model, more time-delay variables to an autoregressive moving average (ARMA) model, more hidden-layer neurons to a neural network, or change the transfer function within a neuron, then you are changing the model structure. For phenomenological models, if you change a

Nonlinear Regression Modeling for Engineering Applications: Modeling, Model Validation, and Enabling Design of Experiments, First Edition. R. Russell Rhinehart.

constitutive relation (equation of state, kinetic expression) or add new phenomena then you are changing the model structure.

Validation means that the model structure is logically consistent with the understanding of the phenomena and consistent with the data (or more precisely that the data does not provide evidence to confidently reject the model). In order to validate the model, you must (i) understand the phenomena, (ii) collect data that constitute a legitimate, relevant, and comprehensive test, and (iii) use statistics to accept or reject the model. To do so, you need to acknowledge what the model is supposed to predict or mimic.

Deterministic and *stochastic* models need to be validated differently. The key characteristic of a deterministic model is that each time the same question is asked the answer is exactly the same. "What is 3 times 5?" Today or tomorrow, Person A or Person B, a legitimate procedure will always arrive at the same answer. By contrast, the key characteristic of a stochastic model is that the replicated answer is not the same. "If I roll this die, what number is on top?"

Apply both logical and data-based validation to either deterministic or stochastic models.

Model *verification* is different from *validation.* Validation is a comparison of the model result to expected behaviors and data to determine if the model is consistent with the understanding and data. By contrast, *verification* is the confirmation that the model equations, code, or procedure have no errors in the conversion from model-concept to algebra or calculus, and then to computer code. There are multiple stages in a model development from concept to implementation, and errors could arise in each stage. For example, in the concept stage, if part of the concept was inadvertently overlooked and not included in the differential equation, the model will not be true to the concept. In the mathematical derivation stage, if a sign change error is made in an algebraic transition from one equation to another in the analytical derivation, the model will not be true to the concept. In the implementation stage, the computer code may encounter an error such as:

- in a variable name (H0G instead of HOG, TI instead of T1),
- uses one variable for two separate functions (such as both an iteration counter and an index),
- has an error in an operation ("*" where a "+" or a "**" should be, or where parentheses are incorrectly grouped),
- has an initialization oversight in a variable value (SUM = 0, Xold = X, for example),
- has a subscript index error,
- includes a misuse of an index (using the index value after a loop is exited and the value had been incremented by 1),
- has a careless oversight in translating algebraic time-series equations to computer assignment statements,
- has a shift in input or output value (from or to a data base, in parsing strings), or
- if the numerical approximation used in the code does not have the required precision (step size in finite difference is too large, convergence criteria in iterative operations is too coarse, single precision variable leads to significant truncation when large numbers are differenced).

Then the computer may provide an error-free execution of the procedure, but the output results will not be true to the model concept. Model verification is important, but not the topic of this textbook.

The terms validation and verification, often combined to V&V, have meanings in manufactured products and software that are similar to those developed above for modeling.

16.1.2 Deterministic Models

Deterministic models are of two categories. The most common type calculates a single value for an outcome. Examples of these include models that answer the following questions. "What is the steady-state temperature after these two substances are mixed?" "What is the pressure, given volume, moles, and temperature?" "How fast does it go when the pedal is at 50%?" These models are either expressed as analytical equations or as their equivalent computer programming assignment statements.

The second type will determine a coefficient value for a probability distribution. Examples include the following. "What is the probable error at the high value for input variable x?" "What is the expected variance if N items are averaged?" "What is the lower 95 percentile temperature in a Maxwell model of gas at a particular nominal T?" "What is the average residence time in a chemical reactor?" "What is the average particle size distribution after grinding?" Even though these models are about a distribution, the questions return a single deterministic value. Each time you ask the model, "What is the lower 95 percentile temperature in a Maxwell model of gas at a particular nominal T?" you get the same answer. These models are also expressed as analytical equations or the equivalent assignment statements.

For deterministic models, we will use data-based concepts for comparing model to data – we will inspect the residuals for trends that are not random, for patterns such as bias and curvature. Residual is the difference between model-predicted and data values. Bias means that the model does not on average represent the data. For linear regression the bias or the average residual should be zero. Although this may not be exactly true for nonlinear regression, the bias should be close to zero.

When the residuals are sorted by either a dependent or independent variable, if the residuals show a skew or curved trend, then the model architecture is likely wrong, indicating that the model is not capturing the functionality of Nature.

If the residuals show a trend when sorted with experimental run number (chronological order), then this is evidence that an uncontrolled experimental variable is influencing the data (for instance, ambient temperature, pressure, or humidity, or operator experience, or raw material composition drifts).

Since the experimental data contains uncertainty (is noisy), the tests for trends in the residuals need to be statistical tests. These are of two types. *Parametric tests* presume that the residuals are distributed in a known fashion – usually Gaussian (bell-shaped, normal) with a uniform variance. With a distribution presumed, fewer data points are needed to determine if the statistics have extreme values. However, the underlying assumptions of the distribution and uniform variance need to be affirmed. Alternately, *nonparametric tests* make few assumptions about the underlying distribution. Although nonparametric tests are more general, they need more data to make confident statistical claims.

Figure 16.1 illustrates data where the outcome y is a power law response to input x. In the real world, we often do not know what the true relation is, and here the unknowable truth about Nature is shown as a dashed line. Also illustrated, experimental variability does not place the measurements exactly on the curve. In this figure, regression uses a linear model, a straight line, and the best linear relation is also shown on the graph. Note that the envelope of the data is curved, but the model is linear. If a model matched the underlying phenomena, then we would expect data distribution above and below the model to reveal independent placement. Note that data in the middle of the x-range are predominantly above the linear model and data in the extremes are predominately below the model. Obviously the model does not match the relation

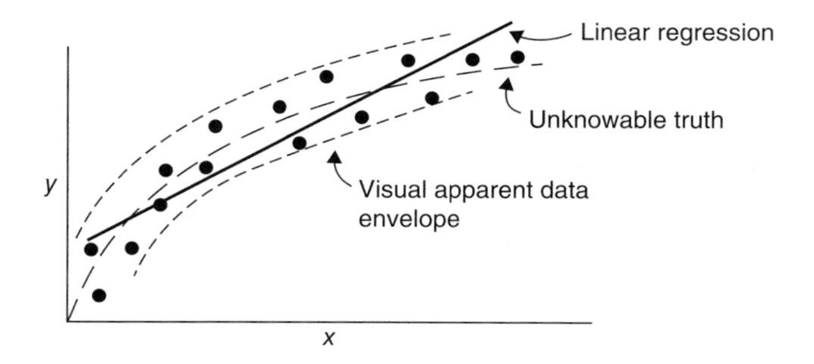

Figure 16.1 Linear model: good r-square does not mean a good model

expressed in the data, even though it is the best possible linear model. The conclusion is that the best linear model, from a least squares fit, does not necessarily represent the phenomena to be modeled.

The value of the r-squared correlation coefficient may be high, which reveals that the model significantly removed variance from the data. However, again, a statistically significant r-square value does not mean that the model matches the $y = f(x)$ behavior expressed in the data. A high r-square value simply means that there is a $y(x)$ relation and that the chosen model somewhat captures it.

The concept of validation is that a "right" model is one that correctly represents the underlying $y = f(x)$ behavior.

In comparing models to data, the methods offered here are grounded in three conditions: first, that the experiment produces accurate data, but with random, independent experimental variability, which leads to an independent and equivalent distribution in the magnitude and direction of "+" and "−" errors in the experimental data.

Let us explore a quadratic model for the data in Figure 16.1 as a stage toward revealing the other two conditions. Figure 16.2 reveals a best quadratic model for the same data. It seems to match the curvature of the data, has an excellent r-square value, and may be acceptable within the x-range of the data. However, since it does not express the power law relation (unknowable) of the data, it will not provide a good model when extrapolated, as the extended trends

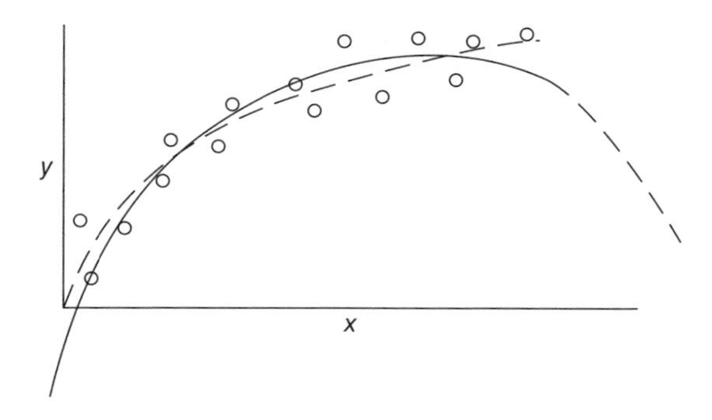

Figure 16.2 Quadratic model: good fit to data, but does not extrapolate

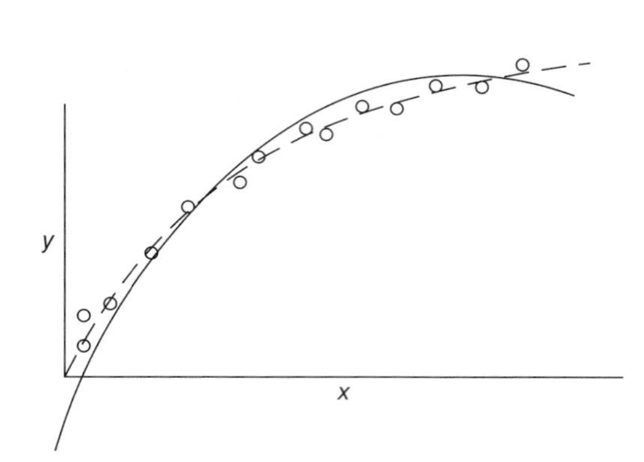

Figure 16.3 Quadratic model inadequacy is exposed by data with less variability

reveal. If the model seems to fit the data, it may be useful within the range, but does not necessarily reveal the underlying mechanistic $y = f(x)$ behavior. It may poorly represent the phenomenon when the range is extrapolated. A good fit does not necessarily indicate that you have discovered the right model architecture.

Accordingly, a second basis for testing is that, in order to be able to confidently claim that a model properly represents the true $y = f(x)$, a large experimental range is needed.

Figure 16.3 reveals the same quadratic model best-fit to the data in the original data range of Figure 16.2, but this time the experimental variability has reduced the "noise" on the data and there are more experimental data points. Since the power law relation is more clearly visible in the less noisy and greater number of data, it becomes obvious that the best quadratic model has extended sections in which it is either above or below the data. A right model will represent the underlying $y = f(x)$ phenomena and will go through the middle of the data envelope. Bad models might not be rejected if there are too few data points or high experimental variability.

Summarizing, to test deterministic models, the range must be large w.r.t. residual variability, N must be large enough to provide confidence that a bad model would be detected, and the experimental data need to represent the unbiased truth about Nature.

16.1.3 Stochastic Models

In contrast to deterministic models, *stochastic models* produce a unique (differing, independent) value at each calculation. An example is: "What number will appear on a die if I roll it?" Right outcomes are the integers 1, 2, 3, 4, 5, or 6. Other examples include: "What is a possible daily production rate, given failure probabilities for process units?" "What might be the water level in a reservoir given probable consumer choices and rainfall amounts?" These values are often the outcomes of what is termed Monte Carlo simulations. These models are usually expressed as computer code (or equivalents such as flow charts or procedure lists) with operations based on random numbers. Each time the procedure is run, it returns a unique number. When the procedure simulation is run thousands of times, each run will be based on a particular (and independent) realization of random numbers. Each run will produce a unique number. The output values for the many runs will create a distribution of values and the histogram of the many individual simulated outcomes reveals the expected natural or experimental distribution. This is similar to sampling from Nature, thousands of times, and creating

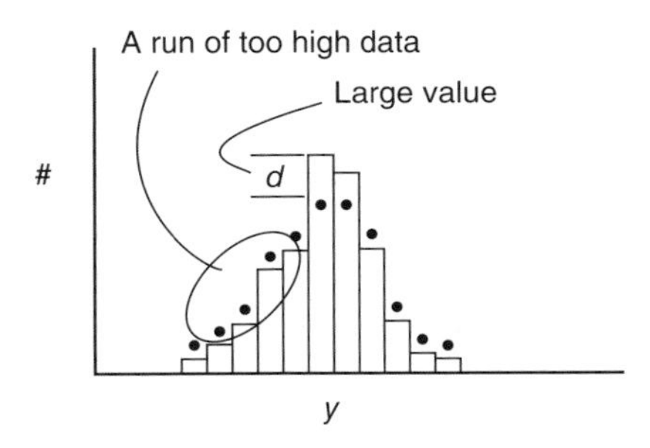

Figure 16.4 Comparison of stochastic model to data

a histogram. Stochastic models create simulated histograms, or equivalently PDF or CDF. To validate stochastic models, you need to compare the simulated and actual data histograms.

The model is based on a concept of how a process/device/product/procedure responds to vagaries of natural influences, and the model seeks to mimic the probability distribution of natural events – adult height, local rainfall, product time to failure, and so on. However, each run of the model represents one realization – one possible result of sampling. Many realizations are needed to represent Nature. Since the model is not perfect, the distribution of outcomes will not exactly match those of Nature. Hopefully, the difference will be undetectable.

One way to validate stochastic models is to compare the histogram of experimental results to that simulated by many model realizations, as illustrated in Figure 16.4. Here the range of results is divided into bins and the experimental and model frequency (number) of values within bins is normalized by the total number. Ideally, if the model distribution matched the data distribution, the model and data will exactly coincide. However, with a finite number of sampling, both experimental and simulation vagaries will generate data that do not exactly represent the truth about either; therefore we expect differences between the model and data bin heights. However, if the model is true to the experiment, we expect the residuals to be randomly distributed and no deviations to be excessively large. As illustrated here, two features reveal that the model does not match the data. First, there is a trend (a set of adjacent bins) of residuals with the same sign. Second, there is an unusual large difference, d, in the bin population fraction.

One basis in validating stochastic models is that there should not be a trend in residuals. A second is that the deviations in population fraction should not be too large.

A valid comparison of data to model requires a sufficient number of bins to see the underlying distribution. In the extreme case of a single bin that encompasses all of the experimental range, Figure 16.5 reveals that the answer is a perfect and identical 100% for each. A third basis for validating stochastic models is that there needs to be a large enough number of bins to be able to reveal the distribution shape.

Finally, Figure 16.6 reveals the impact of too many bins. With too many bins there will only be 0 or 1 or 2 hits in each, which provides too few data for any statistical comparison. The fourth concept is that there needs to be enough numbers of hits in each bin to be able to discriminate differences.

To quantify aspects related to the number of bins and number of data needed, analyze the probability distribution. The probability of a trial result landing in a particular bin is binomially distributed. It either lands in the particular bin or it lands in a different bin. If p_i is the probability

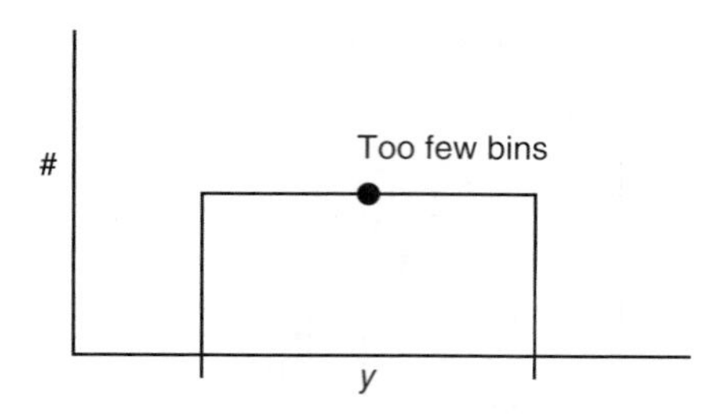

Figure 16.5 Comparison of stochastic model to data – too few bins

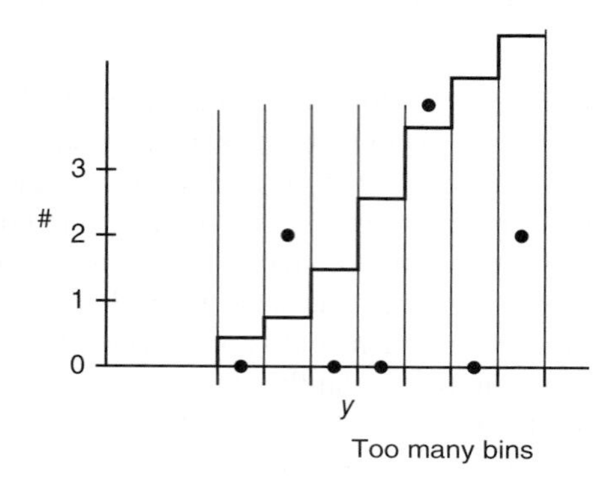

Figure 16.6 Comparison of stochastic model to data – too many bins

of it landing in the ith bin, p_i can be estimated as the ratio of the number of events in that bin, n_i, divided by the total number of trials, N:

$$p_i = n_i/N \tag{16.1}$$

From the binomial distribution, the standard deviation for the number of hits in the ith bin is

$$\sigma_{n_i} = \sqrt{Np_i(1 - p_i)} \tag{16.2}$$

Then the 95% confidence interval (CI) on p_i is approximately the two-sigma interval

$$CI_{pi} \cong 2\sqrt{p_i(1 - p_i)/N} \tag{16.3}$$

Equation 16.3 indicates that the confidence interval diminishes with the square root of the number of trials. More trials provide better precision. If you decide that you want the confidence interval to be 5% of the probability, then solve Equation 16.3 for the total number of trials, based on a particular bin probability. This provides

$$N_i = 1600(1 - p_i)/p_i \tag{16.4}$$

To ensure adequate precision of the histogram for all bins,

$$N = \max\{N_i\} = \max\left\{\frac{1600(1 - p_i)}{p_i}\right\} \tag{16.5}$$

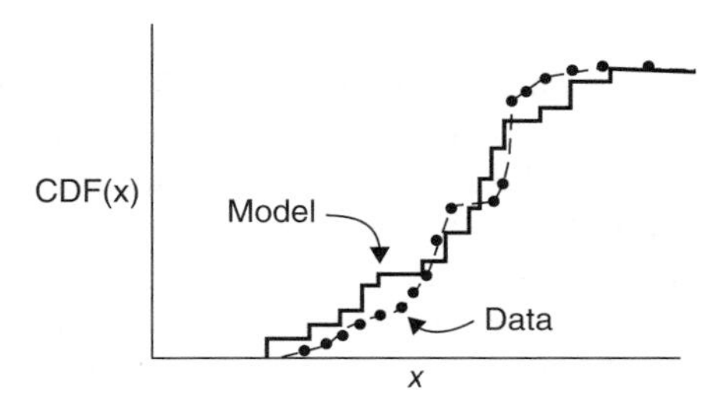

Figure 16.7 Comparison of stochastic model to data – CDF approach

This directs that the bin with the lowest probability of receiving hits is the one that determines the total number of trials required.

Since p_i will be calculated after N trials, one cannot estimate the N to provide an adequate number of data until after some data is sampled. You could use a theoretical expectation of p_i or an estimate from similar data as a first estimate to establish an initial estimate for N.

In general, use 10 bins within the range and at least 1000 samples. Structure the bin widths so that there are no fewer than five hits in a bin. This means possibly combining bins in the tails.

When the bin widths are uniform, as illustrated in Figure 16.4, the shape of the histogram reproduces the underlying PDF. However, uniform bin widths are not necessary. It is possible to choose bin widths to have a uniform fraction of either the experimental or the simulated values. In this case the histogram will have variable bin widths, but each will have a uniform height.

The large N and human involvement in determining bin widths when using the histogram or PDF representation of the data for validation are both reasons that I prefer the CDF approach (see Figure 16.7). In a CDF, sort the data by the outcome (the y value, the simulated measurement), smallest to largest, and assign each its fractional value in the ordered list. Plot the fractional value, CDF, w.r.t. the outcome. This represents the data. Add the simulation results or theoretical CDF to the graph. Use the Kolmogorov–Smirnov test to compare the two CDF curves.

Note that a model that produces statistics of a distribution (such as mean and variance) is producing deterministic numbers. Therefore, even though it is about a distribution, it is a deterministic model. Both stochastic and deterministic model types are important in engineering. Deterministic models are more commonly encountered.

16.1.4 Reality!

There are two key concepts that you need to understand. First, your model is wrong. We can never derive a perfect model, one that truly represents the natural $y = f(x)$ phenomena. Nature always appears to be one step more complicated than our mental constructs of the phenomena, or perhaps Nature is simple, but our confounded minds are looking in the wrong direction for the answer. Look at the history of modeling the atomic structure of Nature, of gravity, of gas equations of state, of heat flow, the nature of light, the nature of mass, and so on. Human history reveals that we should not claim that our model is right. At best we can hope to derive models that seem consistent with the data. We should not claim "It is right!" We can only claim "The model is consistent with the data" or, equivalently, "The data does not reject the model."

The second key concept is that Nature lies. The data has variability, often termed measurement error. Even if not a mistake (in reading, transcription, calibration, etc.), measurement error is the result of natural variability due to incomplete mixing, variable sample aging because of time variation in analysis, precision in chemical analysis, electronic or mechanically induced fluctuation, and so on. Measurement error means that the data do not exactly represent the process y-value. In one view, Nature torments us with imprecise data and our best recourse is to sample enough times so that we have a less fuzzy location of the truth.

The concepts for this section are that the model is imperfect, but it may be close enough to Nature's natural misrepresentation of itself so that we cannot say the model is wrong. (You know the model is wrong, but we do not promote that. Nature lies, but it is not nice to say bad things about Mother Nature.) We will use the data to reject or not-reject our models. We will tentatively accept the not-rejected model until we have sufficient evidence to be able to reject it.

There are two general approaches to model validation. First, does the model behave as logically expected? Second, does the data support the model?

16.2 Logic-Based Validation Criteria

This is often called a parametric or functional analysis. It is a set of sanity checks to explore the question, "Does the model make sense with what we know or expect?" This validity testing decision is based on qualitative reasoning. Following are several guides for functional analysis and for logic-based model validation:

1. Take time to its limits and look at the initial and final values. Are asymptotic values and their rates-of-change as expected? If the expectation is that the process should approach an equilibrium condition then, at large time values, rates of change should be zero. However, the asymptotic rate of change might be constant if it is an integrating process.
2. Take coefficient values to their extreme limits of 0, or 1, or infinity, and look at the asymptotic limits of the terms. Do they reduce to ideal conditions or functionalities? For example, if coefficients a and b in Equation 2.2 are set to zero, then, happily, it returns the ideal gas law of Equation 2.1.
3. Take variable values to extreme conditions (dilute or concentrated, hot or cold, high or low flow rate, short or long tube, early or long time) and look at asymptotic limits of the model terms. Do they reduce to ideal conditions?
4. Consider sensitivities of the model output to either input variable or coefficient values. Does the model response change in a logically, expected manner? Is the sensitivity as expected? This needs to be true throughout the ranges of all variables and coefficients. For instance, when mixing hot and cold water, the mixed temperature should fall with increasing flow rate of cold. The model derivative of T w.r.t. F_c should be negative. Further, in the limit of very large cold flow rates, the temperature should approach that of the cold water and the derivative of T w.r.t. F_c should approach zero with extreme F_c values.
5. Segregate data from disparate experimental conditions (Instruments 1 and 2, batch and continuous, high and low flow rates) and perform the regression on the segregated data. Are coefficient values independent of experimental conditions? If a coefficient is supposed to

represent an inherent material property, then it should not change with experimental conditions. For example, reaction activation energy should not change with flow rate through the reactor.

6. Compare coefficient values to expectations. Are magnitudes and signs of coefficient values consistent with the expected values from related or similar situations? For example, the diffusivity of H_2 gas should be similar to He gas. The overall heat transfer coefficient for one condenser should be similar for another. Distillation tray efficiency should be between 80 and 110% (internal reflux could make it above 100%). Is a coefficient value consistent with the trend in homolog materials? For example, since the specific heat of propane is larger than that of ethane, it would be expected that the specific heat of butane is larger still.
7. Some coefficient values should change with experimental conditions. When phenomena are lumped into one coefficient, does it change as expected? For example, does increasing turbulence and tortuosity of the fluid path increase the dispersion coefficient? Does in-process transport delay increase with lower flow rates?

Be your own Devil's Advocate. You will have a tendency to want to accept your model. You created it. It is your progeny. It may be difficult for you to see its inadequacies.

16.3 Data-Based Validation Criteria and Statistical Tests

By contrast to the logical and asymptotic functional considerations above, which are "fuzzy" because they are grounded in human expectations, the data-based statistical tests of this section will be quantitative. However, considering the vagaries of experimental measurements, these quantitative measures need to be statistically based, making them fuzzy also.

These tests need to be appropriate for the several classifications of data: continuous valued w.r.t. discrete or category, deterministic w.r.t. stochastic, and steady state w.r.t. transient.

16.3.1 Continuous-Valued, Deterministic, Steady State, or End-of-Batch

What patterns in the data might we observe to determine if a model is consistent with the data? Ideally, the model-predicted value will exactly match the measured value, and in a plot of y-model w.r.t. y-measured (a parity plot), data will fall on the 1:1 line. However, experimental uncertainty will make the model not exactly equal to the measurement, so the data will be scattered about the ideal 1:1 line. However, from a properly conducted experiment the measurement uncertainty should be independent for each data point, meaning that, for a true model, the data should be independently scattered above and below the 1:1 line. The above data (or below data) will not be grouped sequentially. Further, the expectation is that half of the data will be above and half below, and the two groups will have equivalent deviation distributions, some close and some far from the line.

However, not exactly half will be above or below in any particular collection of experiments. Flip a coin six times; must you get exactly three heads and the other half tails? Accordingly, if there is an improbably lopsided trend, then reject the model. If there is just a mildly lopsided trend, which might be probable, do not reject.

The general procedure for a statistical test is:

- Start with the null hypothesis. The two treatments are identical and the model matches the data; there is no difference. For the subject of this book, the null hypothesis will be "the model is true to the phenomena."
- Define a quantifiable metric, the statistic that would distinguish a difference if it existed. This could be the average difference, it could be the number of data in a run, or it could be a variance ratio. Depending on the metric, it might be that too small a value or too large a value would indicate the difference. There would be an ideal value for the statistic if the null hypothesis is true. For instance, the difference in averages should be zero, the ratio of variances should be unity, and the number of "+" residuals should be equal to the number of "−" residuals.
- Because of the inherent variability in the data, the experimental value of the statistic will not have exactly the ideal value, but it will express normal variation about the ideal value.
- Define the normal variation for the statistic and limit(s) or critical values that define a confidence interval. Typically, this is the 95% limit.
- Compute the value of the statistic from your experimental data.
- If the value of the data-based statistic exceeds the critical value, reject the null hypothesis. Alternately, accept the null hypothesis.

Note that rejecting the null hypothesis does not necessarily mean that the model was wrong. It is possible to flip a fair coin 10 times and get 10 heads. However, if I lost 10 times in a row, I would reject the fair-coin hypothesis. It is possible for experimental data to generate an extreme confluence of results, to produce a possible but very rare pattern that makes a true model appear to be untrue. Therefore, rejecting the null hypothesis is not a definitive action. Rejection would be at the confidence limit that was used to define the critical value.

As a result, in spite of the definitive quantitative procedure, like a qualitative logic-based evaluation, the statistical evaluation is also fuzzy.

A complete statement would be "If the model were a true representation of ___ (fill in the blank with the phenomena) then the ___ statistic would have an ideal value of ___. The actual, data-based value of ___ is beyond the 95% critical value for the statistic of ___. Accordingly, if the model is true, there is only a 100% − 95% = 5% chance that this could happen. On the other hand, if the model is a wrong representation then there is a very high chance that the extreme value of the statistic would be encountered. We will bet on the situation with the higher probability and reject the model."

The short form is "Using data, reject the model at the 95% level of confidence."

Conversely, accepting the null hypothesis means the model was not rejected by the data. It does not mean that the model is true. A "not guilty" verdict does not mean innocence. It means that there was not enough evidence to confidently make the guilty verdict. Flip a trick coin once and it shows an H. Well, a fair coin could have done that. Therefore, one flip does not provide enough experimental evidence to claim it was rigged. A weighted die might have a 0.2 probability of showing a 1, which could not be detected by counting the number of times it shows a 1 in 10 rolls. Similarly, a model that does not represent the data might appear acceptable. It might take a ton of data to see that it is not right. Therefore, not rejecting a model does not mean that it is true.

A complete statement would be "If the model were a true representation of the phenomena then the ___ statistic would have an ideal value of ___. The actual value of ____ is not beyond the 95% critical value. Accordingly, if the model is true, there is a 95% chance that this could

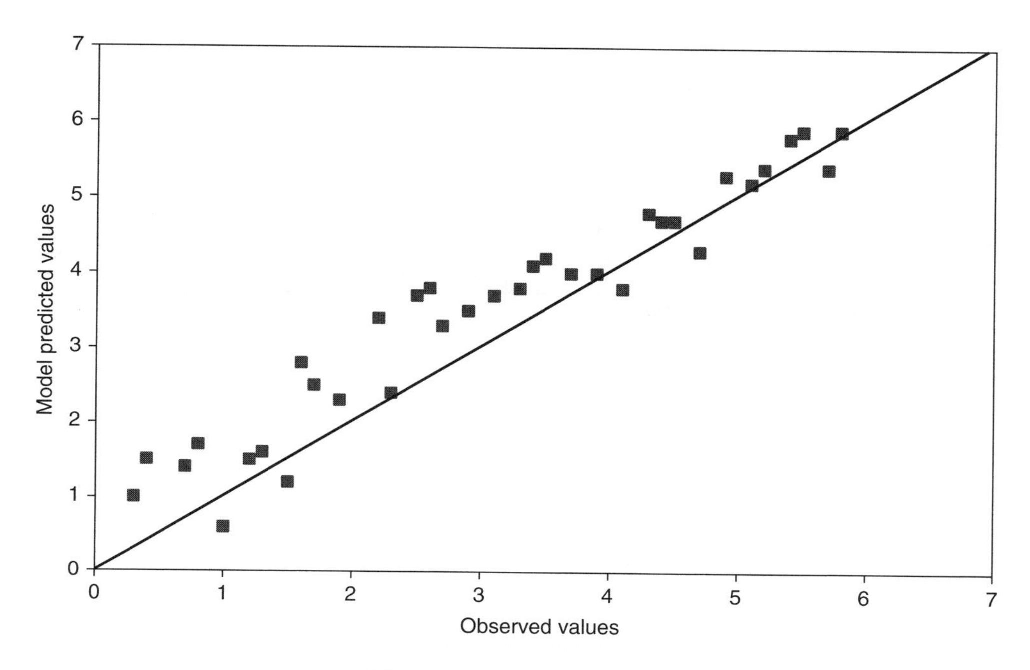

Figure 16.8 Bias illustration

happen. There is inadequate evidence to confidently reject the model. However, I know my model is wrong."

The short form is "The model appears to match the data, within a 95% level of confidence."

16.3.1.1 Data Patterns that Lead to Rejecting a Model

Figures 16.8 to 16.10 reveal three lopsided trends, which would indicate that the model is inconsistent with the data. Figure 16.8 plots a model predicted value w.r.t. the measurement, a parity graph, and reveals bias. The data seem to be predominately on one side of the 1:1 line. You expect half above and half below. You can count the number of heads (+ deviations) and tails (− deviations), and if the difference is suspiciously large, claim the model does not match the data. You also expect the above differences to have a similar distribution as the below differences. In the figure the "above" deviations have +1 as a commonly occurring value; however, the largest of the "below" deviations is about −0.3. The distribution of "+" residuals is not equivalent to that of the "−" residuals. Figure 16.8 would indicate that something is not right.

What could result in a bias of the *y*-model? In linear regression, minimizing the sum of squared deviations (SSD) means that the average residual is zero. If there is a bias on a parity plot such as Figure 16.8, then the average residual is not zero. This would mean that the *y*-model is generally too high (or too low). There are several ways this could happen:

1. In nonlinear regression, minimizing the SSD is not necessarily equivalent to generating a residual average value of zero. However, if the curve fit is good, the bias should be imperceptibly small relative to normal residual variability and not statistically detectable.

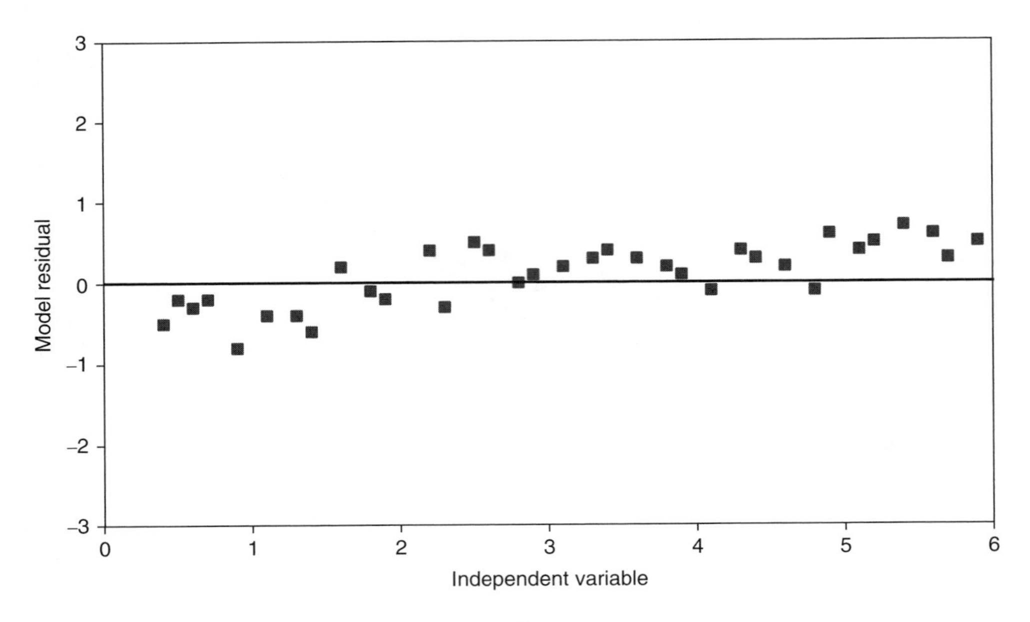

Figure 16.9 Skew illustration

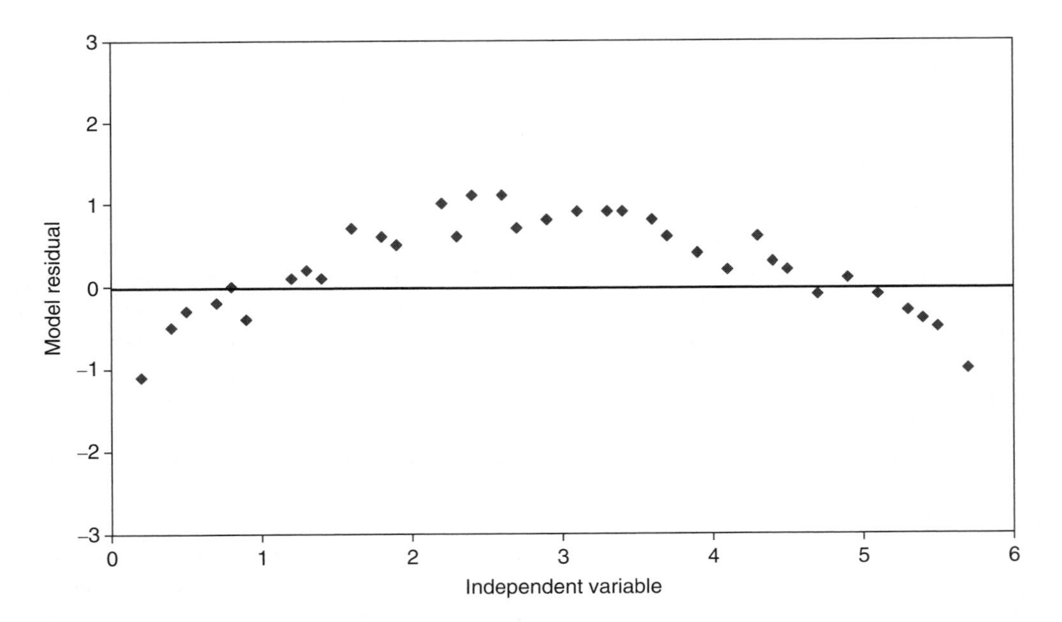

Figure 16.10 Curvature illustration

2. If the convergence of the nonlinear regression was too coarse, then the optimization will stop prior to reaching the optimum, and the SSD will not have been minimized. If this is the problem, reducing the convergence threshold will let the optimizer continue working and find the minimum.
3. The too-much-work override might have stopped the optimizer prior to finding the minimum. It could be that the optimization is structured with an undesired parameter correlation, or it could be that the optimizer type is being confounded by surface aberrations (extended valleys, discontinuities, constraints, etc.), or it could be that the time-out override needs to be extended. Either of these events will be easily detected by observing what stopped the optimizer (time-out, excessive function evaluations, etc.). The situation could be fixed by extending the too-much-work criterion, by restructuring the parameter choices, or by choosing an alternate optimizer that can handle the surface topology.
4. The optimizer may be seeking model coefficient values that make several dependent variables best match the data. If it cannot get all values to best-fit the data, minimizing SSD balances the deviations with the Lagrange multiplier or equal concern weighting. If the weighting makes one variable dominant, the optimizer will be seeking to fit one variable at the expense of another. Then the other variable would have a bias. If the weighting is balanced, both will have a bias. This situation would suggest that the model has not captured Nature or that some of the experimental data have a bias.

Figure 16.9 reveals skew. In this figure the residual is plotted w.r.t. the independent variable. Ideally, the residuals have a value of zero, but expectedly for a true model the residuals fluctuate about a value of zero and show no correlation to an experimental condition. Although there are equal numbers of data on one side and the other, and the magnitude of the "+" deviations is matched by the "−" deviations, and the average residual is zero, the envelope of the data is not parallel to the x-axis. The envelope is skewed across the zero value. This indicates that there is a detectable residual feature, not described by the model, which is correlated with the model variable.

What could cause skew?

1. The model might not adequately capture the phenomena. If so, reconsider the model development.
2. The measurements might have a scale bias. Be sure that your readings are internally consistent (degrees Celsius or degrees Fahrenheit for instance). Reconsider calibration of the data measurement device.

Figure 16.10 reveals curvature. Although there is neither bias (half are above and half are equivalently below, and the average residual is zero) nor skew (a regression line will have zero slope), the envelope of the data has curvature. Curvature is just a higher-order skew. A good model will be expected to predict the data at all values and the only variation should be the independent fluctuations on the noisy experimental data. Again, a definitive pattern in the residuals reveals that the model does not properly capture all of the phenomena.

You expect random experimental uncertainty to distribute the sequence of "+" and "−" residuals randomly. Even if half the flips are head and half tail, a correlation of results to a particular person, time, or any variable is suspicious. Either curvature, skew, or bias results in unexpected sections (regions, periods, treatments, or locations) of data that are either above or below the model when the data are organized by some variable.

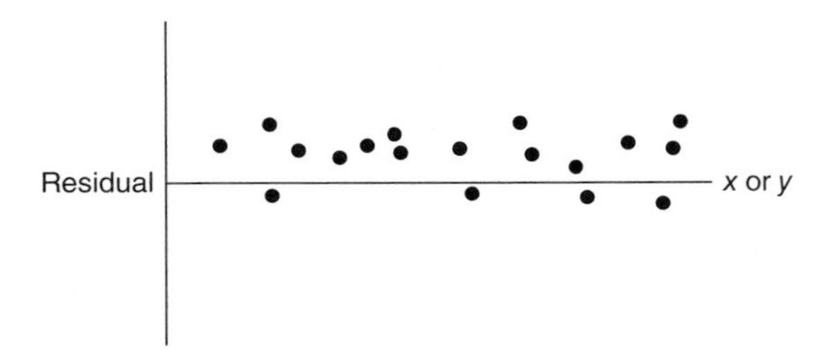

Figure 16.11 Illustration of bias but no curvature or skew

One measure of the clustering is the number of runs in the data. A *data run* is a sequence of data for which the residuals have no zero-crossing. Often this is stated as a sequence with no sign change (a datum with a residual of zero is included in the prior run). The *residual* is the deviation between the model and data.

Where there are data clusters of above or below residuals, this is often called autocorrelation. *Autocorrelation* means that if one data point has a particular value and sign, the next data point in the sequence is likely to have a similar value and sign.

Autocorrelation, or the nonparametric runs test, is better for detecting slope trends than a slope test. To illustrate this, consider Figure 16.10, which reveals curvature, or Figure 16.11, which reveals bias. In either, a regression line slope of the residuals would have a value of nearly zero, yet the figures indicate a definitive trend in the data.

Although Figure 16.8 presents model-predicted w.r.t. observed values, revealing the 1:1 line as ideality, a more convenient presentation of the data will be to look at the trend in residuals, the process-model mismatch, $r = y_{measured} - y_{model}$, with respect to the measurement value *and* with respect to *each* input value *and* for chronological run order for a static model, or with respect to measurement *and* time for a dynamic model, *and* when segregated with respect to treatment if there are suspicions that differences in testing device, operator, lab, raw material supply, data work-up procedures, and so on, may possibly affect results. Ideally, the envelope of the residuals will be centered about the value of zero and it will show a horizontal trend with no evidence of bias, skew, or curvature. If y is a static function of four input variables, x_1 through x_4, then there would be six graphs for the residuals (residuals w.r.t. y, x_1, x_2, x_3, x_4, and run number). The values of the residuals on each graph will be the same; however, they will be ordered differently, which will provide a test about the overall model (residuals versus y) and each independent variable (residuals versus x_i).

Why perform tests on all of the different residual arrangements? The graph in the top half of Figure 16.12 illustrates residuals with curvature but no bias and provides evidence that the model can be rejected. However, the lower graph shows how the same residuals might be rearranged when organized by a different variable, which reveals neither curvature nor bias. In Figure 16.12, the model properly accounts for the one variable impact on the process, but not on the first, suggesting that the modeled mechanism is acceptable for one but not for the other variable. If the model were tested with only one view of the residuals, then a bad model might be accepted.

It is possible that techniques change with experience gained at each experimental run, or with redundancy-induced sloppiness, or with an analytical technician, or with experimental devices that could change during the experimental sequence. Therefore, look at the residuals

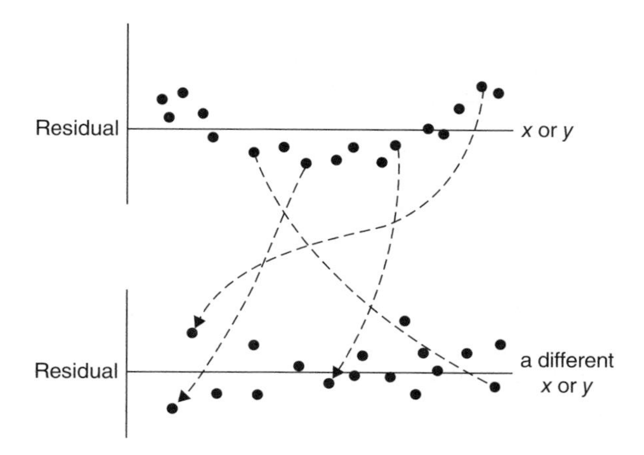

Figure 16.12 Illustration of curvature but not bias and the impact of data order

with respect to the experimental outcome, each independent variable, and the chronological order in which experiments are performed, and, possibly, w.r.t. each of the treatments.

There are statistical tests to measure bias and autocorrelation to be able to reject a model relation with a statistical level of confidence. You can use them. However, if you can clearly observe bias or autocorrelation in a trend, then the statistical test will confirm it and reject the null hypothesis with 95% confidence. If you can clearly see bias or autocorrelation (skew or curvature), then the formal test is unnecessary. Similarly, if you see no evidence of bias or curvature then the test will confirm that. If there is some suspicion, but not strong enough evidence to claim "foul," the statistical test will confirm that (the value of the statistic will be near the critical value but will not exceed it).

Then the question is: "Why use a statistical test?" The answer is: "Because humans often have a prejudice." Consider this situation: Person A was paid by his company for a month to develop process knowledge. He hypothesized mechanisms and derived a model and wants to show his boss that he is a competent and valuable employee. The model must be right! He looks at the data, sees that it captures overall trends, and claims that the model is right and the new knowledge is confirmed. Alternately, consider Person B, who does not like simple models and who wants the personal stature associated with rigorous models. He will look at the residuals from any simple model and will "confidently" see the bias or curvature patterns that indicate that the model is inferior and needs to be improved.

Therefore, to remove personal prejudice, let the statistical methods determine whether the evidence for bias or autocorrelation is strong enough to reject or not-reject the model.

16.3.1.2 Test for Bias

There are several tests that can be used for bias. Unless there is a very unusual distribution of experimental variability, the number of above points should equal the number of below points. Accordingly, a count of the number of residuals of like sign should be half of the number of data points, and you can use the binomial distribution to test the number. However, this does not test for bias, because the above residuals may be further from the model than the below residuals. Accordingly, it is better to test for bias. The most familiar test to see if the bias is significantly different from zero is probably a simple, traditional t-test on the residuals. Test the null hypothesis that the average of the residuals is zero:

$$t = \frac{\bar{r} - 0}{s/\sqrt{N}} \tag{16.6}$$

If the absolute value of the calculated t-statistic from Equation 16.6 is greater than the critical value, reject the null hypothesis, and claim with confidence that the average is not zero and that there is a bias. In Equation 16.6, $\bar{r}$ is the average of the residuals, s is the estimated standard deviation of the residuals, and N is the number of data points. Introductory statistics texts will typically have a table of critical values for “Student’s” t-statistic. You need to choose a level of confidence to select the value. Usually a 95% level of confidence is appropriate for decisions with an economic consequence. However, where team morale and ethics are involved, 99 or 99.9% are common, and where life and safety are involved, 99.999% is often chosen.

If ABS(t-calc) > t-crit then you can reject the model at a chosen percentage confidence. Alternately, you cannot reject the model as there is not adequate evidence to reject the model. The term “cannot reject” does not mean that the model is proven to be right. It simply means that the data are not adequate to confidently reject the model.

The t-test assumes that the residuals have a uniform variance (independent of the influence or response variable values) and that they are normally distributed. It is a parametric test. Before using the t-test, you should check to see if the variance is uniform and Gaussian.

Alternately, a nonparametric test that I like is the Wilcoxon signed rank test, which tests for the median of the residuals. Also relatively simple would be to test the hypothesis that the distribution of above residuals is the same as the distribution of below residuals, and either the chi-square contingency test or the Kolmogrov–Smirnov distribution test can be used. These tests are explained in standard statistical references.

Since the residuals are the same for each graph, just arranged in a different order, the test for the bias needs only be performed once on any one of the graphs, perhaps residual w.r.t. y. However, if the model is predicting several y values, then the tests for residuals must be done for each y value.

16.3.1.3 Test for Skew or Curvature

There are also several statistical tests that can be used to measure sequences in the data that would reveal curvature or skew. Perhaps the simplest is the autocorrelation test of lag 1, with the statistic defined in the following equation:

$$r_1 = \frac{\sum_{i=2}^{N} r_i r_{i-1}}{\sum_{i=1}^{N} r_i^2} \tag{16.7}$$

The subscript on r_1 means a lag-of-one, one time step, one stage, and one place in a sequence or ordered list. It means that each residual is compared to the previous residual in the sequence. The numerator product of $r_i \times r_{i-1}$ is the basis for the sequential comparison. If there is no bias in the model, the r-lag-one statistic can be interpreted as a ratio of variances. Since the residual is the y deviation from the average, the denominator is $(N - 1)$ times the conventional variance. The numerator is $(N-1)$ times the sum of products of sequential deviations, instead of like deviations.

There are $N-1$ terms in the numerator and N in the denominator. If all residuals had equivalent values then r_1 is bounded between $\pm(N-1)/N$; in the limit of large N, the asymptotic limits of r_1 is ± 1.

Unfortunately, nomenclature has developed multiple uses for the same symbol, r. Here, r_i *and* r_{i-1} indicate the residual or process-to-model mismatch, and r_1 is a statistic to quantify autocorrelation in the residuals. Neither of those r concepts is the r-square statistic used to measure the reduction in SSD due to the model nor the ratio statistic for the steady-state convergence criterion.

The denominator represents the variance in the residuals, but not scaled by $N-1$. Regardless of the sign of the residual, squared, the residual makes a positive contribution to the denominator sum and the sum will increase with N. If the residuals are not autocorrelated, if each sequential residual is independent of the previous residual, then the product in the numerator will have positive values as often and as large as it has negative values, and the sum will tend to remain near to zero. The numerator and denominator have the same units and the ratio is normalized to be independent of either the magnitude or the number of the residuals. Consequently, the value of the ratio will be within the extreme limits of $\pm(N-1)/N$, or effectively $-1 < r_1 < +1$.

If the value of r-lag-1 is near zero, the model is consistent with the data. Alternately, if the value of r-lag-1 is too near the ± 1 extremes, the model is rejected by the data. The critical value of r-lag-1 depends on the number of residuals. Multiple sources report that the one-sided 95% confidence limits for the autocorrelation coefficient of lag 1 is given by

$$r_{1.095} = \frac{-1 + 1.645\sqrt{N-2}}{N-1} \tag{16.8}$$

and the two-sided 95% CL is

$$r_{1.095} = \frac{-1 + 1.96\sqrt{N-2}}{N-1} \tag{16.9}$$

Note that both of the coefficient values in Equations 16.8 and 16.9 are the standard normal z-statistic values for $\alpha = 0.05$.

Certainly, those values are illogical for small N, so I will surmise that the critical values are only valid if $N \geq \sim 10$.

If only interested in positive correlation (the expected indication of a bad model, successive residuals have the same sign) then use Equation 16.8 for the critical value. If anticipating either positive or negative correlations (successive residuals have unexpectedly alternating signs), then use Equation 16.9. The coefficient values represent the standard normal statistic, z, for confidence $= (1-\alpha)$ or confidence $= (1-\alpha/2)$. If the absolute value of r_1 from Equation 16.7 is greater than $r_{critical}$ from Equation 16.8 you can claim, at a 95% confidence level, that the data rejects that portion of the model.

The autocorrelation test presumes both zero bias and uniform variance throughout the data sequence, which may not be true. If the data fails the test for bias, then the model is rejected, and there is no reason to consider autocorrelation. If the model is not rejected for bias, then accept that the bias is zero and inspect the data (as they are independently arranged on each graph) for uniform variance.

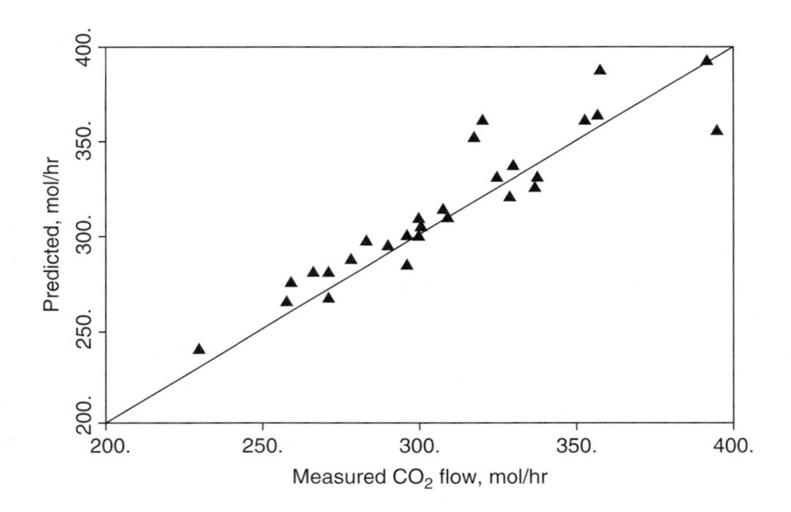

Figure 16.13 Nonuniform variance

If the variance is one region is high, then the data in that region dominate the numerator and denominator terms in Equation 16.7, making the r_1 value focus on that region. Therefore, only use the autocorrelation test if variance is uniform throughout the range of the y and x variables. Visual inspection may be an adequate test for uniform variance.

Figure 16.13, from my PhD thesis (modeling a coal gasification process), illustrates a case of variance not being uniform. Note the collective nearness to the 1:1 line of the data on the left side (less than about 300 mol/h) compared to the collective variability of the data on the right. With such a data pattern, when calculating an autocorrelation r-lag-1, the data in the upper right will dominate the data in the center and lower left, and the r-lag-1 value will not be representative of the whole.

If you suspect that the variance is not uniform, the runs test can be used to detect the presence of curvature or skew. Again, see a standard statistics text. The runs test is nonparametric; it does not presume distribution features. Although this is an advantage in applicability, it makes it less efficient, requiring a greater number of data points to be able to confidently reject the model based on curvature or skew.

Note that the r-lag-1 test is a parametric test and the runs test is nonparametric. If the residuals can be considered homoscedastic (constant variance over time and conditions) and Gaussian (normal and independent) then the parametric r-lag-1 test can be used. Parametric tests require less data to make confident indications than do nonparametric tests.

Since the autocorrelation and runs both depend on the particular sequence arrangement of the residuals, the values derived from each graph (residual versus y or x_i) is likely to be different. It may well be that there is no cause to reject the model when looking at residuals w.r.t. one variable, but there may be evidence to reject the model when looking at residuals w.r.t. another variable (see Figure 16.12). This would mean that the influence of one variable is adequately described by the model, but the other is not. This understanding can direct effort in improving the model.

Also from my thesis, Figure 16.14 illustrates skew. The data in the upper right are clustered about the 1:1 line, but the data in the left are all below the line.

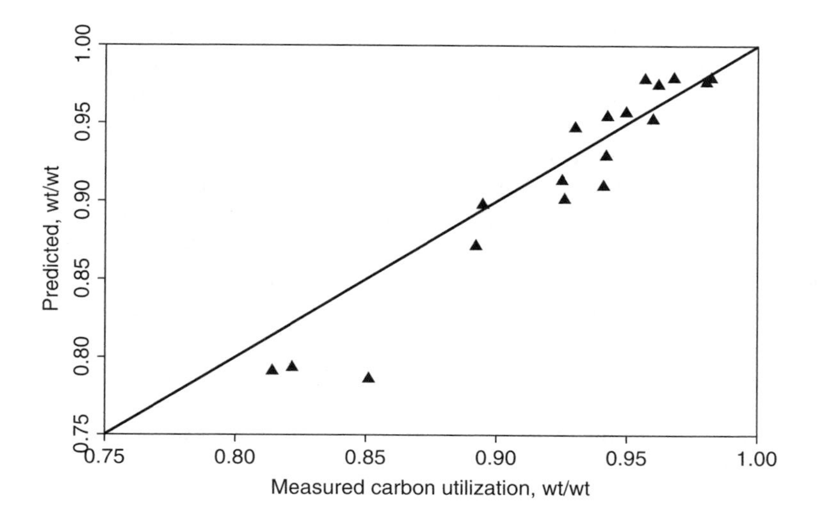

Figure 16.14 Data with skew

In Bray and Rhinehart (2001a, 2001b) the Wilcoxon signed rank test for bias and the autocorrelation test of lag 1 are recommended and demonstrated on the discrimination of reaction kinetic models for carbon tetrafluoride plasma etching of silicone dioxide and of oxygen plasma ashing of photoresist.

As a note, some people prefer the Durban–Watson, Spearman rank, or other similar tests for autocorrelation in data. All are good. I think the r-lag-1 is simple and effective.

16.3.1.4 Test for Variance Expectations

There are two essential comparisons for variance.

1. The variance in the residuals should match that expected from propagation of uncertainty in the data. One could compare the range of residuals to either replicate trials or the propagation of probable error on the data model (the method used to calculate a data value from measurements). If the residual range is much larger than expected, this could imply that the model is wrong or that the experiment was not controlled or understood as expected. If the residual range is much smaller than expected, this could mean that the experimental uncertainty was much smaller than expected. In any case, something was not properly represented.
2. Permission to use a parametric test is that the variance of the residuals should be uniform (independent of the value of the y or x values). If not homoscedastic, use a nonparametric test.

A statistically proper way to compare variances is with an F-statistic. You should set up an F-statistic for both cases, Items 1 and 2 above. However, for the comparison of residuals to those expected from a propagation of variance, Item 1 above, the propagation of variance is just an estimate based on linearization, independence, and human estimates of several of the component uncertainties. It is a reasonable estimate, not the truth. For the homoscedastic test of Item 2, my experience has usually been that the number of data in question is too low (too few residuals) to lead to a definitive test and the human choice as to how to partition the data could lead to challenges. Consider the data in Figure 16.13. There are only seven data points in the upper right that seem to have larger variability. That is a small number to make definitive claims about variance. Further, if they are segregated by the value of 340 on the vertical axis, they are separate from the other apparently low variability data. However, if the data are segregated by a value on the horizontal axis then, with 340 as the boundary, two of the high-variation data are included in the low variation group or, with 310 as the boundary, four low-variability data are included in the high group. How does one defend which axis and which value should be chosen to properly segregate the data?

Although use of an F-statistic to test variances might be "should," either test would be infected with human choices. Accordingly, I think that a human judgment as to whether the residual variance matches the expectation and whether the data are homoscedastic best balances the perfection/sufficiency values.

16.3.1.5 Test Outline

Here is an outline of the procedure for statistical validation of continuous-valued, deterministic, SS or batch processes (line wraps are indicated by the margin dots).

Use rational human interpretation to determine if the range in the residuals is compatible with the propagation of uncertainty on the data. If not investigate why.

```
FOR each response variable
•  IF homoscedastic and Normally distributed residuals (based on visual
                              observation) THEN use parametric tests
      t-test for bias of residuals on any presentation order
      IF t-data > t-critical THEN
            Reject
            Stop
      ELSE
•           FOR ordering data w.r.t. each input variable and run
                                                          number
                  r-lag-1 test for autocorrelation
                  IF reject THEN
                        Record which response and input variable
                  END IF
            NEXT variable
      END IF
ELSE need to use nonparametric tests
•     Wilcoxon Signed Rank test for bias of residuals on any
                                          presentation order
      IF sum-data > sum-critical THEN
            Reject
            Stop
      ELSE
•           FOR ordering data w.r.t. each input variable and run
                                                          number
```

```
                runs test for autocorrelation
                IF reject THEN
                     Record which response and input variable
                END IF
          NEXT input variable
     END IF
END IF
NEXT Response Variable
```

16.3.2 Continuous-Valued, Deterministic, Transient

The similarities to steady-state models of transient models that are also deterministic with continuous-valued variables are strong. The difference is that the experimental run-time defines the sequence that orders residuals.

Here is an outline of the procedure for statistical validation of continuous-valued, deterministic, transient processes (line wraps are indicated by the margin dots)

If the data has steady-state periods, use these to test the steady-state asymptotic limits of the model. Apply the tests of Section 16.3.1.

Then test the transient response with data organized sequentially in time.

```
FOR each response variable
IF homoscedastic and Normally distributed residuals THEN use
                                               parametric tests
      t-test for bias of residuals in chronological order
      IF t-data > t-critical THEN
            Reject
            Stop
      ELSE
            r-lag-1 test for autocorrelation of residuals in
                                            chronological order
            IF reject THEN
                  Record which response variable
            END IF
      END IF
ELSE need to use nonparametric tests
     Wilcoxon Signed Rank test for bias of residuals of residuals in
                                                     chronological order
      IF sum-data > sum-critical THEN
            Reject
            Stop
     ELSE
            runs test for autocorrelation of residuals in
                                         chronological order
            IF reject THEN
                  Record which response variable
            END IF
     END IF
END IF
NEXT Response Variable
```

Table 16.1 A contingency table of outcomes

	Number in category 1	Number in category 2	...	Number in category i	...	Number in category N
Observed experimental outcome	O_1	O_2	...	O_i	...	O_N
Modeled outcome, expected	E_1	E_2	...	E_i	...	E_N

16.3.3 Class/Discrete/Rank-Valued, Deterministic, Batch, or Steady State

When models predict a category, rank, or class the model fidelity to the data is often based on the number of correct categorizations. Here, perform tests with various features or influences and obtain the results of data falling in category 1, 2, 3, ... Create a contingency table, such as Table 16.1, to compare the counts from the model or simulator results to the respective counts for experimental outcomes.

Ideally $O_i = E_i$. Then the measure of discrepancy between the model and data would be the difference squared. This needs to be normalized by the count. The chi-square statistic is

$$\chi^2 = \sum_{i=1}^{N} \frac{(E_i - O_i)^2}{O_i} \tag{16.10}$$

If χ^2 from Equation 16.10 exceeds the critical value for the degrees of freedom, which depends on how you choose to set up rows and columns in the table, then reject the model. Otherwise there is insufficient evidence to reject the model.

The procedure is as follows.

Develop a contingency table and calculate the χ^2 contingency test value based on the number of counts in each category.

```
IF χ²-data > χ²-critical THEN
Reject
END IF
```

Depending on your source of critical values the data might, or might not, be normalized by the degrees of freedom. Be sure that you match a calculated statistic with the tabulated values.

If two models are being compared, the one with the lower χ^2 value would be the better model, but the degree of improvement may not be statistically significant. Only claim one model is better than another if an *F*-test reveals that the χ^2 values are significantly different.

Table 16.1 indicates that all categories are included in one table. However, if the model is right in all but one classification (but definitely wrong in that one classification) and there are many classifications, then the data might not provide a large enough composite chi-square value to reject the model. Alternately, the one bad feature might lead to a statistically significant (extreme) χ^2 value and reject the model (meaning all of the results). However, this is based on the overall model and does not indicate which portion is wrong. To see details, you can compare individual portions of the data or you can group data into broader classifications.

Unfortunately, this involves user choices, which can be challenged. The fewer number of data that often results reduces the ability of statistical tests to make definitive conclusions. Often human judgment is as good as anything for identifying model strengths and weaknesses within the individual categories.

Further, the measure of quality might not be an exact count in a category. For models that assign a rank to outcomes, the user's measure of quality might be related to the degree to which rank is missed or the degree might be scaled by the importance of the rank.

16.3.4 Continuous-Valued, Stochastic, Batch, or Steady State

A stochastic model, a Monte Carlo simulation, simulates the vagaries of Nature. One outcome of a stochastic model is like one roll of the dice or one experimental sampling. It requires many runs of a stochastic model to reveal the range of possible outcomes. The data would simulate a possible realization of experimental samplings, and from the simulated data one can calculate an average expectation, the 95% extremes, and so on.

To compare the model to experimental, the data can be either presented as a CDF (recommended) or PDF.

If a CDF, create the experimental data CDF. Sort the N data in increasing order. Assign a CDF value of i/N to the data. It is desirable to plot i/N w.r.t. data value to have a visual indication, but this is not necessary. Then compare each experimental CDF value (i/n) to the CDF value predicted by the model. State the largest deviation. This is the Kolmogrov–Smirnov d-statistic.

If the number of data is finite, the CDF will have a stair-step appearance. If the number of experimental and simulated data are different, the steps will not match. One trend may take a step up when another is still level. Accordingly, look at the difference between distributions both just before and just after the step in both distributions.

If a PDF, segregate the data into bins. Preferably, choose bin ranges so that the number of data in each is approximately equal. Create a contingency table of the experimental and modeled number of items in each bin and compute the χ^2 statistic. This is not preferred because the user choices of the number of bins and the bin intervals are subjective and could lead to controversy, and the grouping of data together in bins loses the ability to discriminate. I prefer the CDF test.

Two characteristics need to be tested. First, the deviations between the two distributions should be small. The K-S test inspects for this.

However, if one distribution were slightly skewed, then the deviations could be small but display local abundances of "+" and "−" signs. Therefore, as a second test, the distribution of deviation signs should be randomly and independently scattered throughout the range of the CDF variable. Use the autocorrelation (if uniform variance) or runs test to see if this behavior is followed.

There are two cases.

Case 1 – The model could be a deterministic model predicting the distribution or it could be a stochastic model, but with a large enough number of samplings (perhaps 1000 or more) so that the model distribution appears as a continuous curve, effectively deterministic, and not as a stair-step. Here the model is known with certainty. The K-S test and critical values are developed for this case.

Case 2 – The model could be a stochastic model, with its own variability, and with a fewer number of samplings, making the CDF of the model also reveal the stair-step nature. Here both stair-step CDF curves have their own uncertainty. The double variability is expected to lead to larger expected CDF differences. The K-S critical values can be adjusted for such a case.

Here is an outline of the procedure for statistical validation of continuous-valued, stochastic, SS, or batch processes:

```
Compute Kolmogorov-Smirnov d on the CDF
IF d-data > d-critical THEN
      Reject
      Exit IF
ELSE
      Compute the r-lag-1 statistic
      IF r1> critical value THEN
            Reject
      END IF
END IF
```

16.3.5 Test for Normally Distributed Residuals

Parametric tests (t, *r*-lag-1, etc.) on residuals assume that the data are normally distributed. They have the advantage over nonparametric tests that they can make definitive rejections with a lower number of data. If the data are not normally distributed, you should use a nonparametric test.

To visually see if the residuals are normal (Gaussian distributed), create a histogram of the residuals. It should appear bell-shaped. If it appears to be bimodal (if it has two peaks or a hump on one tail) then the data were probably generated by two distinct mechanisms. They are not internally consistent. They are not normal. If the histogram shows skew, a long tail on one side compared to the other, then the data are not normal. If values are coarsely discretized it is not normal, which presumes continuum-valued data. If any of these conditions are obvious, use nonparametric tests.

You can use a K-S test to compare the CDF of residuals to the Gaussian distribution with a mean of zero and a variance expected from data replications or propagation of uncertainty.

If the data reveal bimodal or repeated values when the residual should be from a single, continuous-valued population, then rethink the experimental conditions.

16.3.6 Experimental Procedure Validation

It could be that some uncontrolled variable was progressively changing as the experimental sequence progressed. An example is barometric pressure, which could be changing over a work-day period as a storm is coming in. If the process or a measurement has a dependence on barometric pressure (perhaps a gas density within bubbles or pipes, exit gas velocity, Reynolds number, gas density within measurement devices) then, as the experiment progresses, this uncontrolled variable will impact the results, increasing experimental error.

Other uncontrolled experimental effects that change in time, which would create a correlation with experimental sequence, include: operator fatigue, operator expertise/understanding/skill, device temperature, process fouling, catalyst deactivation, process erosion, spring stiffness, metal/plastic fatigue, calibration drift, battery aging or replacement, relative humidity, sun/cloud/wind impact on feed material or process temperature, atmospheric CO_2 absorption by caustic solutions, concentration of reagents from evaporation of water or solvents, degradation of reagents due to light, shift in personnel who are operating equipment or lab analysis, and shift in measurement devices. You can probably name many other chronologically-dependent wild influences.

Such influences create an experimental bias (a drift or skew) that is correlated to experimental run number. Since the experimental sequence should be random, when residuals are organized by an experimental variable, the residuals will show no pattern. However, when residuals are organized by the experimental run number or experimental chronological sequence, excessive autocorrelation (or too few runs) would reveal the presence of a significant uncontrolled experimental bias.

Sort the residuals by experimental sequence (time). If the autocorrelation (or runs) tests reveal an undesired pattern, the experimental results are infected by uncontrolled events. Use the techniques in Section 16.3.1.

16.4 Model Discrimination

Model discrimination refers to claiming that one model is better than another.

16.4.1 Mechanistic Models

When there are several proposed models, each representing a separate mechanism or derived from alternate assumptions, how do you choose the best one? If you are seeking to understand a mechanism, each model will reflect its mechanism and associated assumptions in the model derivation. If the tests with data reject one model, then that model is wrong. It might be wrong because mechanism or assumptions are wrong, mathematical derivations are wrong, or computer execution is wrong. It might be rejected because the experimental data are wrong. If you are confident in the data and mathematical developments, then you can reject both the hypothesized mechanisms/assumptions. If the bias and curvature tests reject all models, then you have not yet discovered the mechanism/assumptions.

If the tests do not reject a model or do not reject several models, then the mechanisms are possibly "right" (not the truth about Nature, but not rejected by the data). When multiple models are not rejected by the data, the plots of residuals w.r.t. y and each x may reveal where to place additional data that might lead to model discrimination. Then design new experiments, placing data in critical regions, to better discriminate which model is "right."

If two separate models remain not-rejected by the data, then other aspects would lead to a choice of which model is best. The model with fewer empirical coefficients would have the advantage of less influence by noise. The model that is easier to implement and use for process analysis, design, or control would have an advantage. The model that best matches the Section 16.2 criteria for logical tests would have an advantage. However, when models are functionally equivalent, perhaps the primary basis would be to choose the model that your customer (or boss) prefers.

16.4.2 *Purely Empirical Models*

Often we are not attempting to arrive at a phenomenological understanding and simply use a power-series model such as Equation 2.16 or a time-series model such as Equation 2.24 for the regression. Those examples use linear or power series functionality for the input variables. However, the model regressors may have any user-defined functionality, as indicated in Equation 2.25 or 2.27.

Although these models can be derived from a phenomenological model, the use of these models is not grounded in fundamental phenomena. Model flexibility increases with the number of terms, and the question is how many terms to use. The issue is similar for any of the number of empirical model types.

In this case, each additional term in the series makes the model better in the sense that the *r*-square gets better and the sum-of-squared-errors (SSE) gets lower. In the extreme of using an *N*-coefficient power series model for *N* data sets, the model will perfectly fit the data. However, this will give a ridiculous model that fits the noise, not the underlying mechanism. Therefore, while the SSE and *r*-squared get better with increasing model order, the representation of the $y = f(x)$ truth gets worse. Additionally, the greater the model complexity becomes, the greater is the transgression of the engineering KISS principle. Further, the greater is the number of terms in the model, the greater is the number of experiments and associated cost required to determine coefficient values. Whether it is a power series or a neural network model, or any of hundreds of purely empirical mathematical forms (wavelets, Laguerre polynomials, TSK fuzzy model, etc.), the question is the same. At what point does an increase in the number of model terms begin to add undesired aspects more than it adds improvement? Where is the point of diminishing returns with increasing complexity?

If the underlying truth or the $y = f(x)$ relation were known, then we could know the answer to the question, “What is the right model architecture?” However, since the truth is not known, the only nonlogical evidence we have to evaluate empirical models is the SSE and the complexity results. Many people have attempted to find measures of the point of diminishing returns (see Section 13.7 for a variety of methods). My choice is the equal concern criterion of Equation 13.29, but when there is no basis to determine the equal concern factors, I recommend the FPE method of Equation 13.21.

Such procedures do not identify the model that represents the y = f(x) mechanistic truth. Such procedures do not identify the best model for application convenience. Such procedures cannot reveal when model complexity exceeds that of Nature and begins to fit noise. Such procedures do not require a model to pass logic-based validation criteria. The criteria may permit the acceptance of high-order models with inflections that would not represent our understanding of Nature. The criteria are just some of many approaches for balancing complexity and model closeness to data.

You should still apply both logic-based and data-based validation criteria.

16.5 Procedure Summary

The following summary is largely from Bethea and Rhinehart (1991):

1. Ensure that a good experimental technique was used and assume that the data are correct. It is imperative that the data accurately reflect the process if any sort of conclusions (not just model validation) are to be drawn about the process.

2. Assume that the model is correct and that the data are independently perturbed. This is the null hypothesis.
3. Anticipate what the model/experimental data plots should look like if both model and data are correct. Use the guides and outlines in Section 16.3. This step identifies the hypothesis to be tested. Validation requires a conscientious and critical attempt to find ways to prove that the model is wrong by a knowledgeable person. Unfortunately, the most knowledgeable person is the one who developed the model (who also is likely to want it to be a good model) and who may suffer from a conflict of interest. Your particular model may require a comparison not suggested here. Be your own Devil's Advocate. If you are not critical, someone else will be, and the validity of your entire work can be questioned if one point is questioned.
4. Choose a level of confidence. Normally 95% is used when the consequences are economic, but where life, ethics, and safety are at risk, often 99.9% or 99.999% confidence may be appropriate.
5. Choose your experimental program or design (see Chapter 18).
6. Apply the appropriate tests that you selected in Step 3. If the model-to-data mismatch in any of the tests is worse than that expected at your level of confidence, invalidate that portion of the model.
7. Report each of the four items (reject/accept, level of confidence, test used, and the null hypothesis).

16.6 Alternate Validation Approaches

The tests that I prefer have been grounded in classic statistics and their application to test for particular expected features of the residuals. Here validation means that the model meets data-based pattern expectations and cannot be statistically rejected. However, there are other definitions for the term validation.

Since the 1970s, arising with artificial neural network modeling, an alternate approach to model development has arisen, with an alternate meaning for validation. The methods for neural network (NN) training are grounded in human learning and education procedures. The concept is that NNs learn. Therefore, consider the education classroom model: The teacher presents information to the students, showing them how to do examples. Then the teacher observes the student's independent performance on assignments. If the students are learning from the lecture examples, then their performance on the independent work will improve. If, on the other hand, students are memorizing the lectures without understanding the procedure, then their independent work will be very poor. Therefore, the NN training approach is to separate the data (examples) into a training set and a validation set. The NN coefficients (weights) are iteratively adjusted to make the NN output match the training set data. This is called training, but it is really iterative nonlinear optimization seeking to best fit the model to the training data set. With each training iteration (epoch), the NN output is also compared to the validation data set. If the performance on the validation set (the independent set not used for training) is improving, then the NN learning is getting better and optimization continues. However, when the NN performance on the validation set begins to get worse, this indicates that the NN is memorizing the training set data and is not learning underlying concepts, and this is used as the criterion to stop optimization. In this use, validation is a convergence criterion for the optimization, not a test for desirable patterns in the residuals to determine whether the model has captured the underlying mechanisms.

As a perspective, when there are many more adjustable coefficients than data, this might be an acceptable method for regression and determining convergence. However, I do not recommend that approach. First, to have confidence that a model captures the underlying mechanism, the classic rule for regression is that there should be three or more data sets for each adjustable coefficient. Second, all of the test data is valuable and conveys some information about the process being modeled. Removing a portion of it from its use in regression removes some information from directing model development. Third, there was probably some cost associated with acquiring the data. If you want $N = 3M$ sets to shape regression, but decide to use 15% for "independent validation" as a convergence criterion, then you need to pay for $1.15N$ sets. Fourth, in spite of the terminology, the procedure does not validate the model; it is a criterion to stop the regression optimization.

16.7 Takeaway

Your model is not perfect and it cannot possibly exactly and properly account for every mechanism affecting the process. Validation is a procedure to see if the process/model mismatch is so bad that the model must be rejected.

Use both logic-based and data-based criteria to accept or reject both models and the experimental procedure. In data-based validation, first check for bias in the residuals, then organize residuals by y, *and* each x, *and* experimental sequence, and test for autocorrelation (or runs). Although r-square and ANOVA are appropriate for screening models, neither test addresses mechanistic validation. The t-test for bias and r_1 (autocorrelation of lag1) test for skew or curvature are preferred, because these parametric tests are more sensitive (need less data) than the nonparametric tests. However, if there is evidence or suspicion that the residuals are not normally distributed with a uniform variance, then use the Wilcoxon signed rank test for bias and the runs test for skew or curvature.

While the tests either "reject the model at a certain level of confidence" or "cannot reject the model at a certain level of confidence," the corresponding colloquial statements are "reject" or "accept." "Accept" does not mean that the model is right. "Accept" means that the model is not so bad that it can be confidently rejected by the data.

Exercises

16.1 Model in-line mixing of hot and cold water and show that dT/dF_h is always + and approaches zero for large F_h.

16.2 Model in-line mixing of hot and cold water and show that dT/dF_c is always − and approaches zero for large F_c.

16.3 Show that the van der Waals equation of state (Equation 2.2) approaches the ideal gas law in the asymptotic limits of coefficients a and b.

16.4 Show that, in linear regression with the model $\widetilde{y} = a + bx$, after regression the average residual is zero:

$$0 = \frac{1}{N}\sum_{i=1}^{N}(y_i - \widetilde{y}_i).$$

16.5 Repeat Exercise 4, but with a quadratic model, linear in coefficients but nonlinear in x-functionality.

16.6 Show that, in nonlinear regression, after regression the average residual is not zero. Choose a simple model, for instance, $\widetilde{y} = ax^b$.

17

Model Prediction Uncertainty

17.1 Introduction

Uncertainty in experimental data leads to uncertainty on the regression model coefficient values, which leads to uncertainty in the model-calculated outcomes. We need a procedure to propagate the uncertainty in experimental data to the modeled values so that aspects of model uncertainty can be appropriately reported.

In linear regression, this is relatively straight-forward, and mathematical analysis leads to methods for calculating the standard error of the estimate and the 95% confidence limits on the model. If the following conditions are true, (i) the model functional form matches the experimental phenomena, (ii) the residuals are normally distributed because the experimental vagaries are the confluence of many, small, independent, equivalent sources of variation, (iii) model coefficients are linearly expressed in the model, and (iv) experimental variance is uniform over all of the range (homoscedastic), then analytical statistical techniques have been developed to propagate experimental uncertainty to provide estimates of uncertainty on model coefficient values, and on the model.

However, if the variation is not normally distributed, if the model is nonlinear in coefficients, if variance is not homoscedastic, or the model does not exactly match the underlying phenomena, then the analytical techniques are not applicable. In this case numerical techniques are needed to estimate model uncertainty. Bootstrapping is the one I prefer. It seems to be understandable, legitimate, simple, and is widely accepted.

Bootstrapping is a numerical, Monte Carlo approach that can be used to estimate the confidence limits on a model prediction. In Section 3.8, it was presented as a technique to develop the confidence interval on coefficient values. However, because of coefficient correlation, there was a caution not to use propagation of uncertainty on individual model coefficients to estimate the collective impact of model coefficient uncertainty on model prediction. This chapter reveals how bootstrapping can be used to estimate the collective uncertainty of model coefficient values on the model prediction.

Nonlinear Regression Modeling for Engineering Applications: Modeling, Model Validation, and Enabling Design of Experiments, First Edition. R. Russell Rhinehart.

17.2 Bootstrapping

One assumption in bootstrapping is that the experimental data that you have represents the entire population of all data realizations, including all nuances in relative proportion. It is not the entire possible population of infinite experimental runs, but it is a surrogate of the population. A sampling of that data then represents what might be found in an experiment. Another assumption is that the model cannot be rejected by the data, the model expresses the underlying phenomena. In bootstrapping:

1. Sample the experimental data with replacement (retaining all data in the draw from the original set) to create a new set of data. The new set should have the same number of items in the original, but some items in the new set are likely to be duplicates and some of the original data will be missing. This represents an experimental realization from the surrogate population.
2. Using your preferred nonlinear optimization technique, determine the model coefficient values that best fit the data set realization from Step 1. This represents the model that could have been realized.
3. Record the model coefficients.
4. For independent variable values of interest, determine the modeled response. You might determine the y value for each experimental input x-set. If the model is needed for a range of independent variable values, you might choose ten x values within the range and calculate the model y for each.
5. Record the modeled y values for each of the desired x values.
6. Repeat Steps 1 to 5 many times (perhaps over 1000).
7. For each x value, create a histogram of the 1000 (or so) model predictions. This will reflect the distribution of model prediction values due to the vagaries in the data sample realizations. The variability of the prediction will indicate the model uncertainty due to the vagaries within the data.
8. Choose a desired confidence interval value. The 95% range is commonly used.
9. Use the cumulative distribution of model predictions to estimate the confidence interval on the model prediction. If the 95% interval is desired, then the confidence interval will include 95% of the models or 5% of the modeled y values will be outside the confidence interval. As with common practice, split the too-high and too-low values into equal probabilities of 2.5% each and use the 0.025 and 0.975 cumulative distribution function (CDF) values to determine the y values for the confidence interval.

This bootstrapping approach presumes that the original data has enough samples covering all situations so that it represents the entire possible population of data. Then the new sets (sampled with replacement) represent legitimate realizations of sample populations. Accordingly, the distribution of model prediction values from each re-sampled set represent the distribution that would arise if the true population were independently sampled.

If you are seeking to model transient data, you might have set up one situation and then taken 100 measurements over time. However, this is one realization, not 100. All ensuing 100 samples are related to the same initialization. If the set-up has mixing, composition, temperature, and so on, variation from one set-up to another, then the 100 samples do not represent that true variation. They are simply one realization. You should run replicate trials to include the set-up to set-up variability.

Similarly, if you create a large sample, or set up a situation, and then perform multiple tests on the same material (or set-up), all of the tests reflect a single set-up. These replicate part of the source of variability, but not all of it. The data exclude the variation due to the set-up and is in effect a single realization. For example, in rheology we make a mixture, place a sample of it in a device, and measure shear stress for a variety of shear rates (perhaps 20 different shear rates). Then we plot stress w.r.t. shear rate and use regression to match a rheology model to the 20 sets of data. However, the data represent one sample preparation realization, not the population. To make the data reflect the population, perform the tests on replicate set-ups.

Bootstrapping assumes that the limited data represents the entire population of possible data, that the experimental errors are naturally distributed (there are no outliers or mistakes, not necessarily Gaussian distributed, but the distribution represents random natural influences), and that the functional form of the model matches the process mechanism. Then a random sample from your data would represent a sampling from the population and for each realization the model would be right.

If there are N number of original data, then sample N times with replacement. Since the central limit theorem indicates that variability reduces with the square root of N, using the same number keeps the variability between the bootstrapping samples consistent with the original data. In Step 1, the assumption is that the sample still represents a possible realization of a legitimate experimental test of the same N. If you use a lower number of data in the sample, M, for instance, then you increase the variability on the model coefficient values. You could accept the central limit theorem and rescale the resulting variability by the square root of M/N. However, the practice is to use the same sample size as the "population" in order to reflect the population uncertainty on the model.

In Step 6, with only a few re-samplings, there are too few results to be able to claim what the variability is with certainty. As the number of Step 6 re-samplings increases the Step 9 results will asymptotically approach the representative 95% values. However, the exact value after infinite re-samplings is not the truth, because it simply reflects the features captured in the surrogate population of the original N data, which is not actually the entire population. Therefore, balance effort with precision. Perhaps 20 re-samplings will provide consistency in the results. On the other hand, it is not unusual to have to run 100 000 trials to have Monte Carlo results converge.

One can estimate the number of re-samplings, n, needed in Step 6 for the results in Step 9 to converge from the statistics of proportion. From a binomial distribution the standard deviation on the proportion, p, is based on the proportion value and the number of data:

$$\sigma_p = \sqrt{p(1-p)/n} \tag{17.1}$$

Desirably, the uncertainty on the proportion will be a fraction of the proportion:

$$\sigma_p = fp \tag{17.2}$$

where the desired value of f might be 0.1.

Solving Equation 17.1 for the number of data required to satisfy Equation 17.2 gives

$$n = \left(\frac{1}{p} - 1\right) / f^2 \tag{17.3}$$

If $p = 0.025$ and $f = 0.1$, then $n \approx 4000$.

Although $n = 10\,000$ trials is not uncommon and $n = 4000$ has just been determined, I think. for most engineering applications, that 100 re-samplings will provide an appropriate balance between computational time and precision. Alternately, you might calculate the 95% confidence limits on the y values after each re-sampling and stop computing new realizations when there is no meaningful progression in its value, when the confidence limits seem to be approaching a noisy steady-state value.

In Step 9, if you assume that the distribution of the $\widetilde{y}$ -predictions are normally distributed, then you could calculate the standard deviation of the $\widetilde{y}$ values and use 1.96 times the standard deviation on each model prediction to estimate the 95% probable error on the model at that point due to errors in the data. Here, the term error does not mean mistake; it means random experimental normal fluctuation. The upper and lower 95% limits for the model would be the model value plus/minus the probable error. This is a parametric approach.

By contrast, searching through the $n = 4000$ or $n = 10\,000$ results to determine the upper and lower 97.5 and 2.5% values is a nonparametric approach. The parametric approach has the advantage that it uses values of all results to compute the standard deviation of the $\widetilde{y}$ -prediction realizations and can get relatively accurate numbers with a much fewer number of samples. Perhaps $n = 20$. However, the parametric approach presumes that the variability in $\widetilde{y}$ -predictions is Gaussian. It might not be. The nonparametric approach does not make assumptions about the underlying distribution, but only uses two samples to interpolate each of the $\widetilde{y}_{0.025}$ and $\widetilde{y}_{0.975}$ values. Therefore, it requires many trials to generate truly representative confidence interval values.

Unfortunately, the model coefficient values are likely to be correlated. This means that if one value needs to be higher to best fit a data sample, then the other will have to be lower. If you plot one coefficient w.r.t. another for the 100 re-samplings and see a trend then they are correlated. When the variability on input data values are correlated, the classical methods

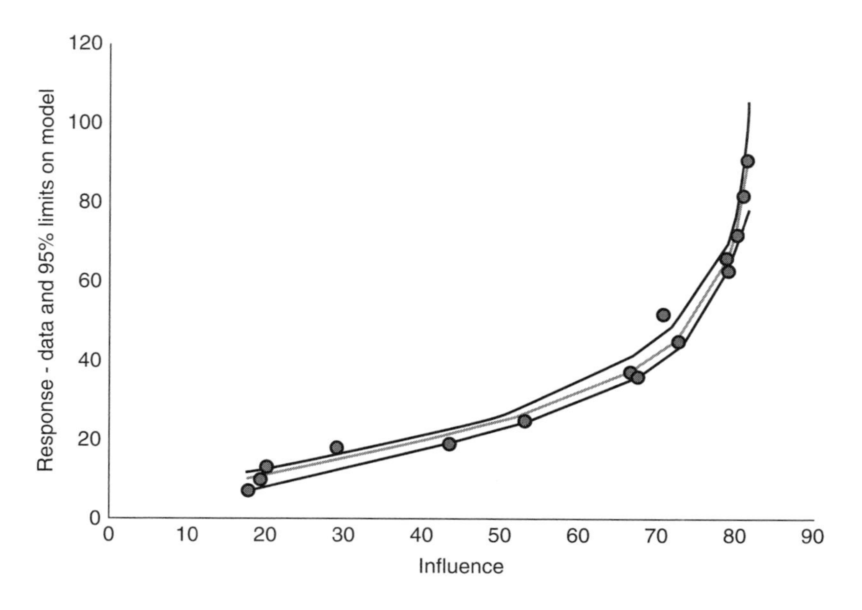

Figure 17.1 Bootstrapping estimate of model uncertainty due to data.

for propagation of uncertainty are not valid. They assume no correlation in the independent variables in the propagation of uncertainty.

Also, Step 9 has the implicit assumption that the model matches the data, that the model cannot be rejected by the data, and that the model expresses the underlying phenomena. If the model does not match the data, then bootstrapping still will provide a 95% confidence interval on the model, but you cannot expect that interval to include the 95% of the data. As a caution, note that if the model does not match the data (if the data rejects the model) then bootstrapping does not indicate the range about your bad model that encompasses the data, the uncertainty of your model predicting the true values.

Figure 17.1 reveals the results of a bootstrapping analysis on a model. The circles represent data, the inside thin line is the modeled value, and the darker lines indicate the 95% limits of the model based on 100 realizations of the data set of 15 elements.

17.3 Takeaway

Bootstrapping is predicated on having enough experimental data to represent all of the aspects of the population and uses a finite number of subsampling to estimate the confidence interval. It assumes that the model is a true representation of the phenomena. It is neither a definitive statement about the confidence limits nor a validation of the model. Bootstrapping provides a reasonable estimate of the uncertainty of the model-calculated values due to the uncertainty in the experimental data.

18

Design of Experiments for Model Development and Validation

18.1 Concept – Plan and Data

To run an experiment, you need to choose values for influence variables. You will implement a specific set of input conditions and then wait until conditions are right to collect response data. This might be to wait until the process settles to a new steady state (SS). Then collect data. Then move to the next set of input values. However, for dynamic models, data from the transient period is of interest. There, you will collect data at time intervals chosen to reveal the time-response.

Whether SS or transient, batch or continuous, the *experimental plan* deals with the sequence that you choose for your experimental conditions, the time you wait until collecting data (or the trigger you choose to initiate data collection), the range of conditions that you explore, the number of runs, replications, and so on. These issues should be selected to support validation of the model, the critical testing of the model.

Additionally, you need to define the method of gathering data, of obtaining response measurements, and the number and placement of data sets so that you minimize the impact of experimental uncertainty on the model coefficient values. The *experimental measurement methodology* should be selected to minimize the impact of data uncertainty. Desirably, the uncertainty in the data will justify a choice of vertical least-squares regression as opposed to the more complicated maximum likelihood or total least-squares approaches.

All of those issues should shape the design of experiments (DoE). The measurement methodology will be addressed first.

18.2 Sufficiently Small Experimental Uncertainty – Methodology

There is uncertainty in both the measured dependent variable response, y, and the independent variables, x, representing the experimental inputs. Ideally, you want no uncertainty in your experimental data. However, this is not possible. At best you can design your experiment so

Nonlinear Regression Modeling for Engineering Applications: Modeling, Model Validation, and Enabling Design of Experiments, First Edition. R. Russell Rhinehart.

that the impact of the uncertainty is inconsequential. Design to achieve a sufficiently small experimental uncertainty so that it is inconsequential to the in-use application of the model.

Choices in the experimental design include the number of data points, their location, the quantity of material sampled, the number of replicates, the time to wait to ensure SS, the measurement device, procedures for sample preparation, device calibration, and so on.

Within the context of regression, the objective of an experiment is to generate data so that you can determine precise coefficient values in a model. However, coefficient values will reflect data uncertainty. Subsequently, when the model is used, the calculated outcome will have an associated uncertainty, which depends on uncertainty in both the coefficient values and the input variable values. Consider Equation 1.1. The coefficient values are a, b, and c, and the input variable is x. Accepting that there is uncertainty on all coefficient values and the input value, propagation of maximum error yields

$$\epsilon_y = \epsilon_a + |x|\epsilon_b + |x^2|\epsilon_c + |b + 2cx|\epsilon_x \tag{18.1}$$

which clearly reveals that uncertainty on the value of y-calculated depends on the uncertainty associated with coefficients (ϵ_a, ϵ_b, and ϵ_c) and the input (ϵ_x).

In general a model equation can be written as

$$\tilde{y} = f(\underline{x}, \underline{p}) \tag{18.2}$$

where $\underline{p}$ represents the vector of coefficient values and $\underline{x}$ the vector of input values.

Propagation of maximum error (uncertainty) on Equation 18.2 results in

$$\epsilon_{\tilde{y}} = \left|\frac{\partial \tilde{y}}{\partial x_1}\right| \epsilon_{x_1} + \left|\frac{\partial \tilde{y}}{\partial x_2}\right| \epsilon_{x_2} + \ldots + \left|\frac{\partial \tilde{y}}{\partial p_1}\right| \epsilon_{p_1} + \left|\frac{\partial \tilde{y}}{\partial p_2}\right| \epsilon_{p_2} + \ldots \tag{18.3}$$

Desirably, convergence criteria in the nonlinear regression was tight enough and the number of data were large enough that we have such low uncertainty on the coefficient values that they contribute inconsequentially to the uncertainty of the model-calculated value. In this case, the uncertainty on the model-calculated value would be

$$\epsilon_{\tilde{y}\ Use} = \left|\frac{\partial \tilde{y}\ Use}{\partial x_1}\right| \epsilon_{x_1} + \left|\frac{\partial \tilde{y}\ Use}{\partial x_2}\right| \epsilon_{x_2} + \ldots = \sum \left|\frac{\partial \tilde{y}\ Use}{\partial x_i}\right| \epsilon_{x_i} \tag{18.4}$$

The postfix "Use" indicates that this value of $\tilde{y}$ uncertainty represents what will be calculated from the equation for the x values of use. This is an unavoidable uncertainty on the calculated value as a result of uncertainty on the input situation.

When you are planning an experiment, choose experimental conditions so that the error contribution on the experimental value of y_{Exptl} is much less than the uncertainty on the calculated value of y_{Use} from Equation 18.4. To begin the design, first consider that all of the measurement uncertainty on the y value is due to errors on constitutive elements of calculating y. In experiments we usually do not measure y, we use a data model for $y_{Exptl.}$ For example, if y represents the volumetric flow rate and we are using a bucket-and-stopwatch method, we calculate y_{Exptl} from experimental measurements of tared weight, time interval to collect the liquid, liquid density, and perhaps an adjustment for the gravitational acceleration where the material was weighed:

$$\dot{Q} = \frac{W_{full} - W_{tare}}{\rho(t_{end} - t_{begin})} \left(\frac{g_{nominal}}{g_{actual}}\right) \tag{18.5}$$

As another example, consider the "measurement" of an acid composition by titration. It is actually calculated from the titration volumes and titrant composition:

$$c_{sample} = c_{titrant} \frac{V_{titrant}}{V_{sample}} \tag{18.6}$$

As a last example, consider "measuring" the number of moles in a gas sample from *P–V–T* data:

$$n = \frac{PV}{RTz} \tag{18.7}$$

In general *y*-experimental is calculated from an equation in which m_i represents the several measurements. These are termed *data models*, to differentiate them from *process models*.

From this experimental calculation, using m to represent a measurement in the data model equation, propagation of uncertainty reveals how measurement uncertainty affects experimental "measurement" uncertainty:

$$\epsilon_{y_{Exptl}} = \sum \left| \frac{\partial y_{Exptl}}{\partial m_i} \right| \epsilon_{mi} \tag{18.8}$$

Choose experimental conditions such that $\epsilon_{y_{Exptl}} \ll \epsilon_{y_{Use}}$, where $\epsilon_{y_{Exptl}}$ is from Equation 18.8 and $\epsilon_{y_{Use}}$ is from Equation 18.4. "Much, much less" usually means at least an order of magnitude less, or $\epsilon_{y_{Exptl}} < 0.1\epsilon_{y_{Use}} a^2 + b^2$. In the bucket and stopwatch method, choose a collection interval long enough so that uncertainty on weight and time interval is inconsequential relative to the in-use uncertainty. If making samples, or titrating, use larger volumes or more precise measurements, or both, to reduce the impact of volume uncertainty, and so on.

This analysis used propagation of maximum error. You could also use propagation of probable error.

In either case, the uncertainty on the input and measured values has to be estimated. Experience in the experimentation and in the conditions of use of the resulting equation need to be understood.

It may well be that you cannot change the measurement procedure to obtain an adequately small value for $\epsilon_{y_{Exptl}}$. In this case use multiple measurements. Ensure independence of the measurements by randomly placing them throughout the experimental procedure. Applying the central limit theorem, variance is reduced by a factor of *N*, the number of trials. Choose experimental conditions and N such that:

$$\epsilon_{y_{Exptl}} = \left(\sum \left| \frac{\partial y_{Exptl}}{\partial m_i} \right| \epsilon_{mi} \right) / \sqrt{N} < 0.1\epsilon_{\tilde{y}_{Use}} = 0.1 \sum \left| \frac{\partial y_{Use}}{\partial x_i} \right| \epsilon_{xi} \tag{18.9}$$

If it is believed that there is no uncertainty on the *x* values, then one could specify a desired $\epsilon_{\tilde{y}_{Use}}$ from understanding how the model would be used and then find conditions such that

$$\left(\sum \left| \frac{\partial y_{Exptl}}{\partial m_i} \right| \epsilon_{mi} \right) / \sqrt{N} < 0.1\epsilon_{\tilde{y}_{Use}} \tag{18.10}$$

There may be inherent sources of variability in the *y*-measurement that are in addition to the measurement calculation from the *m* values. Including this unquantifiable term, $\sigma_{other\ sources}$

in the propagation of maximum error relation, and equating maximum uncertainty to 2.5σ, $\epsilon_{yExptl} \cong 2.5\sigma_{yExptl}$:

$$\left(\sum\left|\frac{\partial y_{Exptl}}{\partial m_i}\right|\epsilon_{mi} + 2.5\sigma_{other}\right)/\sqrt{N} < 0.1\epsilon_{\tilde{y}_{Use}} \tag{18.11}$$

r, including it in a propagation of variance analysis,

$$2.5\sqrt{\left(\sum\left(\left|\frac{\partial y_{Exptl}}{\partial m_i}\right|\frac{\epsilon_{mi}}{2.5}\right)^2 + \sigma^2{}_{other}\right)/N} < 0.1\epsilon_{\tilde{y}_{Use}} \tag{18.12}$$

Since all of these are based on estimates of uncertainties, the simpler propagation of maximum error is probably the best practice to guide pre-experimental planning.

Example 18.1 Orifice Calibration An orifice is a restriction in a flow line and the differential pressure, dP, across the orifice is, ideally, proportional the flow rate squared. Therefore, measuring the dP can be used to infer flow rate, F. In an instrumented control system, the dP signal is transmitted as an electrical milliamp, scaled signal i as a deviation from the zero flow value i_0. Assume that all of the experimental uncertainty on the flow rate is due to the measurement errors on collection mass and time. In an orifice measurement calibration experiment, (i) set a valve position and collect water in a bucket and read the electric current signal sent from transducer to controller, (ii) measure mass and collection time, and calculate the flow rate, (iii) repeat for about 10 conditions, and (iv) determine coefficient values in the model $F = a(i - i_0)^b$ to minimize the vertical *sum of squared deviations* (SSD). The relevant equations are:

Calibration flow rate:

$$F_{Experimental} = M/(\rho\Delta t)$$

Controller calculated flow rate:

$$F_{Orifice} = a(i - i_0)^b$$

Generic propagation of probable error:

$$\varepsilon_{95\%} = 1.96\sqrt{\sum\left(\frac{\partial F}{\partial x_i}\sigma_{x_i}\right)^2}$$

Applied to the experimental data model, assuming inconsequential uncertainty on the density:

$$\varepsilon_{95\%\ Experimental} = 1.96\sqrt{\left(\frac{1}{\rho\Delta t}\sigma_M\right)^2 + \left(-\frac{M}{\rho\Delta t^2}\sigma_{\Delta t}\right)^2}$$

Applied to the orifice process model assuming that there is no uncertainty on model coefficient values:

$$\varepsilon_{95\%\ Orifice} = 1.96\sqrt{\left(ab(i - i_0)^{b-1}\sigma_i\right)^2 + \left(-ab(i - i_0)^{b-1}\sigma_{i_0}\right)^2}$$

Desiring that the experimental uncertainty in the measurements defining flow rate (M and Δt) are much less than the inherent uncertainty from the mA measurements, design conditions (choices of Δt, M, and N) so that

$$\left[\left(\frac{1}{\rho\Delta t}\sigma_M\right)^2 + \left(-\frac{M}{\rho\Delta t^2}\sigma_{\Delta t}\right)^2\right] / \sqrt{N} < 0.1\left[\left(ab(i-i_0)^{b-1}\sigma_i\right)^2 + \left(-ab(i-i_0)^{b-1}\sigma_{i_0}\right)^2\right]$$

The standard deviations for M, Δt, i, and i_0 are estimated from the reasonable range that was inferred by human observation; then use $\sigma \cong \varepsilon/2.5 = R/5$.These will not be rigorous values. These are reasonable estimates, which if off by 10 or 20% do not change the bottom-line analysis. The equation for propagation of variance has many assumptions: locally linear relations, independence of perturbations, and Gaussian distribution, and leads to a 95% probable error, not an absolute truth. This being so, the estimates for the sigmas need only be reasonable, not exact. In an application once, I used $\sigma_M = 0.01$ kg, $\sigma_{\Delta t} = 0.2$ second, $\sigma_i = 0.2$ mA, and $\sigma_{i0} = 0.02$ mA.

Until values are known for model coefficients a and b, one cannot calculate the combinations of Δt, M, and N needed to satisfy the relation. Therefore, collect a few data points, work up the data to determine the early estimate of model coefficient values, and use this to determine the experimental conditions, Δt, M, and N. It might become realized that a desired Δt for N = 10 samples results in an M that exceeds the full capacity of a bucket and that N needs to be increased to 25. Let the developing knowledge define this.

18.3 Screening Designs – A Good Plan for an Alternate Purpose

There is a large body of knowledge related to statistical DoE. Unfortunately, it is substantially irrelevant to the model validation and discrimination objectives presented in this textbook. Classical DoE is grounded in the use of a power series representation of the phenomena, as given in Equation 2.16, and seeks the best placement of experimental conditions to determine values for the model coefficients *a*, *b*, *c*, and so on, with greatest precision with the fewest number of experimental trials. The assumptions are that the order you choose for the Equation 2.16 model matches the true $y = f(x)$ behavior, that the experimental variability is solely due to *y*-variability (no uncertainty on *x*), that variability is uniform throughout the entire range of values (homoscedastic), and that any combination of *x* values within their ranges can be experimentally obtained (one *x* value does not impose constraints on another). This leads to box-designs, star-designs, Latin-square designs, and so on, which are good in screening experiments in which you want a rough description of the surface to be able to hone in on a local region of interest.

However, here our models (usually nonlinear), our assumptions about variance (not necessarily homoscedastic), and our objectives (to validate a mechanistic representation of Nature) are different from those in classic DoE. In order to perform the tests for bias and curvature there need to be enough data throughout the entire *x*-range (not just at high, low, and medium levels), and the *x*-ranges need to be large enough so that the influence on *y* is large relative to the experimental variability. If the data are located at only three levels or if there is a large intermediate void in the data, those aspects do not permit investigation of the intermediate regions. The model may be wrong, but the absence of data hides the fact. Data need to be taken so that the model can be critically evaluated. Data location is one difference between classic DoE and the experimental plan recommended here.

Screening designs and classical DoE procedures serve an alternate purpose than validation of mechanistic models, but they do have the convenience of pre-experimental formulation of a definitive plan and simplicity of communicating with others. To validate mechanistic models, you need to be the Devil's Advocate. Choose the experimental conditions to critically test the model. As data evolves, your understanding of the model will evolve. After obtaining each data point, work-up the data, update the model, and consider where new data should be located to best test the model. Do not wait until the end of the experimental plan to work-up the data. The evolution of the experimental plan is a second difference w.r.t. DoE.

Additionally, the range of x needs to create a range of y that is not masked by the experimental variability, so that the influence on y can be definitely assessed, relative to the experimental uncertainty.

18.4 Experimental Design – A Plan for Validation and Discrimination

18.4.1 Continually Redesign

This diversion may be a useful and entertaining revelation. It is captured by the sentence, "You don't know what you don't know." It is a seeming silly statement, but what it means is that you don't even know that you don't know it.

Here is an analogy. If you plan a path though a never-visited forest to find the Elves' treasure, your plan is based on what you think lies in the forest. However, if you think the treasure is near the ridge in the North, but it is really in the South, your pre-plan will lead you to explore an unimportant area. Worse, if you don't know there is a lake or cliff or hill along the way, then your pre-plan path may take you to barriers to completing the project. Accordingly, you don't resolutely follow the pre-plan, the imagined reality, the pre-conceived notion. Every step into the forest gives you knowledge and you use the developing clues to better understand the trickery of the Elves and their cleverly worded misdirection in the Legend of the Treasure. As you become aware of what you did not know, you revise the forest exploration plan accordingly.

Your map app even does it. It first maps out a driving plan to get from A to B, but if you encounter a detour, or if you decide to take a side diversion, it does not yell at you, "Get back on the original plan. Don't deviate!" It pleasantly voices, "Recalculating" and redefines a new best path to the destination.

You know that.

Do this with your experiments. After each data set is collected, after each run, totally work up all the to-date data. See the implications for the model coefficients. See the data locations on the final graphs. Compare results to expectations on variability, data patterns, and so on. This will provide insight on your understanding of the process, model, experimental procedure, and so on. Use this developing knowledge to understand the "secrets of the Elves" and the misdirection they placed the messages you initially had about the location of the Treasure.

Do not take all of your data and then afterwards begin work-up. Once you have three experiments you can begin to see the relations on the graph. This reveals clues to the Mystery and what to do next. It may seem like an interruption to the mission of running experiments, but the mission is not to follow the pre-plan, it is to develop knowledge.

You don't know what you don't know. Don't let your naivety about the process, model, and experimental procedures misdirect your search. The Elves will watch, smiling, as you waste time searching in fruitless regions. However, your boss will prefer that you get useful results in the most efficient manner.

18.4.2 Experimental Plan

What are the attributes of useful results and efficient manner? We want:

- Enough data to be able to critically test the model, but not an excessive amount.
- Data over a wide range and with no void spots to be able to critically test the model.
- The impact of uncertainty on input values or sampling time to be insignificant relative to the uncertainty on the experimental response value, so that vertical least squares can be used as the regression objective function.
- A randomized sequence of trials so that response to controlled variables (inputs, designed conditions) do not correlate with changes in environmental influences (experience, warm-up, ambient pressure, etc.).

Accordingly, here are rules I suggest to guide experimental planning for model development:

1. Take enough data to be able to make strong (confident, defensible) evaluations. Use the maximum of:
   (a) Fifteen or more independent experimental runs, the minimum required to have enough substance for the bias and sequential autocorrelation tests.
   (b) Three or more independent experimental runs per each coefficient value to be adjusted by regression, enough to create confidence on coefficient values.
   (c) Large enough N to obtain desired model precision – based on your choice of Equations 18.9 to 18.12.
2. Place data throughout the range of each variable (y and each x) not just at the high, middle, and low x values.
   (a) In thinking this you will probably select x values uniformly within the x-range. It does not have to be a uniform interval. Values could be at randomized intervals.
   (b) An exhaustive experimental design is not required. For example, if there are 3 independent x variables and it is desired to have 20 levels for each, it is not necessary to have 20^3 experimental runs. Randomly pair the x values. For example, place 2 of each of the 20 values in Hat #1 for x_1, 2 of each of the 20 in Hat #2 for x_2, and so on. Then draw one value from each hat, the set representing one experimental condition. This randomly defines 40 experimental conditions (trials or runs) covering 20 values in each variable. Of course, you do not need hats. Random number generators can generate the pairing. Nor must it be 20 values and 40 trials.
3. The random pairing might miss combinations of interest, such as high values of all variables, so feel free to shape the randomization with certain selections. Unexpected nonlinearities in the process model and data model may cluster all points in the high (or low) range and leave sparsely populated y values in the other range or in a region of rapid

change. Therefore, as the experimental results evolve, add additional sets of input conditions designed to fill in the empty spaces, questionable spaces, or locations of a high rate of change, on the y-variable values.

4. Seek a range on each x that produces a great enough influence on y so that the resulting range on y is much greater than the experimental uncertainty on y, permitting a critical inspection of the x-functionality in the model. Exactly what this range is may not be known until some of the experimental data reveals the sensitivity of y to each x and uncertainty on y. Equation 18.13 reveals a desire that the range on an independent variable produces a variation in the dependent variable that is 10 times the standard error. Note that both the range on x and the number of trials are DoE choices:

$$R_x \geq 10(3\sigma_y/\sqrt{N})/\left|\frac{\partial \tilde{y}}{\partial x}\right| \tag{18.13}$$

5. Randomly sequence the individual experimental runs. Preferentially, do not let experimental sequence (chronological order, run-time) be correlated with one of the variables. Consider this example: the height of an able person has no effect on their ability to measure length. Suppose people were organized by height, shortest to tallest, and asked to measure the length of the same piece of wood with a metal ruler. There should be no trend of a measured length w.r.t. the person's height. However, suppose the relative humidity was changing and the wood was progressively shrinking during the measurement sessions. Then a graph of measured length w.r.t. an individual's height would show a decreasing trend, an erroneous artifact, a misleading finding, a misrepresentation due to the correlation of an experimental condition (height of the individual) with an uncontrolled variable (relative humidity). If the selection was tallest to shortest, then the apparent trend would have the opposite slope. Randomization prevents changes in experimental technique, drifts in uncontrolled variables, progressive improvement of technique, and so on, to be correlated with input variables, which could make a right model appear to have a skew trend.
   (a) Of course, this preference may have to be overridden if changing one variable takes significant time, effort, cost, or waste generation. Then cost, speed, and convenience could override the random sequence desire.
   (b) I would select the initial trials to represent data near to the extremes to best develop an initial understanding and models for subsequent reshaping of the experimental plan.
   (c) At the end of the planned runs, the data might be redirected to regions that have become of high interest.
6. Calibrate measurement devices that you will use to obtain both the y and the x data.
   (a) Calibration is a small cost relative to the cost of the experiments. If the measurements are wrong, then the experimental investment is devalued.
   (b) Be sure that the measurement technique minimizes both bias and variability (accuracy and precision) of the experimental error. You may want to use a more precise measurement device or increase the sampling rate or sample size than those devices or procedures used for normal plant monitoring and control. This also would include measurement, y and input, x, variability.
   (c) If there are several instruments, labs, technicians, or historical periods in generating or analyzing data, be sure they are internally consistent. Perhaps test controls and bias one set of data by the difference in results.

7. Choose measurement conditions so that the uncertainty on the measured y values are inconsequential to an in-use desire.
   (a) Use Equations such as (18.9) to (18.12) to define the number of trials and conditions.
   (b) Be sure that steady state is achieved for SS models.
   (c) Run long enough at SS so that averaging of noisy measurements provides the desired precision.
   (d) Run long enough at SS so that the averaging window includes the expected uncontrolled influence vagaries.
   (e) Do not let samples spoil (oxidize, agglomerate, cool, react, evaporate) between sampling and analysis or between preparation and use.
8. As experimental evidence evolves, adjust the experimental plan.
   (a) As constraints are revealed that prohibit particular pairing of input conditions, choose new constraint-free parings. Attempt to retain the desire to have no voids and a large range-to-uncertainty in the y and x data.
   (b) As empty spaces are revealed in data plots, add experimental conditions to fill them in. Look at data organized by y and each x. Even if data are uniformly distributed in the x-range, nonlinear relations could leave voids within the y-range.
   (c) As regions are revealed in which there is much change in y relative to x, shift the experimental plan or add runs to place conditions preferentially in the regions of higher change.
   (d) As regions are revealed in which there is little change in y relative to x, shift the experimental plan to add diversity of results or delete runs to lessen effective duplication.
   (e) If developing results seem to indicate that there might be a transition where y behaves differently before and after (a slope or level discontinuity in the y w.r.t. the x plot), then add points near the transition point.
   (f) As the sensitivity of x on y is revealed, change the range of any x-variable to ensure its influence on y can be definitely observed relative to the noise on y.
   (g) As data variability (noise, uncertainty) is revealed, change (increase or decrease) the number of experimental runs. Increase the number if variability seems high enough to mask trends. Decrease if the trends can be clearly seen with fewer data.
   (h) As developing data patterns in the residual w.r.t. y and x graphs raise suspicions of autocorrelation of the residuals (indicating mismatch of model functionality to data), place additional experiments in that range.
   (i) As regions appear where the residual values are large (relative to the other model-to-data differences), place additional experiments in that range.
   (j) Use bootstrapping to analyze the post-data model uncertainty. If uncertainty on the model is adequately low, stop the experiments. If not, add new experiments.
9. Do not attempt to get the input conditions exactly at the planned values. If the experimental implementation has approximately the right value, accept it and move on. Experiments take time and consume resources. If you are selecting a dozen x values from a range of low to high, it does not matter if a slightly different set were chosen. Set the knobs and if the resulting conditions are reasonable, use that data. Do not progressively fine-tune the knobs to get the input conditions to exactly match the planned values.
10. However, control the input conditions to a value that remains constant in time. For example, if you are interested in the flow rate response to a heat transfer coefficient and set the valve to a fixed position, the upstream pressure might change during a run,

which would have the flow rate change in the middle of the run. Since the flow rate is the process influence, control it, not the valve position, not something indirectly related to the process influence. As another example, consider the water temperature in a water heater. It rises when the heater is on and falls when cold inflow mixes with the in-tank water when the hot water is removed. If the heater thermostat is set to a single value, the water used in your experiment might actually be cycling by several degrees during a run.

11. Be sensitive to other humans. As you adjust the experimental plan, do not upset the humans in the enterprise by doing so. This includes operators who made preparations for the initial plan and managers who authorized the initial experimental design. Preview the how and why of the plan revisions.
12. Always be considering environmental, health, safety, and loss prevention (EHS&LP) issues. Although this chapter has a focus on experimental design to determine valid models, this cannot come at the expense of EHS&LP.

18.5 EHS&LP

The acronym represents environmental, health, safety, and loss prevention.

E – Environmental. Minimize resource consumption, waste generation, and pollution. Use small quantities. Turn off the process when in a thinking/discussion mode to conserve chemicals, materials, electricity, and so on. Do not let the process run (idle) between runs; start the next run as soon as possible. Wait for SS, but no longer. Choose materials with minimal environmental impact. Re-use material, do not discard after first use. Neutralize prior to putting in the drain. Cover samples to prevent evaporation. Leave equipment in a sealed, empty, and totally off state.

H – Health. In your handling and preparation of materials, take care to choose procedures that prevent chemical injury – ingestion, breathing, eye irritation, skin contact. Monitor the wellness of others – attitude, infection, cuts, bruises, behaviors, and so on. Know the life cycle of materials. (After you wipe a methanol smear with a tissue and toss it in the trash can, where does the methanol go? After the CO_2 leaves the column, where does it go? What are the health consequences?)

S – Safety. Work in a manner that is safe. Enjoy the activity and friendships, but not practical jokes or horseplay. No risky climbing, reaching, transfers, sitting, and so on. Eliminate clutter. Know where to find exits, showers, eyewash, fire extinguishers, and so on. Take a break when tired or unfocused. Remain alert to the lab surroundings (no distractions by texting or personal entertainment). Take care when moving to not bump things or other people. Give ample space to those carrying materials. Label all chemicals and materials – even if it is just water.

LP – Loss Prevention. Work so that things do not get broken accidentally or lost because they are misplaced. Keep data organized and traceable, so experimental effort is not "lost." Put tools and equipment back in their place. Take care to not drop, break, bend, tear, waste. Clean it when you dirty it. Keep the lab secure. Do not place glass or full containers near the edge where they can get bumped. Things that can roll, like thermometers, will, and will try to break themselves. Do not let them. Be sure that you have a visual/sound assurance that things are ready to go to the next step (for example, that cooling water is on before starting to heat the reboiler, that there is no liquid in the line before you turn on the compressors, etc.) so that equipment is not damaged by your operation. Cover samples to prevent contamination by CO_2 or dust or microbes. Keep degradable chemicals refrigerated. Use minimal amounts of materials and numbers of replicate testing to reduce costs. When running low on a consumable, write it on the shopping list, so that others can maintain the inventory needed to keep operating.

18.6 Visual Examples of Undesired Designs

Figures 18.1 to 18.3, from my PhD thesis, and Figure 18.4, from work reported by Bray and Rhinehart (2001a), illustrate some of the problems associated with experimental design. First, Figure 18.1 is what we hope a model-to-data comparison looks like. There is a high range of the response variable relative to variation and data are centered on the 1:1 line without evidence of skew or curvature (the distribution of "+" deviations match those of "−" distributions, which are independently distributed) and that variability seems independent of range.

Contrasting that good example is an undesirable pattern of Figure 18.2. Because of the missing data in the region of about 230–260 mol/h, there is no data to test for either curvature or model fit in that region. In looking at the figure, you might imagine that the data points in the 300 mol/h region suggest a horizontal trend, which curves downward to intercept the data in the 200 mol/h range. Having insufficient data to reject the model does not affirm the model. A good experimental design will not have voids in the data; it will provide data in patterns that would permit critical model evaluation.

In Figure 18.3 the vertical variability is about 0.5 mol ratio. However, the range (2.5−1.5) is about 1.0 mole ratio. This makes the range to variability only have a 2:1 ratio. We would like to have a 10:1 ratio to be able to critically test any trends. Here, the range is not enough to critically test the model. One could report that the model predicts the H_2/CO ratio on average, but there is no evidence that it can properly describe the trend in the ratio.

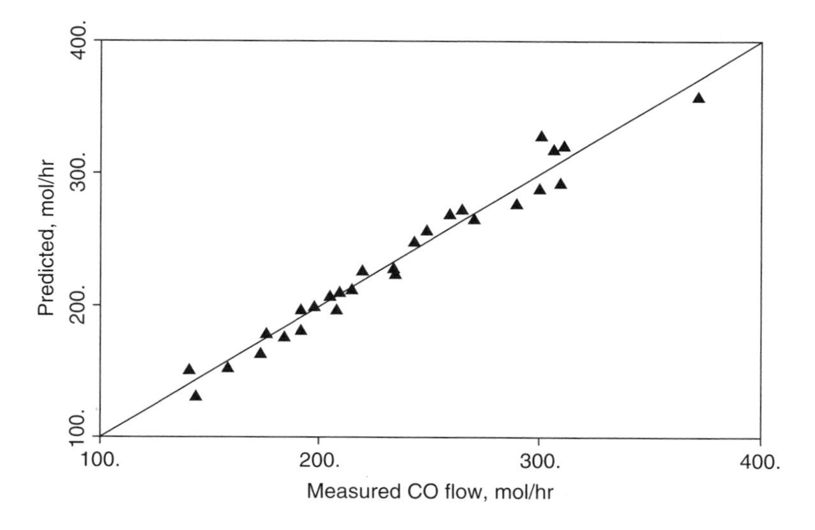

Figure 18.1 Desirable data patterns.

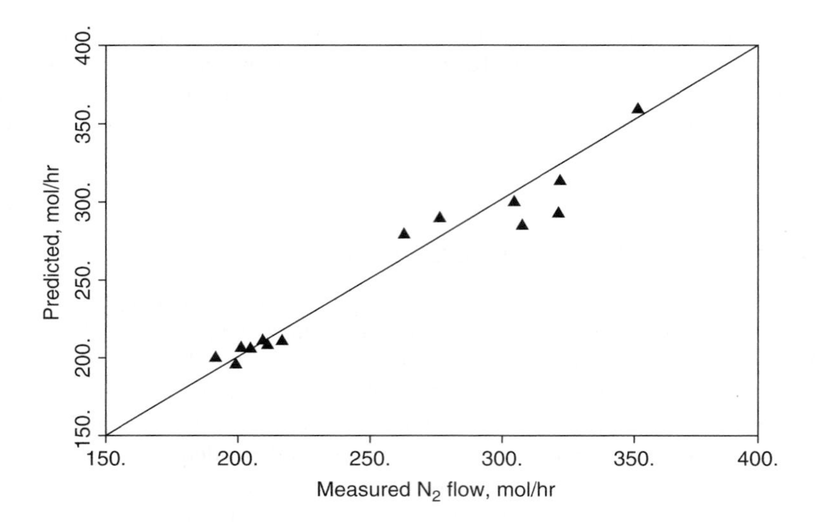

Figure 18.2 Regions of missing data.

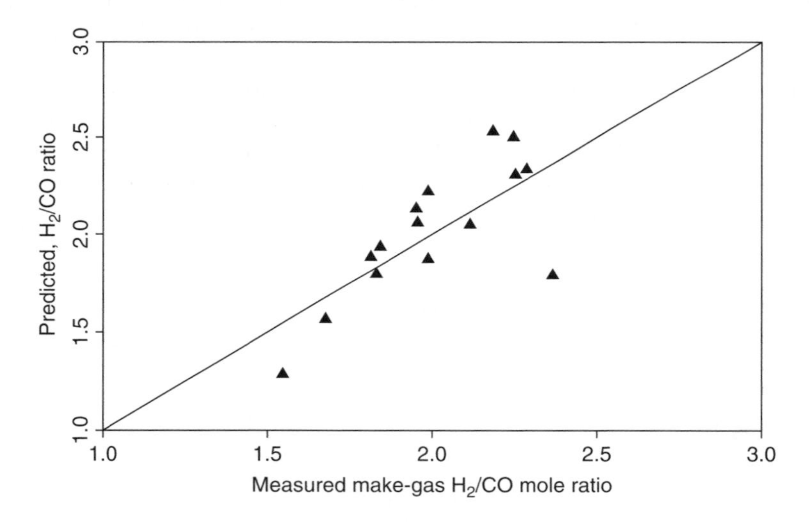

Figure 18.3 Range is not large relative to variability.

Figure 18.4 is from a Bray and Rhinehart study (2001a) of the influence of radio frequency (RF) power to a gas, creating an oxygen plasma used in the etch stage of semiconductor manufacturing. There were three distinct power settings on the production reactor, undesirably limiting research investigation. It prevents disclosure of the etch rate trends in between.

Although the variability seems large relative to the range, notice that the residual and top power scales are different. Experimental evidence indicates that the standard deviation on the etch rate is about 250 Å (angstrom)/min, and the model indicated that sensitivity of the etch

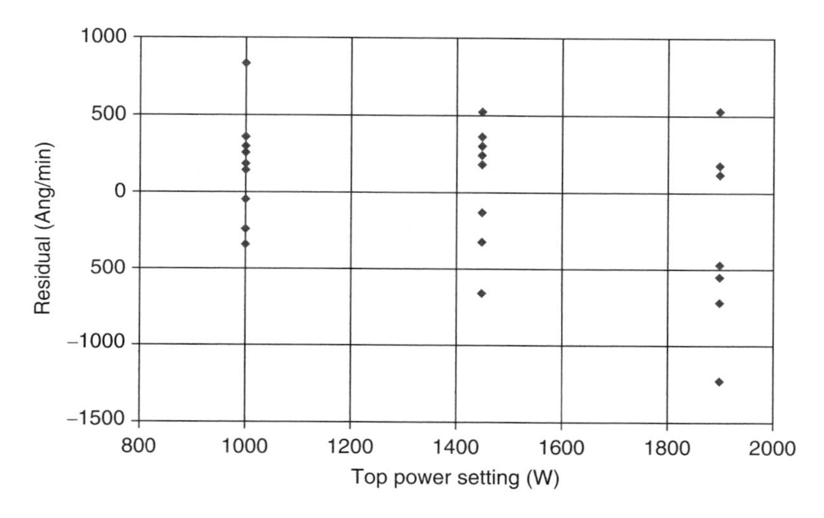

Figure 18.4 Levels in experimentation prevent evaluation of in between behavior.

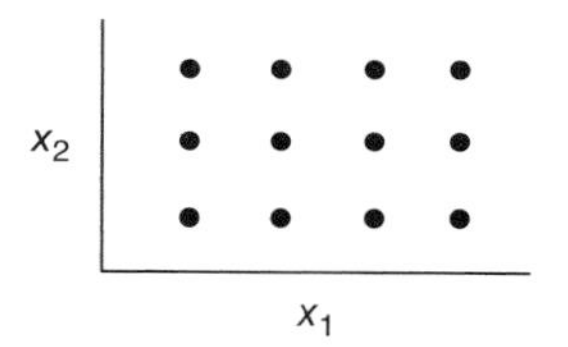

Figure 18.5 Array.

rate to RF power is 2.969 A/min/W. Accordingly, Equation 18.13 indicates that the desired range on RF power should be at least about 850 W, which it is.

18.7 Example for an Experimental Plan

Consider that you may be changing two variables as experimental influences. For instance, flow rate, x_1, and temperature, x_2, of water have an effect on the heat transfer coefficient in a heat exchanger, y, and you want a model of $y(x_1, x_2)$. You will adjust the flow rate by setting the valve position (Knob 1) and temperature by setting the rheostat position (Knob 2) of an electrical heater. You might plan for a dozen experiments and initially have them organized with four x_1 values and three x_2 values, and your plan might be to operate at the conditions shown in Figure 18.5 that cover the entire operating range.

However, what sequence do you run them? Figure 18.6 indicates several choices. In (a) the runs are sequentially, logically changed. One variable, x_2, is held constant, while just one variable, x_1, is changed, but this means that between Runs 4 and 5 and 8 and 9 two variables are changed. Panel (b) is more efficient in that only one variable is changed at a time, and is better than (a) because the forward and backward sequence on x_1 will tend to eliminate correlation

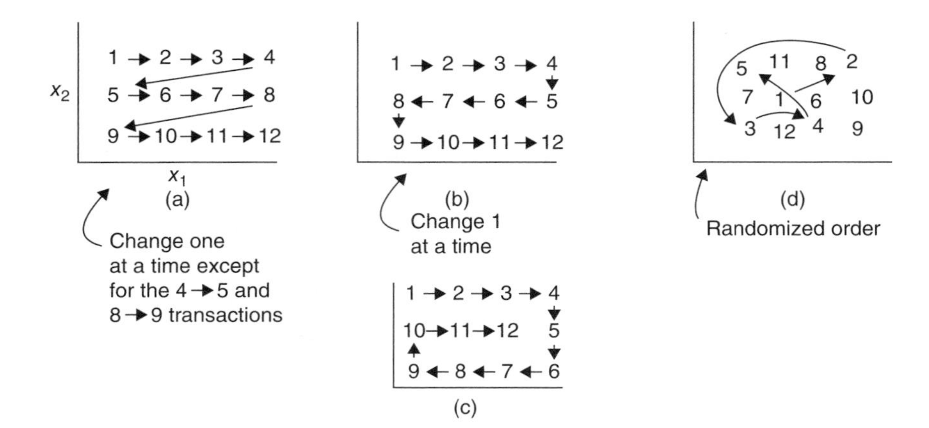

Figure 18.6 Array run order.

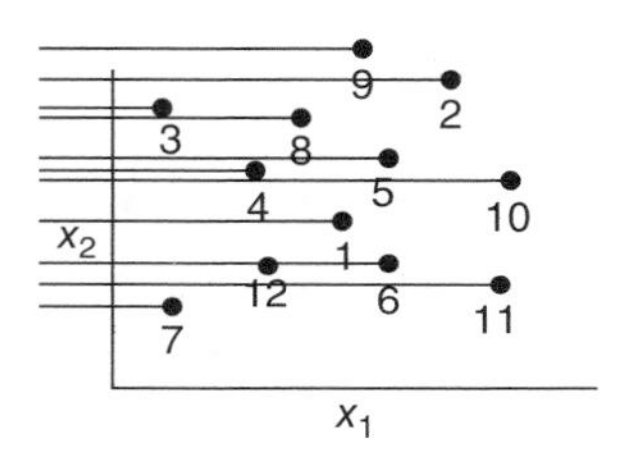

Figure 18.7 A better plan.

to uncontrolled variables. However, the value of x_2 is correlated with time, so plan (c) is a step better, maintaining a change one at a time sequence but somewhat overall preventing a correlation to time except for the 1–2–3–4 and 4–5–6, and so on, runs. A better plan for removing time-dependent correlations from confounding the results is (d), which randomizes the run sequence. However, this might not be preferred for human tracking of the progress or considering the transition time and cost.

However, that plan only permits three x_2 values, which does not provide critical evidence for testing the model. Figure 18.7 randomizes the data throughout the range, providing 12 x_1 values and 12 x_2 values for testing, in a randomized order, with only 12 experimental runs. This is a better plan.

The next level of consideration would be to override purely random x_1 and x_2 locations to ensure that the vagaries of randomization do not create voids, or combinations that are uninteresting or constrained, and to sequence the selection of the initial half of the runs in order to be sure that the entire region of interest is covered in the first few runs. Then let progressive insight redirect the location of the remaining trials. Figure 18.8a indicates the first six runs (solid circles), the original tentative plans for the remaining runs (open circles), and the revised plan (solid triangles) to place data to eliminate voids in the y-response range are not shown. Figure 18.8b indicates that the original plan would have placed very few data points

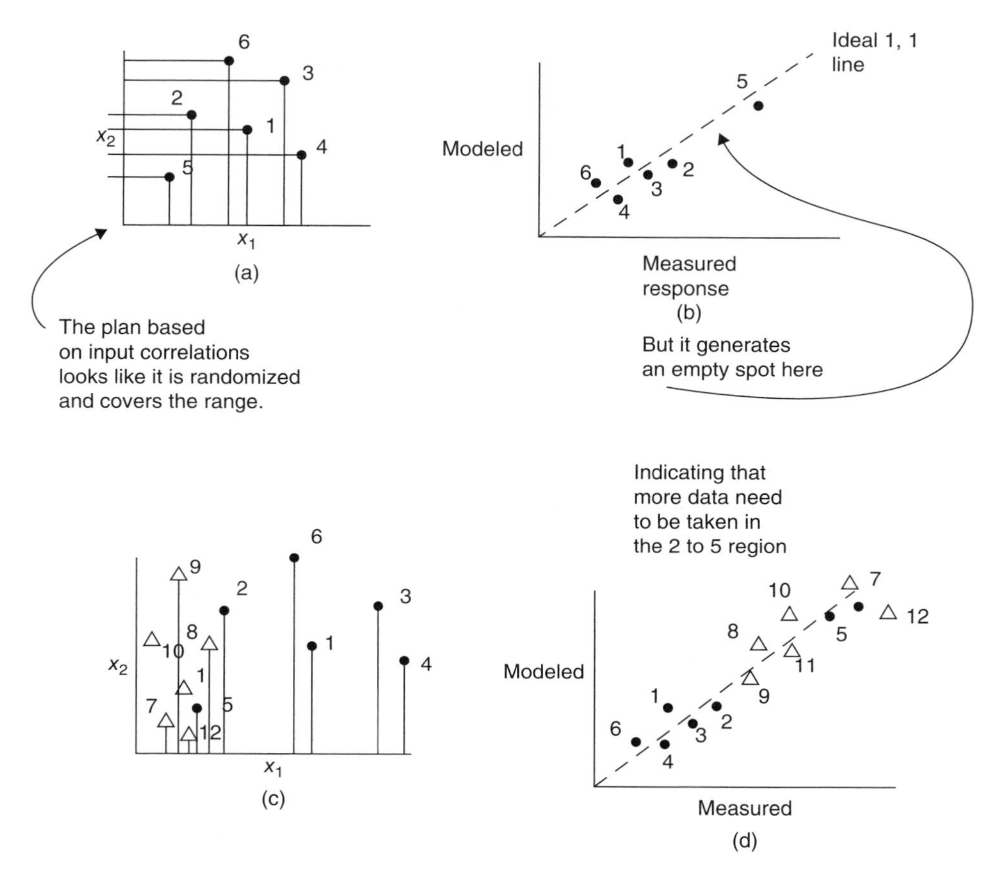

Figure 18.8 Experimental plan evolution: (a) the initially desired plan; (b) parity plot reveals voids; (c) revised plan adds data (triangles) to fill the gaps; (d) parity plot with added data shows full coverage.

in the upper y-range and that the revised plan fills in the voids even though it has a preference for low x_1 values. The experimental sequence remains randomized.

Finally, the experiment is run by adjusting the valve position and potentiometer to get the desired flow rate and temperature values. Some knobs are turned to achieve the planned x_1 and x_2 values. Perhaps the valve and potentiometer are set to 19 and 92% to attempt to get a particular planned situation of a low flow rate of 2.2 gpm and a hot temperature of 143 °F. However, suppose the resulting conditions were 2.1 gpm and 146 °F? That is as good a point as the target 2.2 and 143. Take it and move on. Do not try to fine-tune the knobs to get the data point spot on.

18.8 Takeaway

When seeking to validate models, design experiments that provide data intended to critically test the models. Use the direction in Section 18.4.2. Classical DoE for screening trials may not provide useful data for validation. Use the direction in Section 18.2 to choose the measurement type and number so that measurement uncertainty from data models has an inconsequential influence on regression results. Work-up data as each point is obtained and let the progressive understanding reshape the remainder of the experimental plan.

Exercises

18.1 Describe what is good and bad about this experimental plan. Hold x_1 at a central value and do seven x_2 values; then hold x_2 at a central value and do seven x_1 values.

18.2 The model is $y = a/(x + b)$ and the regression parameters are a and b. The actual values of both the independent variable, x and the dependent variable, y, have uncertainty. Values will be experimentally measured and those data pairs used to determine the best a and b values and uncertainty on a and b will compound the in-use uncertainty on y due to x. Determine the balance between the number of independent data sets and the value of experimental uncertainty on a measurement of y so that the uncertainty in a and b from the calibration data is inconsequential to in-use y-uncertainty due to x.

18.3 Where should the next data point go? Choose a model that is nonlinear in the $y(x)$ response and a plan that has x values uniformly distributed throughout the x-range. Show the impact on the y values and revise the x-data plan to provide more uniform y-coverage.

19

Utility versus Perfection

19.1 Competing and Conflicting Measures of Excellence

One view of a measure of model excellence could be its ability to describe Nature. A *true* or *right* model represents the unknowable truth about Nature. However, looking at the history of human attempts to understand Nature, I do not believe that it is possible to ever be sure we have perfectly understood Nature. I do not think we can ever claim that we have truly identified the model. I do not think we ever understand Nature well enough to get perfect knowledge.

With using closeness to perfection as the measure of excellence there are two categories for a model. A *wrong* or *incorrect* model would be rejected by either a data-based or logical validation. By contrast, an *accepted* model would not be rejected. Acceptance does not mean that the model is right. It just means that it is close enough to the truth that there is not enough evidence in the data or logical analysis to reject it. With additional data, or with data of finer precision, or with additional tests of the model, we might find that it can be rejected.

However, our aim in developing and validating a *true* model is to obtain that elusive perfection. Our model will be wrong, I think, but the data might not reject the model when the model error is masked by experimental variability. Therefore, we tentatively accept that the model is right until we collect enough data with enough precision to finally reject it.

By contrast, one could use utility as a measure of excellence. A *useful* model will sufficiently describe the data so that the model meets all criteria for functionality in use. The functionality criteria may include rapid computation, no possible execution errors, simple (to understand, transfer, code, and adapt), robust, and accurate enough to be accepted for design or control.

A linear model could very well be chosen as a good model to use even though it may be acknowledged as not representing the underlying $y = f(x)$ behavior. Utility aspects of simplicity and adequacy of the answer may override perfection. Due to its having an explicit inverse, the ideal gas law may be selected over a Ping–Robinson equation of state in the preliminary process design stage.

Nonlinear Regression Modeling for Engineering Applications: Modeling, Model Validation, and Enabling Design of Experiments, First Edition. R. Russell Rhinehart.

19.2 Attributes for Model Utility Evaluation

There are many aspects of utility in use. Here are eight:

1. *Simple*. The perfect model may require iterative procedures to solve implicit relations, making it difficult to solve. We often prefer a simpler model with explicit relations where we can guarantee that a solution will be possible and that the solution will be obtained within an upper limit of computational time. Will the model be simple enough to be implementable?
2. *Cost effective*. Rigorous models, those seeking perfection by including all phenomena with the most comprehensive of constitutive relations, take a lot of effort to develop and to determine coefficient values. The cost of model development may not be justified by the benefit of approaching perfection. Is the model cost effective? Will the benefit it provides have an acceptable future pay-out?
3. *Relevance*. Each model coefficient must be important. If the uncertainty on a model coefficient includes zero, then that model coefficient should (probably) be set to zero or its functionality (its role, the concept it represents) be deleted. If changing the model coefficient value over a reasonable range has an inconsequential impact on the model result, then that term should (probably) be dropped. (However, if a significant human, such as a customer, boss, or representative of the institutional way wants the term to be included, it is probably best to keep it in.) If a term is dropped, you should re-do the regression optimization to re-evaluate the remaining coefficient values.
4. *Accuracy*. Although an imperfect model will not match the data throughout the range, the deviation may be inconsequential in light of the use of the model. For example, in process control we typically model high-order nonlinear dynamic processes with linear first-order plus dead-time models. If a model is only 75% right, then it is 25% in error, but at each sampling the controller seeks to correct the error. Therefore, as a very coarse estimate, after 10 control actions the error has been reduced to 25% of 25% of 25% … to about $0.25^{10} \approx 0.000\ 000\ 95$, which is inconsequential to the impact of the unknowable but continually changing environmental disturbances. What level of accuracy is required of the model within its application? In control, nearly always, sufficiency trumps perfection.
5. *Precision*. Consider propagation of uncertainty. The model-predicted value will be uncertain due to both coefficient uncertainty and input data uncertainty. You could apply propagation of maximum error or propagation of probable error as elementary ways to estimate combined uncertainty from both coefficients and inputs. If coefficient uncertainties are not known, use bootstrapping to estimate model uncertainty due to data vagaries and then use that instead of propagation of uncertainty from model coefficients. Use a Monte Carlo approach if the underlying distributions are not Gaussian (normal) or if there are interactions or conditional probabilities. The level of uncertainty would be likely to change with the x values, so explore the entire range. Evaluate the impact of propagated uncertainty on

the model use. Is the uncertainty acceptable within the context of how the model output will be used? Devise experimental plans to sufficiently determine coefficient values, as opposed to seeking the exactly true value for your incorrect model.

6. *Ease of use*. Is the model invertible? We usually develop models to describe how Nature responds to an influence. However, in engineering use, we use the models to determine the operating conditions or device design we must create to make Nature behave the way we wish. We have to solve the model for the influences that produce the desired response. This is the inverse, the opposite of using the model to see how the response responds to the influences. Is there a unique solution to the inverse? Is the solution bounded throughout the range? Is the input space feasible throughout the range? The model design must provide a functional inverse.
7. *Flexible*. Is the model adaptable? Processes change. How will you adjust the model to match the new catalyst reactivity, heat exchanger fouling, ambient heat losses, friction losses, cold-work hardened spring, temperature-affected resistance, air density, and so on?
8. *Multi-purposeful*. Can the model serve alternate objectives? Is it the "ONE" model that can be used in design, control, process performance analysis, inferential sensing, data reconciliation, operator training, and plant-wide optimization? Often, for the same process, control will use first-order plus dead-time (FOPDT) or finite impulse response (FIR) linear dynamic models, while performance analysis and data reconciliation will use first principles steady-state models, inferential sensing may use a neural network (NN) model, and optimization may use one vendor-sourced black-box rigorous model and design another. Each model needs to be understood and maintained (updated when the process changes), and if they represent diverse software environments, modeling approaches, and access by the user, the diversity of styles and codes will confound ease of success in use.

You can probably think of other features of engineering utility that would override scientific perfection. In general ask, Will the model be acceptable to all stakeholders using it as it is implemented? How will your customers use the model? What will they like or not like about it?

The answers to the above questions depend on the application context.

A *useful* model balances perfection with sufficiency. Sufficiency is not defined through mathematical, statistical, or scientific values. Sufficiency represents the aspects of utility of use. The industrial advisors for our ChE program have clearly told us professors, "Teach the students to balance perfection with sufficiency."

19.3 Takeaway

Balance model utility with perfection. Know how your customers will use the model and what attributes they will say supports utility (ease of use and functional adequacy) and perfection (their perception of the truth about Nature).

Exercises

19.1 Discuss the balance of sufficiency and perfection, and explain the reasons that diverse users would argue to explain why Model A is best or Model B is best.

(a) Neural network versus first principles.
(b) Vendor software versus autoregressive moving average (ARMA)
(c) Differential equation versus FOPDT.
(d) Rigorous versus first principles.
(e) Fuzzy logic versus power series.

20

Troubleshooting

20.1 Introduction

You may encounter any number of diverse issues in nonlinear regression that may be related to either regression (optimization), modeling, or experiments. This chapter summarizes many and provides advice to solve them.

20.2 Bimodal and Multimodal Residuals

When the data reflect disparate procedures, operators, data sources, test protocols, calibration standard sources, categories, or treatments, which are not an input for the model, then the data may be internally inconsistent. One data source might provide values that are consistently above the other. The model will seek to go through the middle of the data and one set of data will be generally above and the other generally below that model. If the data were internally consistent, then all sets will have the same distribution from the model, which is expectedly Gaussian. However, if the data represent disparate sources then there will be two distributions, each expectedly Gaussian, but combined, will reveal a bimodal distribution.

If the data are internally consistent (same operator, same instruments, same process) then, except for a possible outlier, the data will provide a single representation or the process. However, if the data were generated by operators with different techniques, or different lab instruments that may not be identically calibrated, or from the same process but at time periods before and after a change, or from ideally identical processes that are not actually identical, then the data set from one treatment will not exactly match that from another and there will be several different treatments. If there are several different treatments or combinations of them, then the distribution of residuals may be multimodal – composed of several, not just two, distributions.

The bimodal residual distribution may be visually obvious when either looking at the residuals from the model or at a histogram of the residuals. However, there might not be enough data or enough separation in treatments to make it strikingly visible and it might appear as a somewhat broadened, but otherwise normal, distribution.

Nonlinear Regression Modeling for Engineering Applications: Modeling, Model Validation, and Enabling Design of Experiments, First Edition. R. Russell Rhinehart.

To test for normalcy in the residual distribution use the Kolmogorov–Smirnov test, as explained in Chapter 4. Alternately, to test if the two treatments are identical use a chi-square contingency test. Each treatment should have the same fraction of "+" residuals and "−" residuals. If the null hypothesis (that there is no difference in treatment) can be rejected then you should consider the following actions.

- Segregate the data into the before and after, or treatment categories, and generate separate models for each.
- Correct the rogue data set to values to be consistent with those of the preferred data set and regenerate the model. (Infer a calibration correction that makes the bimodal data appear to be consistent.)
- Add a term to the model to account for the treatments.
- Improve training of operators and technicians.

20.3 Trends in the Residuals

Residuals, the process-to-model differences, should not correlate to the input data, output data, experimental sequence number, experimental run time, or any variable. If there is a trend in the residuals w.r.t. any variable, then it indicates a problem. Use the r-lag-1 test of Chapter 4 to test for trends in the residuals.

- If a trend is noticed w.r.t. the chronological order, then perhaps the data are influenced by an unmodeled, uncontrolled disturbance (such as ambient temperature, humidity, pressure, experience, progressive sloppiness, or instrument warm-up). You can minimize this by running experiments in a randomized order, but if a trend is detected, it means that the unaccounted for uncontrolled variable has a significant impact. Consider including it in the model or changing the experimental procedure to eliminate the problem.
- If it is a steady-state model and a trend is noticed w.r.t. an input variable, this is an indication that the influence is somewhat different from how it was modeled. Consider revising the model.
- If it is a steady-state model and a trend is noticed w.r.t. run wait-time for SS, then perhaps some of the shorter runs did not reach steady state. Consider repeating the experiments using a longer settling time.
- If it is a dynamic model there is not a unique relation between influence and response. Do not investigate residuals w.r.t. input values. Only investigate residuals w.r.t. time and the output variable.

20.4 Parameter Correlation

In parameter correlation, the value of one model coefficient (parameter) is correlated to another. You can only detect this if the regression optimizer is run many times from randomized initial values, leading to many converged solutions. For each solution, plot one coefficient value w.r.t. another coefficient value. If there are N decision variables (DVs) then there are $N(N-1)/2$ possible coefficient pairings that should be explored.

- If there is a distinct trend, then it could be that those model coefficients are redundant. If redundant, many combinations of the coefficient values have the identical net influence on the model and all seemingly different solutions will return equivalent objective function (OF) values.
  - If a reciprocal relation, then the coefficients are effectively multiplied in the model.
  - If a linear relation with an intercept of zero, then the coefficients are effectively divided (used in a ratio) in the model.
  - If a linear relation with a non-zero intercept, then the coefficients are effectively added or subtracted in the model.
- If there is a distinct trend then, alternately, it could be that the OF w.r.t. the DV surface has a steep valley with a small slope to the bottom and the optimizer stops prior to reaching the optimum. If so, consider tightening the convergence criterion or using a different optimizer that is more appropriate for the topology.

20.5 Convergence Criterion – Too Tight, Too Loose

The classic convergence criterion for optimization is to stop iterations when the changes in the DV moves are smaller than some value, the convergence threshold. The user choice for the threshold might be too large or too small.

If the convergence criterion is too loose, meaning that the thresholds to claim convergence are too large, then the optimizer will stop in the vicinity of the optimum, but prior to finding the precise optimum. If run many times, with random initializations, the optimizer will approach the optimum from many directions and a graph of one DV value w.r.t. another will be a scatter plot.

- If parameter correlation graphs reveal a scatter plot, consider making the convergence criterion smaller. However, it might be that it appears to be a scatter plot on the scale that you are viewing it.
- If the multiple trials report similar OF values that seem to be noisily representing one true value, then consider making the convergence criterion smaller. However, it might be that it appears noisy on the scale that you are viewing it.

By contrast, if a convergence criterion is too tight, meaning that the thresholds to claim convergence are too small, then the optimizer will invest excessive work in seeking to excessively fine-tune the trial solution.

- Monitor the OF and DV values with iteration. If, after a period of noticeable OF change, there is no noticeable improvement with optimizer iteration, then consider making the convergence criterion larger.

I like the alternate convergence criterion offered in Section 12.6, to observe the rms of a random subset of data w.r.t. iteration and claim convergence when there is no detectable improvement relative to the inherent variability in the random subset rms values – the steady-state convergence criterion.

20.6 Overfitting (Memorization)

An N-coefficient polynomial can perfectly fit N data points (ideally). However, if there is no trend or just a simple trend (linear perhaps) and the data are noisy, then the N-term model will not be representing the trend; it will be fitting the noise. With a different data realization, the noise will be different and the N-coefficient model will be different, even though the true model should not change. The model will reflect noise and not represent the truth. This is called overfitting or memorization. It is similar to a parlor trick: You can teach a 3-year-old to say "four" then when friends are over ask your kid, "What is the third decimal digit in pi?" and impress your friends with how smart your kid is, but it was just memorization without understanding. The kid might actually be saying "fore" as heard on the golf channel.

Classic measures of regression, such as r-square improve with each additional model term, often tempting a researcher to put excessive terms in the model, to the point of beginning to have the model fit noise, not true trends.

- One traditional heuristic in statistics is to have at least three data sets for each adjustable model parameter.
- Another is to use the final prediction error (FPE) (or any of several similar criteria) to test when the increase in complexity overshadows the improvement in fit (see Section 13.7).

20.7 Solution Procedure Encounters Execution Errors

Sometimes the optimizer will be led to choose DV values that produce any of several sorts of execution errors (divide by zero, overflow, log of a negative number, root of a negative number, subscript out of range, etc.). These would be constraints, but perhaps were not identifiable and not explicitly stated. Therefore, the optimizer might choose a trial solution DV set that encounters the execution errors. Then it stops with no solution. If this happens:

- Change initial values and restart the optimizer. Approaching from a different direction might avoid the excursion into a constrained region.
- Add execution-error anticipation lines to your code to stop the execution to avoid the calculation and to return a flag indicating to the optimizer that a constraint was hit.
- Change the optimization algorithm.
  - Use backward rather than central difference methods.
  - Temper the jump-to distance for the next trial solution.
  - Choose a different optimization algorithm altogether. My preference is for direct search algorithms.

20.8 Not a Sharp CDF (OF)

Run the optimizer many times from randomized initializations and generate a CDF(OF) graph. If the optimizer repeatedly gets to the true minimum, then the cumulative distribution will make a distinct and abrupt jump at the minimum OF. By contrast, if the optimizer stops when in the proximity of the minimum, then the CDF will progressively rise with the OF value. Stopping in the proximity of the minimum could be due to several reasons:

- If the OF surface w.r.t. the DV value is characterized as a sharp valley (steep walls with a gently sloping floor) then the optimizer might be stopping prior to finding its way down the valley.
  - Consider making the convergence criterion tighter to make it keep going.
  - Consider changing the convergence criterion to another variable (on the OF instead of the DV).
  - Consider changing the optimizer algorithm to one that better matches the surface.
- If the optimizer is encountering a constraint, consider changing the optimizer to better accommodate the constraint. For instance, use a constraint-following approach such as the generalized reduced gradient (GRG) algorithm.
- If the optimizer is stopping on the side of a hill, consider making the convergence criterion tighter.

The visual interpretation of whether the CDF(OF) has a sharp or gradual rise at the minimum OF is partly dependent on the scale (range) of the OF axis. If there is a single and easy-to-find optimum, then all trials will stop in the vicinity of it, with variation due to the convergence criterion. In this case the CDF(OF) graph will not appear to be sharp, but the OF range will be small on the order of the convergence criterion.

20.9 Outliers

It is not unusual to encounter experimental data that are irregular, spurious, wrong, unconventional, not representative of the norm, or such. For properly controlled experiments, the results are reproducible and accurate. However, sometimes errors happen, like digit transliteration or sample contamination. Sometimes good people make an experimental mistake. The errors, spurious events, or outliers are a part of the natural process. If you wish to model the whole of Nature, exceptions as well as the rules, then do not discard outliers. However, more often we seek to model the normal or expected behavior and outliers should be discarded from the data.

To identify outliers, look at the residuals, the deviations from model to data.

- Construct a confidence interval using the rms deviation value with all data, and if a residual is beyond the critical value, then consider rejecting the data point. Use Chauvenet's criterion to set up the limits (see Section 14.3.1).

As a caution, a good experimental plan will not generate a lot of outliers. If several data appear to be outliers:

- Reconsider the experimental plan, it could be that there were influential uncontrolled events, or disparate treatments.
- Take more data in the proximity of the outliers.

Be aware that an outlier might indicate that the model is wrong.

- If the outlier is at an extreme value, consider redevelopment of the model.
- If several outliers in one region agree, consider redevelopment of the model.

20.10 Average Residual Not Zero

If the model matches the data, the average residual (data to model deviation) should be zero, but this needs to be zeroish w.r.t. variance in the residuals. Therefore, use a t-test for bias to see whether the null hypothesis, "average residual = zero," can be rejected. If the null hypothesis cannot be rejected, that does not mean all is well, but it is a good sign. However, if it is rejected, then it is likely that something was wrong with either the optimization or the model.

- Consider the optimization:
  - Perhaps convergence criteria was too coarse and it stopped prior to finding the optimum.
  - Perhaps it got stuck on a constraint.
  - Perhaps it was not looking at all of the data.
- If the problem lies in the model, there should be obvious trends in the residuals that may be able to reveal where to focus on model redevelopment.

20.11 Irrelevant Model Coefficients

You might have expected a variable to be an influence on the process and included it in the model with a particular function or architecture. However, that variable might be irrelevant; it might have no influence, or very slight influence, or the functional model might be incorrect. In this case, after regression, the value of the coefficient for the variable (or for the associated term) will be nearly zero relative to the uncertainty on the coefficient.

Bootstrapping can be used to reveal uncertainty on the coefficient. If the realization to realization range on the coefficient is large and encompasses a zero value:

- Perhaps the variable or term is irrelevant and should not be included in the model.
- Perhaps more data are needed over a wider range or with greater precision to be able to properly account for the variable influence.
- Perhaps it is an influence, but a minor one, and can be trimmed to make the model simpler while still adequately useful.

20.12 Data Work-Up after the Trials

Should you plan the experiments, run all of them and then work-up the data analysis? Only do so when the procedure has been routinized, when the method is a standard, or when the procedure and results are well understood.

When exploring, when developing a new model, work in parallel not serial. You don't know what you don't know. If you run the entire experimental plan and use up your allocation time on the equipment, then you are stuck with the data. Post analysis might reveal voids, an inadequate range of variables, excessive variability, and so on. If the trials are complete, you cannot use the insight from data analysis to guide an improved plan. Therefore, completely work up the data after each set is collected – create plots, generate the model, test against expectations, analyze uncertainty, look at patterns in residuals, and so on. See all of the results and let progressive insight guide the model redevelopment and the revised experimental plan (see Section 18.4).

20.13 Too Many *r*s!

There are many *r*-statistics, all expressing some form of sum-of-squared deviations (SSD).

- The regression *r*-squared is a ratio of the variance reduction due to a model. It is the SSD in the original data less the SSD in the residuals, scaled by the SSD in the original data. This is the classic *r*-squared value presented in regression software. It is a measure of whether the model provides some form of correlation to the data, but is not a measure of model validity. A linear model can provide a strong correlation to quadratic data.
- Some products present the square root of the regression *r*-squared value.
- A similar statistic is the *r*-squared-sub-*m*, which uses variance, the SSD scaled by degrees of freedom. It is the $SSD/(N - 1)$ in the original data less the $SSD/(N - m - 1)$ in the residuals, scaled by the $SSD/(N - 1)$ in the original data, where m is the number of model coefficients.
- The *R*-statistic for probable steady and transient state identification is a ratio of variances in a data sequence as measured in two ways: by the deviations from the average and by the sequential data deviations. It is not a measure of model match to data. It does not use the *r*-squared terminology; even though it relates to deviations squared and it is a ratio.
- The autocorrelation statistic, *r*-lag-1, is the sum of the product of sequential residuals, divided by the sum of squared residuals. It does not use the *r*-squared terminology; even though it relates to deviations squared and it is a ratio. I believe this provides a sensitive and revealing indication of patterns in the residuals. If the model expresses the phenomena and if the data are true, then the residuals should be randomly and independently distributed, when sorted by any variable. If there is a nonrandom pattern a clustering of "+" or "−" residuals, then there is evidence that the model is not expressing the data.

20.14 Propagation of Uncertainty Does Not Match Residuals

One can apply propagation of uncertainty to the experimental data to estimate the probable error in the experimental data. Then, after regressing the model to the data, one can get the 95% confidence limits on the residuals. If all goes as desired (if the experiment is well understood, and the model adequately matches the data, and the optimization finds the right solution, and there are no errors in the analysis) then propagation of uncertainty on the experimental data should provide a measure of data variability that matches that found in the residuals.

- If the propagation of uncertainty probable error is much larger than the 95% confidence interval on the residuals, perhaps the experimental procedure was not well understood or the model has too many coefficients and is fitting the noise.
- Alternately, if probable error is much smaller than the residuals, perhaps the experimental procedure was not well understood or the model is not matching the experimental mechanism.
- If the difference is large in one range and small in the other, it may mean that the uncertainty in the givens (the controls, the experimental inputs) needs to be included in the propagation of uncertainty.

20.15 Multiple Optima

There are diverse reasons as to why independent optimizer initializations would lead to different optima.

- If the surface has several local optima then random initializations will find one of the several optima and many independent initializations will repeatedly find the same few optima. In this case there are a few distinct optimum values and the CDF(OF) curve will reveal distinct steps. In my experience with nonlinear models with limited data, it is not uncommon to have a few local optima. To be confident in finding the best fraction of possibilities use the best-of-N equation to determine the number of starts. Also consider adding more data sets, especially at the extreme values or in locations of a high rate of change.
- If the surface has many local optima, the CDF(OF) will again show distinct steps, but there will be many. In my experience, this is not a natural possibility with continuum-valued variables, but likely an artifact of a coarse numerical discretization in the model solution procedure. Consider using a smaller time or space increment in the model solution to better approximate the ideal concept.
- If the CDF(OF) curve does not show distinct steps, it is possible that either the solution gets stuck on a constraint or that the solution is in a gently sloped ridge or valley with steep walls.
  - Check to see if a constraint is active. If so, consider changing it from a hard constraint to a soft constraint, or reducing the penalty, or switching to a constraint-following optimizer.
  - If the issue is a gently sloped ridge or valley with steep walls, then consider switching the optimizer to one that better handles the surface topology.

20.16 Very Slow Progress

If the optimizer requires many iterations and seems to be slowly approaching the optimum:

- Let it run overnight. It just might be a difficult topology.
- Switch to an optimizer that uses a distinctly different search logic.

20.17 All Residuals are Zero

It would be a rare event for the model to both perfectly match the natural phenomena and have zero noise on the data. If all residuals are zero (not just the average) then:

- It may be that the data was generated by the model, not by experiment.
- It may be that there are too many coefficients in the model relative to the data. Good practice has three independent data sets for each adjustable model coefficient. More data sets could be considered excessive and costly. Fewer could be considered as not providing enough characterization of the experimental procedure.
- Alternately, there are too few data sets. Take more data, aiming to have three for each adjustable model coefficient.

20.18 Takeaway

Do not expect the first try of either modeling or experimental design to provide the definitive right answer. Run the optimizer multiple times from independent initializations and use patterns in the results (parameter correlation, bimodal residuals, trends in residuals, inconsistent expectations from propagation of uncertainty, etc.) to reveal possible issues that will guide model and experimentation improvement. Do this total data work-up after each experimental set is obtained, while the experimental plan is being executed, and use progressive insight to refine the model and the experimental plan.

Exercises

20.1 Derive the three parameter correlation relations described in Section 20.4.

20.2 Derive the $N(N-1)/2$ quantity in Section 20.4.

20.3 Prove that, for linear regression, for example, $y = a + bx$, the average residual will be zero.

Part IV

Case Studies and Data

21

Case Studies

21.1 Valve Characterization

Although many power law like models can be easily solved with classic gradient based optimizers, it is not hard (in my experience) to encounter ones that generate problems. Here is a simple one that arose from a desire (in model-based control) to calculate the valve stem position required to produce a desired air flow rate. The model is derived from first principle pressure losses in the pipe and valve, with the valve characteristic (modified equal percentage) modeled as a power law relation:

$$u = a\left(\frac{F}{\sqrt{2783.2 - kF^2}}\right)^b$$

where u is the controller output (nominally the valve position) and F is the flow rate. The three model coefficients are a, b, and k.

Experimentally, we chose the u value (controller output), waited to steady state, and measured the flow rate. In use, however, the equation needs to provide the correct u value to obtain a desired flow rate. The experimental data are:

Run number	Controller output, u (%)	Air flow rate, F (scfm)
1	45	72.67
2	91	81.5
3	7	17.75
4	25	53.1
5	72	80.09
6	36	67.47
7	37	66.5
8	18	29.15
9	13	20.21

Nonlinear Regression Modeling for Engineering Applications: Modeling, Model Validation, and Enabling Design of Experiments, First Edition. R. Russell Rhinehart.

Run number	Controller output, u (%)	Air flow rate, F (scfm)
10	82	80.78
11	63	78.89
12	66	78.85
13	19	43.5
14	10	19.42

The constants in the model, 2783.2 and 2 represent expectations. The value for coefficient b is expected to be between 1/3 and 1, and the values of a and k must be positive. Seems simple!

1. Determine N, the number of regression optimizer trials that you should use.
2. Use any regression optimizer of your choice to determine the model coefficient values. Choose convergence criterion and threshold value.
3. Look at the CDF(OF) results and comment on the number of random optimizer initializations.
4. Test for several model validation aspects. Is the residual average (or median) for the best model nearly zero? Does the data variance seem uniform over the range. If yes, does the model pass the r-lag-1 test (relative to x, y, or run number)? If no, is there a pattern in the runs?
5. Is there an outlier? If yes, how is it determined and, when removed, what are the new model attributes?
6. The flow rate data are calculated from the orifice relation $F = 23.1(i - 4.02)^{0.46}$ with an observed range on the milliampere signal of ± 0.25. What is the uncertainty on the data model values for flow rate?
7. The valve position is not actually set. The u value is the signal to the valve, and after moving to a new value, the actual valve position might actually be off by a random amount (depending on the vagaries of the transition) with a likely maximum error of about 5%. There is no positioner on the valve. What might be the probable y-error in the data presentation of u w.r.t. F?
8. Does the experimental design comply with good practices (enough data points, enough range, sequence prevents environmental correlation, all data range covered)?
9. Does the model pass logical tests?
10. Does the model pass utility tests?
11. Since it seems that there is uncertainty on both the x and y data, explore Akaho's method of normal least squares as an estimate of maximum likelihood. Use that for the regression OF and repeat Exercises 3 to 5. Compare the results to vertical least squares.
12. Use bootstrapping to determine the 95% CL on the model (either OF or with or without the outlier removed). Is it a significant uncertainty relative to the data range?
13. Use propagation of uncertainty on the model to see if the implicit uncertainty in the given constant (2783.2) and nominal error in the ideal square relation (the 2 is often discovered to be better represented with 1.8) are inconsequential to uncertainty in the F data.
14. Use propagation of uncertainty in the orifice data model of Exercise 6 to see if implicit uncertainty in the model coefficients (23.1 and 0.46) is inconsequential to uncertainty in the F data.

21.2 CO_2 Orifice Calibration

An orifice flow meter is a single-point restriction for fluid flowing in a pipeline. When fluid flows, the restriction causes a pressure drop that is, ideally, proportional to the square of the fluid velocity. A transducer responds to the pressure drop and transmits an electrical 4–20 milliampere current to the control room computer that uses a calibration equation to convert the milliampere signal to flow rate. You will then develop the calibration equation, the process model.

In this case, the flowing fluid is gaseous CO_2. The data are from Vidhya Venugopal's calibration of a pilot-scale packed tower absorber unit as she was preparing it for the instructional chemical process lab at Oklahoma State University. Column 1 in the table is the run number, Column 2 is the measured milliampere signal from the transducer eyeball averaged (human observation and mentally averaged) over the measurement time interval of about 30 seconds, and Column 3 has the measured flow rate of the actual gas through a dry-test meter, corrected from actual pressure and temperature to standard conditions. The milliampere signal with zero flow is 3.95 mA, close enough to the ideal 4.0 mA. With a standard transducer, the pressure drop is linearly proportional to the milliampere deviation from the zero flow milliampere. Normally, the high flow rate should produce about a 20 mA signal, but in this equipment, the maximum flow rate only generated about 10.45 mA.

Run	i (mA)	F (scfm)
1	4.25	0.777
2	9.13	4.182
3	4.43	1.060
4	6.07	2.261
5	7.37	2.897
6	4.89	1.555
7	10.45	4.452
8	5.52	2.049

The process model is a power law representation, not the ideal square root model:

$$F = a(i - i_0)^b$$

The data model is the ideal gas law corrected measured volume per collection time interval (the P and T are absolute – Kelvin or Rankine – not deviation temperatures):

$$F_{data} = \frac{Q}{\Delta t} \frac{T_{std}}{T_{actual}} \frac{P_{actual}}{P_{std}}$$

The milliampere value is not uncertain. At the zero flow condition, the eyeball average might be off by ±0.01 mA. At the 10.45 mA value the fluctuation is larger, and has an eyeball uncertainty of about ±0.02 mA. Additionally, the measured data values are not uncertain. The Q-value is observed from a rotating dial at a timer signal, and both start and stop values have an uncertainty of ±0.01 CUFT, dominating the uncertainty in the data model.

Nominally, the model coefficients a and b would be the best fit to minimize the sum of squared deviations (SSD) of the eight data values from the corresponding model values.

However, the power law model can be linearized with a log-transformation. Also acknowledging uncertainty on the milliampere data, perhaps a maximum likelihood OF will be more appropriate.

1. There are eight data points; where should the next several be placed and what order will best randomize the sequence?
2. Propagate uncertainty on the data model to estimate the uncertainty on the "measured" flow rate for calibration. For all practical purposes the T, P, and Δt uncertainties are inconsequential.
3. Propagate uncertainty on the process model to estimate the uncertainty on the "measured" flow rate for the in-use application of the process model. Assume that the uncertainty on the a and b coefficients is negligible.
4. Use the results from Steps 2 or 3 to determine the threshold value for the convergence criterion on the optimizer. You could use any of a number of convergence criteria – change in DVs, change in OF, relative change in OF due to the change in DVs, and so on.
5. Log-transform the process model to linearize it and determine the a and b coefficients from linear regression. This does not need convergence criterion, because it determines an exact solution (disregarding the numerical truncation error in the computation).
6. Use a nonlinear optimizer to determine the a and b coefficients from the nonlinear power law model. As a caution, if you use the Excel chart trend line power law model, the coefficient values will actually be calculated from a log-transformed linearized model and will be exactly the same values as those from Step 5. Do not use the chart trend line. You should have slightly different a and b values from Step 5.
7. Use Akaho's approximation to maximum likelihood to determine the model coefficient values. For convenience use a constant uncertain range of the milliampere difference signal of about ± 0.03 mA. Since the OF is different, you should have slightly different a and b values from either Step 5 or 6.
8. The in-use model will predict flow rate, not the log-of-the-flow rate, and not the flow rate scaled by standard deviation. Therefore, to compare the models from Steps 3, 4, and 5, use the SSD of the flow rate. Which is better?
9. Is the average deviation of each model statistically indistinguishable from zero?
10. Does the average deviation from the model match the deviation expected from propagation of uncertainty on the data?
11. Apply the r-lag-1 test to each of the three models. Is there a trend in residuals w.r.t. chronological order, F or mA? Which approach is best for r-lag-1?

21.3 Enrollment Trend

There appears to be a cyclic trend in time of the national BS graduation rate of Chemical Engineers. The data in the table present the national graduation numbers of BS ChEs per year from 1980 to 2013. One conceptual model for the cycling phenomenon suggests that large BS production saturates the employment opportunity. Then high school graduates who might have chosen ChE choose an alternate degree path. Four or so years later, when the BS ChE graduation rate is lower than the employment demand, the high salary and seemingly large demand attracts a large number of freshmen matriculates, which floods the market four, or so, years later. If this concept is true there should be a strong relation between the freshmen matriculates and the national BS production. However, is it a linear, quadratic, exponential, reciprocal relation? Further, there is also a time period for news about career opportunities to become recognized, published, and digested by high school students to influence their decision. Therefore, there might be a several year delay between the influence (National BS rate) and the response (freshmen matriculates). The cycling phenomenon leads to a 2:1 swing in enrollment, which is important when universities are planning for the future, and a 2:1 swing in the pool of new hires, which is important for industrial workforce planning.

Academic year, ending in May of the date	NCES data, number of ChE BS degrees in the US	Freshmen matriculates at OSU declaring ChE as a major, in the fall of the year
1980	6320	—
1981	6527	—
1982	6740	—
1983	7185	—
1984	7475	47
1985	7146	36
1986	5877	35
1987	4991	29
1988	3917	29
1989	3663	34
1990	3430	43
1991	3444	53
1992	3754	58
1993	4459	73
1994	5136	67
1995	5901	55
1996	6319	57
1997	6564	40
1998	6319	57
1999	6033	54
2000	5807	48
2001	5611	52
2002	5462	59
2003	5109	54
2004	4742	59
2005	4397	67
2006	4326	68
2007	4492	66
2008	4795	80
2009	5036	97
2010	5740	96
2011	6311	103
2012	7027	159
2013	7529	140
2014	—	105
2015	—	130

1. What is the best lag (delay) between the response and influence variable? I would presume a linear model and explore the impact of the delay.
2. What is the best order of a power series model $\tilde{y} = a + bx + cx^2 + \cdots$. Perhaps use the final prediction error (FPE) to make the decision. Perhaps use my equal concern method.
3. Is there a better functional relation over the power series terms? Consider a reciprocal relation, or a noninteger power, or an exponential. How will you defend whether it is better?
4. If the model is true then there should not be a trend in the residuals with respect to either response, influence, or time. Use the r-lag-1 test to see this.
5. It appears that the data in time periods prior to and after 1997 are inconsistent. Segregate the 1997 and prior data and repeat the modeling. Explore a power law relation. Use both nonlinear regression and the log-transformed linearization of the model. The model coefficient should not have ideally identical values. Is there evidence that one model is better than the other?
6. Use bootstrapping to determine the 95% confidence limits on the model projected values.

21.4 Algae Response to Sunlight Intensity

Algae seem to offer a promising approach to renewable fuels and other biomaterials. They grow in ponds and use sunlight to convert CO_2 to biomass that includes oils that can be converted to fuel. A key aspect of modeling algae is how their growth depends on sunlight. Henley (1993) provides data that Jayaraman and Rhinehart (2015) used for modeling. The data are in the following table. The units on the intensity values were changed and the data reveals an excessive number of digits. For very low sunlight conditions, the growth factor is negative, indicating that the total biomass decreases during dark periods.

Sunlight intensity (W/m^2)I	Growth factor $f(I)$
0.000	−1.011
17.608	−0.855
30.967	−0.335
62.548	−0.023
66.679	0.862
80.085	1.278
111.761	1.381
124.955	2.266
142.775	2.943
201.169	4.087
295.723	5.437
499.135	6.576
734.316	7.610
1001.600	7.809
1495.750	7.375

The Jayaraman and Rhinehart empirical model is based on the observed trend that seems to have a classic exponential approach to an asymptotic value but with a non-zero initial value, and used this functional choice:

$$f(I) = a\left(1 - e^{-bI}\right) + c$$

However, the terms $a(1) + c$ can be combined to an alternate model:

$$f(I) = -ae^{-bI} + d$$

A classic empirical model is of the power series type. The data indicates that the order is higher than linear, but what is the right order?

$$f(I) = a + bI$$

$$f(I) = a + bI + cI^2$$

$$f(I) = a + bI + cI^2 + dI^3$$

$$f(I) = a + bI + cI^2 + dI^3 + gI^4$$

Another classic model is a power law model:

$$f(I) = a + bI^c$$

The data also appears to suggest that the slope at very low light has a tail, suggesting that a either a logistic or a second-order model might be appropriate:

$$f(I) = a + b\left[1 + e^{s(I-c)}\right]^{-1}$$

$$f(I) = a + be^{-cI} + de^{-gI}$$

In all of the models the coefficients a, b, c, d, g, and s would be adjusted in regression to make the model best match the data.

1. Reveal the parameter correlation in the first model and that the coefficient rearrangement in the second removes it. Discuss any aspects of the form of the first that makes it preferential over the second.
2. Graph the data and visually assess the trend. Conceptually, consider how light intensity might affect plant growth and what happens in the extreme I values. Now perform logical tests on each model and identify any that pass or fail expectations.
3. Use vertical least squares to fit the first model to the data and comment on the homoscedastic (or not) appearance of the data. Since there are only 15 data pairs, there are too few to use an F-test to distinguish variance differences between regions.
4. There is a significant slope change from the zero light to high-intensity conditions. If there was significant uncertainty on the I values and homoscedastic biomass measurement variability, then the vertical deviations at the low I region would be larger than those in the high I region. Use propagation of uncertainty to support that claim. Then use what you observe in the data to argue that there is insignificant uncertainty in the I values and

that vertical least squares is the appropriate regression objective (as opposed to maximum likelihood, total least squares, or Akaho's method).

5. Perform regression using each model. You may need to scale the I-data in the classic power series models so that the c, d, and g coefficient values are not vanishingly small and appear to have been converged.
6. Perform an r-lag-1 test on each model to see if there is enough autocorrelation in the data residuals to reject the hypothesis that the model represents the data.
7. Use an FPE (or similar) analysis to indicate the best order for the power series model.
8. The first (or the equivalent second) exponential model has one exponential term and three coefficients. The logistic model has four adjustable coefficients and the second-order exponential model has five adjustable coefficients. The fit to the data, as measured by the SSD, improves with the number of model coefficients. However, it is better to include complexity in assessing models. Use FPE (or similar) to choose the model that best balances goodness of fit with complexity.
9. Use the K-S test to determine if the residuals can be considered to be normal (Gaussian). If they were not normal, what action should be taken?
10. Use Chauvenet's criterion to determine if any data could be considered an outlier.
11. Use bootstrapping on the model of your choice to reveal the 95% uncertainty on the model due to the uncertainty in the experimental data.

21.5 Batch Reaction Kinetics

In a batch reaction ingredients are mixed in a vessel and then left to react. The contents of the vessel change in time as the reaction proceeds. The data below represent an enzyme catalyzed conversion of cellobiose to glucose in a water solution. The glucose product concentration is measured in time.

Ideally, an enzyme is not consumed by the reaction. It is like a key that opens a series of locks. The enzyme molecule randomly drifts in the medium, driven by thermal energy, bouncing off diverse molecules until it finds a cellobiose molecule. It links the cellobiose to water, converts it to two glucose molecules, and then returns to drifting while waiting to find another cellobiose molecule. Initially in the reactor, there are many cellobiose and enzyme molecules; the probability of one finding the other is high and the reaction rate is high. However, as the cellobiose is consumed, the chance of an enzyme molecule finding a target is reduced and the reaction rate is reduced.

An elementary, homogeneous type, kinetic expression for the isothermal (constant temperature) rate of generation of glucose is

$$r_G = 2kc_C c_E$$

The variable c indicates concentration of cellobiose or enzyme. The numerical value of 2 indicates that one cellobiose molecule is converted to two glucose molecules. Since water is also a reactant, the right-hand side (RHS) should also be multiplied by c_W; however, since water is in considerable excess, c_W does not change in time and that constant value has been included in the kinetic coefficient k.

That model is based on a concept that all three molecules collide at the same time and place. However, a more sophisticated model is of the Hougan–Watson (or Langmuir–Hinshelwood or Michaelis–Menten) type, which includes a representation of the mechanism for attachment of the enzyme to the substrate:

$$r_G = \frac{2k_1 c_C c_E}{k_2 + c_C}$$

It could be that the cellobiose molecule is the active one that needs to wander. Then

$$r_G = \frac{2k_1 c_C c_E}{k_2 + c_E}$$

There is suspicion that the enzyme population does not remain constant, that the enzyme degrades in time, as described by the following homogeneous reaction model:

$$r_E = k_3 c_E$$

Which combination of models (glucose 1, 2, or 3, and each with and without the enzyme degradation) best fits the data?

Reaction time (min)	Glucose concentration (mmol/l)
0	0
2	2.2
4	3.7
6	4.7
8	6.2
10	7.1
12	7.2
14	8
16	9.1
18	9.2
20	9.5
22	9.4
24	9.8
26	10
28	9.9
30	10.1

Thanks go to Julie Partner, Bron Shoffner, and Brice Stewart for generating the data as part of their undergraduate lab experience in the fall of 2014. The initial charge was 0.5 mM enzyme and 10 mM cellobiose per liter of water. The sampling was on a 2-minute schedule with an estimated uncertainty of ±5 seconds related to sample removal and assay time. The uncertainty of the glucose assay of ±0.6 mM/l was estimated from a range of replicate measurements at steady state.

Using a simple Euler's method (explicit finite difference approach) to numerically solve the differential equations with the two homogeneous models

$$c_{C,i+1} = c_{C,i} - 2(kc_{C,i}c_{E,i})$$

$$c_{E,i+1} = c_{E,i} - 2(k_3 c_{E,i})$$

where the numerical value of 2 represents the 2-minute time interval and the i subscript indicates the sampling time counter (experiment time = $2i$).

Once the cellobiose concentration is modeled, the glucose concentration can be calculated as

$$c_{G,i} = 2(c_{C,0} - c_{C,i})$$

Here the number 2 represents the stoichiometric conversion of one cellobiose to two glucose molecules.

1. Plot the data to see how glucose changes in time and from the data estimate whether the uncertainty on the sampling time is a significant factor or not. If the impact on modeled c_G is insignificant relative to the uncertainty in the c_G measurement, then vertical least squares is acceptable. If not, a method that accounts for both x and y variability (such as Akaho's) is appropriate.
2. Use vertical least squares to generate the three k values for the second cellobiose and the enzyme degradation models. (I think this is the best combination.) Explicitly choose the appropriate convergence criterion and its threshold. Run enough optimization trials from random initializations to be confident that the global best solution has been found. If residuals seem homoscedastic, use the r-lag-1 test to determine whether the data accepts or rejects the model and the t-test of the average of residuals to accept or reject that the model has an average residual near zero.
3. Repeat Step 2 using Akaho's approach for perpendicular least squares. If your conclusion from Step 1 is that the uncertainty in the sampling time is inconsequential, then the results here should essentially match those from Step 2.
4. Repeat Step 2 using the third cellobiose model. Is there enough statistical evidence to claim one model is better?
5. Repeat Step 2 using the first cellobiose model. This has one less coefficient to adjust. Is there enough statistical evidence to claim one model is better? Use FPE or an equivalent to judge the balance of fit and complexity.
6. Repeat Step 2, but exclude the enzyme degradation model and use the first cellobiose reaction model. Keep the original 0.5 mM/l concentration throughout the run. This has two fewer coefficients to adjust. Is there enough statistical evidence to claim one model is better? Use FPE or an equivalent to judge the balance of fit and complexity.
7. Repeat Step 2, but exclude the enzyme degradation model and use the second cellobiose reaction model. Keep the original 0.5 mM/l concentration throughout the run. This also has two (one less) coefficients to adjust. Is there enough statistical evidence to claim one model is better? Use FPE or an equivalent to judge the balance of fit and complexity. Look at the values of the resulting k_1 and k_2 coefficients and comment on parameter correlation.
8. Apply bootstrapping to a model of preference to determine the 95% confidence limit of the model.
9. Compare the variation in model residuals to the experimental uncertainty. Are they consistent?

Appendix A

VBA Primer: Brief on VBA Programming – Excel in Office 2013

A.1 To Start

- Open an Excel Worksheet (and save it).
- Set Macro Security Setting to Medium. Click on the Office Button in the upper left of the menu bar, Excel Options button in bottom of window, Trust Center in the left-hand menu, Trust Center Settings in the middle right of the window, and choose your security preference ("disable with notification", probably). Click OK.
- Press ALT-F11 to open the VBA editor. Alternately, use the "Launch VBA" icon on the Developer Toolbar. (To install the Developer Toolbar, click on the "Office Button" (upper left), click on "Excel Options" (lower border of the new window), in the "Popular" set, check the "Show Developer Tab in the Ribbon" box. Close the "Office Button" window and you should see the "Developer" tab on the upper ribbon.)
- In the new open window "VBA Editor" you should see three windows (Project, Properties, and Immediate) and a grayed space for another window. If the "Project" window does not appear, use the "View" menu item to add "Project Explorer".
- Right-click in the Project window, then click on "Insert" and then "Module". This will open a window in the grayed space for writing the VBA code. Alternately, you can double-click on the spreadsheet name in the Project window to open a code window. Either way allows you to write VBA code. However, code in a code window attached to a spreadsheet (for instance, "Sheet1") can only interface with that sheet.
- There are two categories of code – subroutine and function – with attributes similar to that of any language. To create a subroutine, type "SUB", spacebar, subname, and "()". For example, "Sub Practice()". Then hit Enter and "End Sub" will appear. Your code goes in between these two lines. To create a function, type "FUNCTION", spacebar, functionname, "(", argument list, and")". For example, "Function F_Multiply(x,y)". Then hit enter and

Nonlinear Regression Modeling for Engineering Applications: Modeling, Model Validation, and Enabling Design of Experiments, First Edition. R. Russell Rhinehart.

"End Function" will appear. Your code goes in between these two lines. Subs do not need an argument list, but can take one. There can be no spaces in the name of either the subroutine or the function, but you can use the underscore, such as "Sub Help_Me()".

A.2 General

- There are no rules about the use of columns (unlike Fortran columns 1–6, 7, and 8–72).
- To continue a long statement on the next line, break it with a blank_underscore at the end of the line. " _" acts like a hyphen in English.
- Names for variables, functions, and subroutines cannot include spaces or characters and must start with a letter. However, you can use the underscore to join two_words into a single name.
- Some combinations of letters are reserved for functions or values. These include log, sin, sqr, rand, name, TRUE, single, and so on. You cannot use these for variable names.
- The assignment symbol is simply "=", not ":=".
- There is no end-of-line symbol, like ";".
- Unless the user declares "Option Explicit", variable types do not need to be explicitly declared.
- Comments are indicated by the single quote mark, " ' ", and whether starting a new line or following code on the same line, anything on the line after the " ' " mark is ignored. The term "Rem", short for "Remark", does the same.
- Run a program by pressing the F5 key, or by clicking on the "Run arrow", which looks like a sideways triangle, on the VBA toolbar.
- If an Excel cell is in the edit mode, you cannot run a VBA program. Press the "Esc" key or click on another cell to close the cell in edit mode.

A.3 I/O to Excel Cells

- Cells are addressed by the row and column number, as Cells(4,2) representing the B4 cell. The operator name is "Cells", not "Cell". For your convenience, switch the Excel display from the default "A1" notation (column-as-a-letter row-as-a-number) to "R1C1" notation (row-as-a-number column-as-a-number). To do this, open an Excel workbook, click on the "Office Button" then "Excel Options" at the bottom of the window, then "Formulas" in the left-hand menu, and then check the R1C1 notation box in the "Working with Formulas" group of items. OK-out of the sequence.
- Like an array, "Cells" is the name (not "Cell") and the row and column numbers are the indices. The VBA assignment statement "A = Cells(5,6)" reads the value in cell "F5" (R5C6) and assigns it to the variable "A". "Cells(i,j) = q" assigns the value of "q" for display in the cell in the *i*th row and *j*th column.

A.4 Variable Types and Declarations

- Variables that store text are called "string" variables. Those storing integer values are called "integers". Double precision integers are called "long". Variables storing single precision real values are called "single" and double precision reals are called "double". They do not have to be explicitly declared, but it is good practice. They are declared using a "dimension" statement, "Dim" for short, which sets up storage space as follows:

```
Dim Name As String        'sets a location called "Name" for text
Dim I As Integer
Dim J As Long
Dim a As Single
Dim b As Double
```

- Dimensions of vectors and arrays have to be declared, but the variable type does not. Examples include:

```
Dim List_One(25) as String          'declares both number and type
Dim Table(10, 10)               'only declares the array size
Dim Intensity(20, 20, 20)
Dim Population(100, 3) As Double
```

- Assignment statements for string variables require double quotes. The others are conventional. Examples include:

```
List_One( k ) = "Paul Taylor"
Table( I , J ) = I * J
```

- It is good practice to organize all declarations in the header of the program, above all subroutine and function declarations, but VBA does not require this structure. If you declare variables within each subprogram, they are locally used, but variable values are forgotten when exiting the subprogram. Declaration in the header space above all subprograms makes the variable commonly available, with its value retained when control is switched from one subprogram to another.
- The statement "Option Explicit" requires every variable to be declared with a "Dim" statement. This is good for subsequent users, if you include the definition and units as a comment line. It is also good to be sure that you did not type JELLO for JELL0, or line for 1ine, or dimond2 for diamond2.

A.5 Operations

- Precedence of mathematical operators is the same as Fortran. Parenthesis first, in-to-out, then left-to-right. The parenthesis symbols are "(" and ")", whether nested or not. Exponentiation is second, in-to-out then left-to-right. The VBA symbol for exponentiation is " ^ " and in Fortran it is "**", followed by multiplication and division, left-to-right and then addition and subtraction left-to-right. These symbols are "*", "/", "+", and "–".
- Functions have parenthesis, and return the value that results from operating on their argument value. For example,

```
A = fun_fun(b)
```

- VBA can concatenate. The operator symbol is " & " (space-&-space). For example, the statements

```
Name2 = "Rhinehart"
Name1 = "Russ"
Cells( 5, 9 ) = Name2  &  ",    "  &  Name1
N = 3
Cells( 4, N ) = "R^"  &  N
```

would write "Rhinehart, Russ" in cell I5, and "R ^ 3" in cell C4.

A.6 Loops

The loop type closest in function to the Fortran Do loop is the FOR-NEXT loop. The loop is started with a FOR statement and an initialization of the loop index, an extreme limit, and an optional step increment. The loop ends with a NEXT statement, at which place the loop index is incremented and tested whether it is beyond the extreme to either exit the loop or re-enter. Here is an example of a loop that counts by 2s:

```
For k = 1 to 20 Step 2
      Cells(k, 1) = k
Next k
```

If the "Step" is not defined the default is "Step 1". "Step 5" would decrement.

A.7 Conditionals

- The conditional statement closest in function to the Fortran IF is the IF-THEN-ELSE conditional. It starts with "IF" followed by the condition and the word "THEN". If the condition is true, the code following the THEN is executed, otherwise not. The <, =, and > keyboard symbols are used for the comparison. Here are two examples, one with and one without an "ELSE":

```
IF name < "C" THEN Cells (17, I) = name
IF Abs(x) > = Abs(y) THEN
      Cells(k, 1) = "Great! x is at least as large as y"
ELSE
      Cells(k, 1) = "Sorry, x had a smaller magnitude than y."
END IF
```

- Compound antecedent tests require each comparison to be explicitly stated and joined with either the "AND" or "OR" or other conjunction operator. For example, to test if the value of x is between "a" and "b", as in "Is a < x < b?" you would write:

```
IF  a < x  AND  x < b  THEN …
```

- DO WHILE or DO UNTIL. These are preferred over the GOTO statement. The end of the WHILE or UNTIL range is the statement LOOP. The form is:

```
Constraint = "INDETERMINED"
DO UNTIL Constraint = "PASS"
      x = Rnd()           'Assigns a random value to variable x
      Constraint = Constraint_Test(x)      'Calls function
LOOP
```

A.8 Debugging

- If VBA detects an error during the compile stage (actually an interpreter but it does look at the entire code for syntax errors such as FOR-without-NEXT or a GOTO-Label-not-defined) (whether pre-execution or during your edit stage), or if it encounters an impossible operation during the execution stage, it will make a "BONK" sound (so that all of your friends can hear) and open an error message window. If you click "debug" it will highlight the problem line in yellow. This is termed being in "Break Mode". If the break mode occurs during execution, you can place the cursor on any variable in the VBA editor window and it will display its value. This is a great convenience for debugging.
- Click on the square "Reset" button on the toolbar to exit "Break Mode". You must do this to be able to run.
- If you want execution to stop at a particular line so that you can mouse-over variables to see their values, then click in the slightly gray column just to the left of the programming text area. It will create a red dot. The program will stop, in break mode, when it gets to that line.
- Alternately, you can step through the program during execution, line-by-line, by pressing the F8 key. VBA will highlight each line that is about to be executed. In "Break Mode" you can observe variable values from formerly executed lines by a mouse-over (placing the cursor over) the variable symbol. Use the Escape key or the reset button to exit this "Step Into" mode.
- Instead of printing intermediate values to the Excel cells for display, you can add a watch window to the VBA editor. Use the "Debug" drop-down menu, click on "Add Watch", and follow the directions.

A.9 Run Buttons (Commands)

- You can place a button on the Excel spreadsheet to run a subroutine. In Excel, open the "Developer" toolbar and click on the "Insert" icon in the "Controls" category. Then click on the "Button" icon in the upper left. Move the cursor to the location on the spreadsheet where you want the button (the cursor will have a + shape) and left-click. It will open a window that provides a list of subroutines and create a button in the grayed boundary edit mode. Choose the sub that you want the button to start and edit the button name. If the button is not in the active edit mode, right-click the button.

A.10 Objects and Properties

- The Excel Workbook is an object. Objects contain objects and objects have properties (attributes). For example, consider "The red car has a flat tire." Tire is an object of the car object. The car object has a property (color) and a property value (red). The tire object has a property (air pressure) and a value (0 psig). The Workbook object contains worksheet objects, which contain cell objects, which might contain a number. The number object has a numerical value, but it also has other properties such as font type, font size, and font color.
- You can use VBA commands to do any formatting thing that you can do using the Excel toolbar items – change font, create cell borders, set display characteristics, set font and background colors, set cell dimensions, and so on. Use the VBA help menu for details.

A.11 Keystroke Macros

- Often it is easier to perform an operation such as sort, plot, clear contents, and change font and color, with keystroke/mouse operations in an Excel Worksheet rather than in VBA. You can record the keystroke/mouse sequence in VBA and then call it from VBA programs.
- Click on the "Developer" tool tab on the Excel ribbon; then click on "Record Macro". This opens a window. Fill in the name you want to assign to the keystroke sequence and location to write it. I usually accept the default name and location. Perform your keystroke/mouse operations and then click on "Stop Recording" where you found the "Record Macro". The VBA code that the keystroke/mouse sequence represents will be in a new module in the VBA project window.
- Your VBA code can call that Macro subroutine.
- You can read the Macro code to see what the VBA instructions are.
- You can copy/paste that code, or appropriate sequences from it, into your VBA program.

A.12 External File I/O

To open a file for input or output, state "Open" the file path including the name, the purpose, then the number you choose for the file. If the path is not explicitly defined the default is the directory of the open Excel Workbook. The numbers are your choice. For example,

```
OPEN "C:/My Documents/VBA Programs/filename.txt" FOR OUTPUT AS #4
OPEN "source data.txt" FOR INPUT AS #1
```

You should close a file, when through, with the CLOSE #n, statement, for example, CLOSE #7.

INPUT and PRINT read and write from and to text files. Specify the file number followed by a comma and then the variable list. When the Input or Print list is complete, the next read or write "call" starts at the beginning of the next line in the file. Examples are:

```
INPUT #8, a, b, c(2)
PRINT #2, Name2 & ",  " & Name1, age
```

A.13 Solver Add-In

Solver is an optimization add-in to Excel. I am currently running Office 2013, but it is very similar in earlier versions. To install Solver, open Excel, click on the file, then options. This window should open as follows.

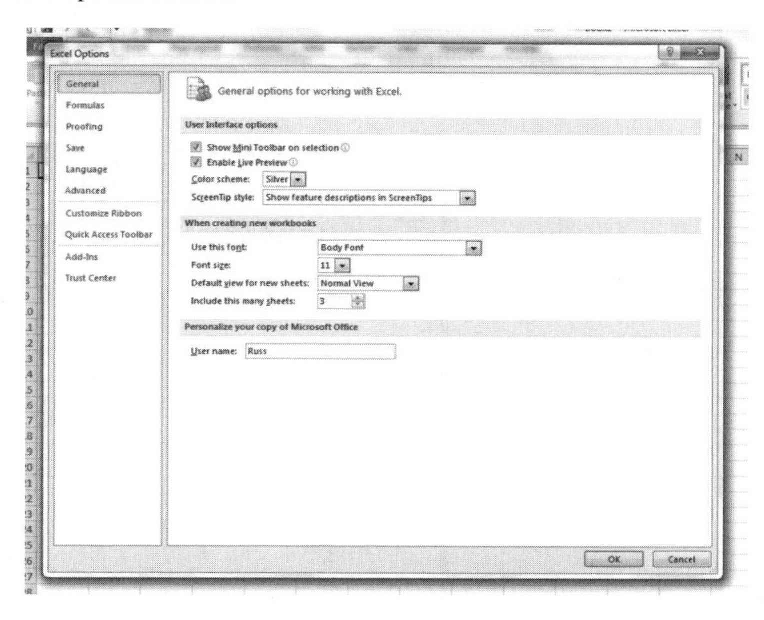

Click on Add-Ins in the left column and the view changes to this

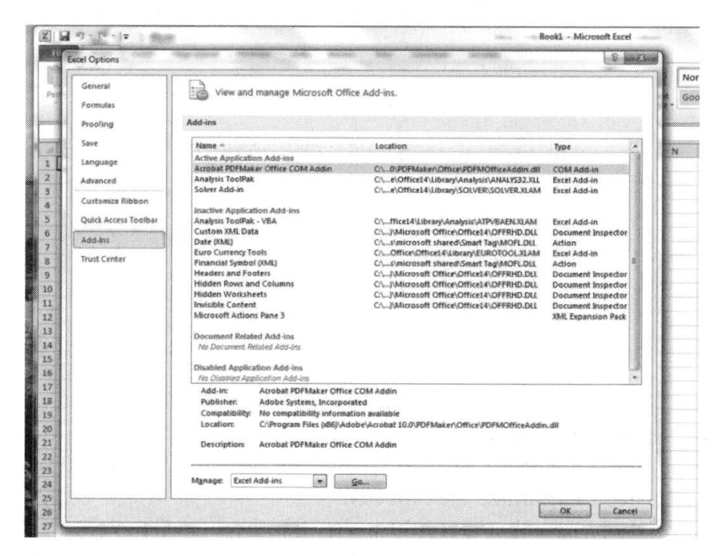

In the lower section, there is a "Manage" window, choose the "Excel Add-Ins" and click on the "Go" button. This opens a selection list.

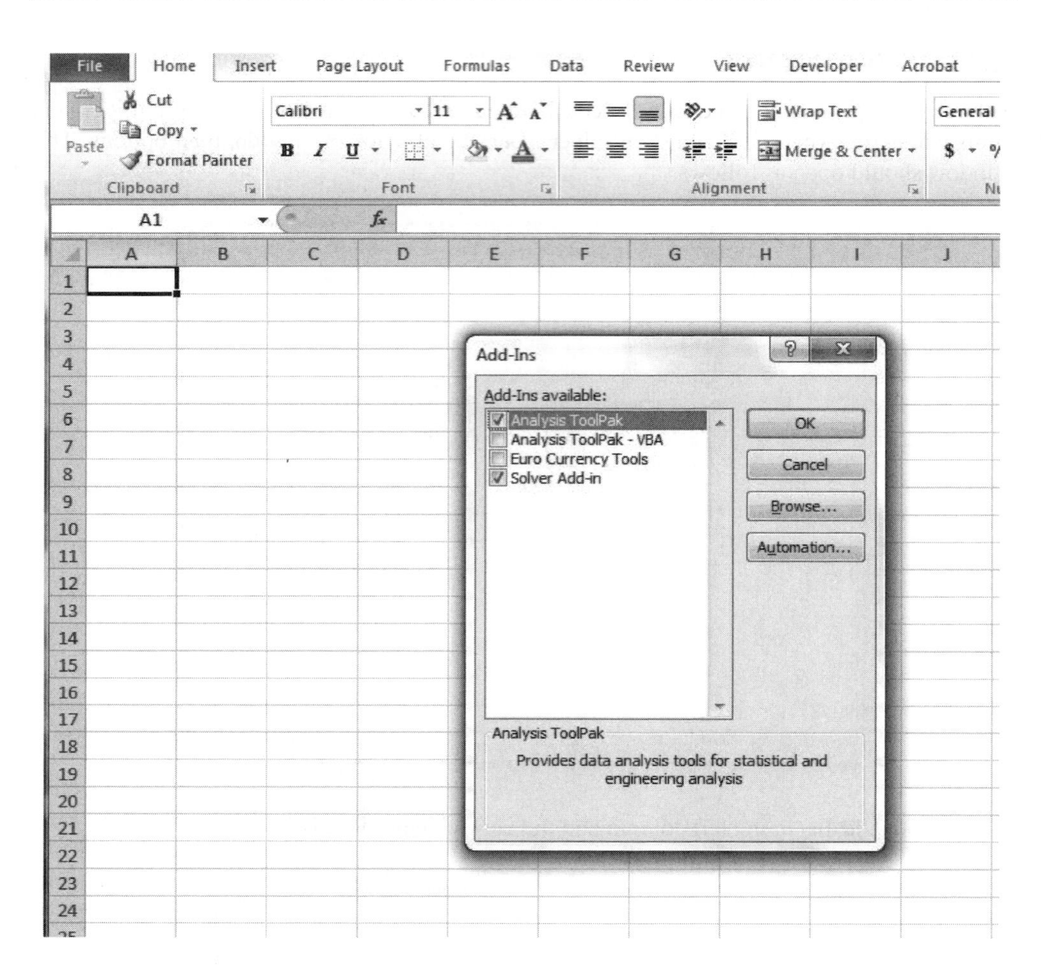

Click on the Solver box to check it, then click "OK". Now the Solver should appear on the far right of the Excel Data menu.

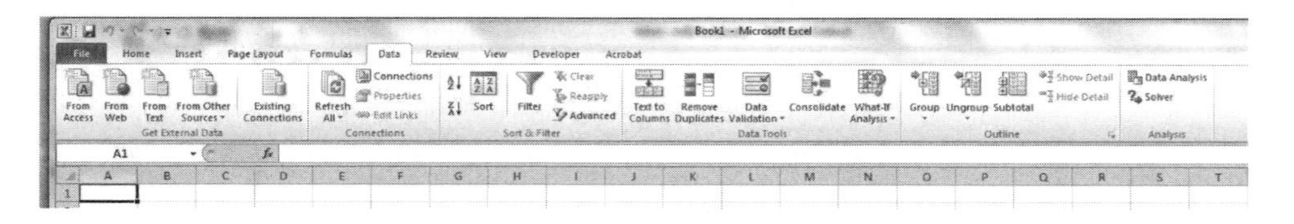

To use Solver, click on the menu icon. Then select the cell to be optimized (choose objective), choose min or max or value of, and choose the decision variable (DV) cells with the values that are adjusted to optimize the objective.

Solver has several optimization algorithms. In 2013 I prefer Generalized Reduced Gradient (GRG) Nonlinear. If you click on options you can change the convergence criterion and choice of forward or central difference derivatives. I think it is a Levenberg–Marquardt (L-M) approach that, when it encounters a DV constraint violation, reduces the number of DVs by one. I think it uses L-M on the reduced number of DVs and calculates the other DV to remain on the constraint. I think if it finds it is free of the constraint, it returns to the original L-M with all DVs. I suspect all of this because it is a conventional approach and because I am guessing at what Excel does. I cannot get answers from MS.

A.14 Calling Solver from VBA

To Call Solver in VBA:

1. Make sure the Solver add-in is installed to VBA.
   In VBA editor, go to Tools, then References Menu, and add solver as a reference to the project.
   Seek Solver help in VBA editor for more details.
2. Record a Solver Macro from a keyboard implementation of Solver.
3. Modify the arguments for Solver.

```
SolverOk SetCell:="$J$" & nI & "", MaxMinVal:=3, ValueOf:=0,_
        ByChange:="$D$" & nI & "", Engine _
        :=1, EngineDesc:="GRG Nonlinear"     'a typical call open for
                  'You'll need to add the following 2 lines to close
                                     solver and to keep the solution.
SolverSolve UserFinish:=True
SolverFinish KeepFinal:=1
```

A.15 Access to Computer Code

Those interested can visit the author's web site, www.r3eda.com, for open access to Excel VBA macros to many of the procedures in this book.

Appendix B

Leapfrogging Optimizer Code for Steady-State Models

This is written for VBA in Excel. Dots in the margin indicate the comment or code wraps around to the next line.

```
' Static Model Parameter Determination by Vertical Least Squares Minimization
'   Copyright R. Russell Rhinehart
'       School of Chemical Engineering, Oklahoma State University
'       rrr@okstate.edu
'       Last Revised 26 October 2015
'   Contact R. Russell Rhinehart, rrr@okstate.edu, for a gratis copy of the
'       Excel .xlsm file
'
'   Leapfrogging Optimization
'   Steady State Stopping Criteria
'   Best of N starts
'   Response Model is Static (Steady State) type. The new modeled value does not
'       depend on the past modeled value. But, this application does include a
'       delay to the influence, so order of the data does matter.
'
•  '   The user needs to change subroutines and some declarations to make the code
                                                          match the application
'       DIM statements - possibly add or change model coefficient names
'       SUB Assign - assigns model coefficient values from LF player values
'       SUB Response_Model - calculates the response
'       SUB ConstraintTest - tests to see if any constraint is violated
•  '   The user needs to enter data in the green cells in the Work Sheets labled Main
                                                                    and DataIN
•  '       This particular version of the code explores models of BS ChE enrollment
cycling.
'       It has two inputs and one response and four model coefficients.
•  '   The user could change the OF calculation from classic vertical SSD to other
                                                                       metrics
'       for goodness of fit in the OFCalculate subroutine
'   Press "Display Model" button to see the model response to user-entered coeffi-
cient values.
```

Nonlinear Regression Modeling for Engineering Applications: Modeling, Model Validation, and Enabling Design of Experiments, First Edition. R. Russell Rhinehart.

```
'       See results in the graphs - response w.r.t. input, and parity plot of
                                                               model w.r.t. actual.
'   Press the "Start Regression Optimizer" to run the N trials from randomized
                                                         initial coefficient values
'       Observe approach to steady state, and developing model progress
'       The CDF(OF) is displayed in the graph.
'   Press the "Display Best Model" to display the best of N models
'
'
'   **************************
Option Explicit     'This irritating statement requires you to define every
                                                  variable in a dim statement
                    'You can comment it out. It is good for being sure there
                                                        are no spellilng erors
                    'in variable names. And it is good for creating a variable
                                                           list to help a reader
                    'understand the program.

'       By defining all variables here, all variable values are common to all
                                                 functions and subroutines, and
'       values remain preserved when control is transferred from one subroutine
                                                          or function to another.

'   Arrays for experimental data

Dim udata(10, 1500) As Double   'experimental input value
Dim ydata(1500) As Double       'experimental response value
Dim Ndata As Integer            'Number of experimental data points in time,
                                        determined from data - maax 1500
Dim NInputs As Integer          'number of input variables to model - max 10
Dim DataNumber As Integer       'data number during iterations
Dim InputNumber As Integer      'input variable number
Dim yhigh As Double             'high value of y
Dim ylow As Double              'low value of y

'   Model arrays and coefficients
Dim ymodel(1500) As Double      'model response value, delayed value of yFO
Dim influence1 As Double        'value of one input to model (and to process)
Dim influence2 As Double        'value of second input to model (and to process)
Dim ModelA As Double            'value of model coefficient - you can add others
Dim ModelB As Double
Dim ModelC As Double
Dim ModelD As Double
Dim ModelE As Double
Dim ModelDelay As Double        'value of model coefficient

'   Constraint testing variable
Dim constraint As String        'returns "PASS" if no constraint violation,
                                                      "FAIL" if a violation

'   Main Program Control Variables
Dim IterationNumber As Integer  'counter for iterations in the LF Optimizer
Dim MaxIterations As Integer    'maximum number of iterations permitted within a
                                                                           trial
Dim TrialNumber As Integer      'counter for independent LF optimization trials
Dim Ntrials As Integer          'number of independent trials of LF optimizer
Dim StartNum As Integer         'start number for trials
Dim EndNum As Integer           'end number for trials
```

```
Dim MaxIter As Integer          'maximum iterations of optimizer to stop before
                                                                 convergence
Dim Iteration As Integer        'Iteration number comprised of DVDimension number
                                                                  of leapovers
Dim SubIteration As Integer     'To get all DVDimension leapovers
Dim Watch As String             'screen updating choice Y or N
Dim Include As String           'Include prior best player in the new trial
                                           (or totally randomize positions)
Dim Show As String              'show players
Dim CoefficientLabel(10) As String  'text name for the decision variable,
                                                             max 10 DVs.
Dim NowTime As Double

'   SS Stopping Criteria Variables
Dim RSdata(1500) As String      'selected data for random SSD 0=not included,
                                                               1=included
Dim status As String            'select status of a data set, taken or open.
Dim SelectedSetNumber As Integer    'Number of randomly selected set
Dim SSDRSS As Double            'sum of squared deviations for the ramdomly
                                                           selected sets
Dim RRMS As Double              'Root-mean-square value of randomly selected sets
Dim RRMSold As Double           'old value of RRMS
Dim RRMSfilter As Double        'Filter value for RRMS
Dim varnum As Double            'numerator measure of RRMS variance
Dim vardeno As Double           'denominator measure of RRMS variance
Dim NRSsets As Integer          'Number of data sets to be used for ran-
dom SSD - set at 20%
Dim RStatistic As Double        'Ratio-Statistic for SSID
Dim RCritical As Double         'Critical value of R fro SSID
Dim RSSIndex As Integer         'index for randomly selected set number

'   Leapfrogging (LF) Optimizer Variables
Dim DVNumber As Integer         'counter for DV dimension
Dim DVDimension As Integer      'number of DVs, here set to a maximum of 20
Dim PlayerNumber As Integer     'player number for LF teammates
Dim NumTeammates As Integer     'number of LF teammates, here set to a maximum
                                                                    of 50
Dim FirstOtherPlayer As Integer '2 if include prior best else 1
Dim LFLowpn As Integer          'player number for player with lowest OF value
Dim LFLowpnold As Integer       'player number for previous player with lowest
                                                                 OF value
Dim LFHighpn As Integer         'player number for player with highest OF value
Dim LFHighpnold As Integer      'player number for previous player with highest
                                                                 OF value
Dim PlayerPosition(10, 100) As Double       'DV position of the pn-th LF player
                                               (DV Dimension, Player Number)
Dim PlayerOFValue(100) As Double            'OF value of the pn-th LF player
Dim HighPlayerPosition(10) As Double        'x position of player with highest
                                                                 OF value
Dim LowPlayerPosition(10) As Double         'x position of player with lowest
                                                                 OF value
Dim PlayerLeapDelta(10) As Double           'change in x position from low to high
Dim OFhigh As Double            'OF value of lowest player in an iteration
Dim OFlow As Double             'OF value of lowest player in an iteration
Dim InitializeLow As Double     'low value of initialization range
Dim InitializeHigh As Double    'high value of initialization range

'       Evaluation and Convergence Criterion Variables
Dim NOFE As Integer             'Number of Function Evaluations
```

```
Dim SSD As Double                  'sum of squared deviations
Dim RMS As Double                  'Root-Mean Square of deviations
'
'    *************************************************
Sub main()

    Call ClearOldResults
    Call InitializeData

    For TrialNumber = StartNum To EndNum
        Cells(4, 6) = TrialNumber
        NOFE = 0
        If Watch = "N" Then Application.ScreenUpdating = False
        Call InitializeOptimizer
        PlayerNumber = LFLowpn
        Call Response_Model
        Call DataOut
        For Iteration = 1 To MaxIter
            Cells(5, 6) = Iteration
            For SubIteration = 1 To DVDimension 'Each iteration permits one
                                                  leapover per DVDimension
                Call Leapfrogging
                If Show = "Y" Then Call Show_Players
            Next SubIteration
            PlayerNumber = LFLowpn
            Cells(12, 6) = OFlow
            Cells(9, 6) = LFLowpn
            Cells(10, 6) = LFHighpn
            Call RandomSubsetSS
            If RStatistic < RCritical And Iteration > 50 Then Exit For
        Next Iteration

        Sheets("DataRESULT").Cells(TrialNumber + 8, 9) = TrialNumber / EndNum
        Sheets("DataRESULT").Cells(TrialNumber + 8, 10) = TrialNumber
        Sheets("DataRESULT").Cells(TrialNumber + 8, 11) = Iteration
        Sheets("DataRESULT").Cells(TrialNumber + 8, 12) = NOFE
        Sheets("DataRESULT").Cells(TrialNumber + 8, 13) = OFlow
        For DVNumber = 1 To DVDimension
            Sheets("DataRESULT").Cells(TrialNumber + 8, DVNumber + 13) = PlayerPo-
sition(DVNumber, LFLowpn)
        Next DVNumber

        If Iteration < MaxIter + 1 Then
            Cells(TrialNumber + 15, 1) = TrialNumber
            Cells(TrialNumber + 15, 2) = Iteration
            Cells(TrialNumber + 15, 3) = NOFE
            Cells(TrialNumber + 15, 4) = OFlow
            For DVNumber = 1 To DVDimension
                Cells(TrialNumber + 15, DVNumber + 4) = PlayerPosition(DVNumber,
                                                                        LFLowpn)
            Next DVNumber
        End If

        Call Sort_Trials

        Application.ScreenUpdating = True
        DoEvents            'Have Excel update graphs and display to catch up with
                                                                        VBA output
```

```
        NowTime = Timer    'Need to give screen time to update graphs in my
                                                                    computer
            Do Until Timer > NowTime + 0.1
        Loop

   Next TrialNumber

End Sub
'
'   ************************************************
Sub InitializeOptimizer()

    DVDimension = Cells(4, 4)
    NumTeammates = Cells(9, 4)

    If Include = "Y" Then
        PlayerNumber = 1
        For DVNumber = 1 To DVDimension
            PlayerPosition(DVNumber, PlayerNumber) = Sheets("DataRESULT").Cells
                                                                (9, DVNumber + 13)
        Next DVNumber
        Call Assign
        Call Response_Model
        Call OFCalculate
        PlayerOFValue(PlayerNumber) = RMS
        FirstOtherPlayer = 2
    Else
        FirstOtherPlayer = 1
    End If

    For PlayerNumber = FirstOtherPlayer To NumTeammates 'initialize player parame-
ter values - yInitial should be zero
        constraint = "Unassessed"
        Do Until constraint = "PASS"          'each must be in a feasible
                                                 location, repeat if not
            For DVNumber = 1 To DVDimension
                InitializeLow = Cells(3 + DVNumber, 9)
                InitializeHigh = Cells(3 + DVNumber, 10)
                PlayerPosition(DVNumber, PlayerNumber) = Initial-
izeLow + Rnd() * (InitializeHigh - InitializeLow) 'randomize km for the player,
                                                                could be negative
            Next DVNumber
            Call ConstraintTest
        Loop
        Call Assign
        Call Response_Model
        Call OFCalculate
        PlayerOFValue(PlayerNumber) = RMS
    Next PlayerNumber

    Call Find_High
    Call Find_Low
End Sub
'
'   ************************************************
Sub Find_High()
'   Search for player with highest OF value
```

```
    LFHighpn = 1 + Int(NumTeammates * Rnd())      'Random Assignment for
                                     initialization in case floor is flat
    OFhigh = PlayerOFValue(LFHighpn)
    For PlayerNumber = 1 To NumTeammates                    'search through all
                                                                     players
        If PlayerOFValue(PlayerNumber) > OFhigh Then            'Reassign if worst
            LFHighpn = PlayerNumber
            OFhigh = PlayerOFValue(PlayerNumber)
        End If
    Next PlayerNumber
    For DVNumber = 1 To DVDimension
        HighPlayerPosition(DVNumber) = PlayerPosition(DVNumber, LFHighpn)
    Next DVNumber
End Sub
'
'   ************************************************
Sub Find_Low()
'   Search for player with lowest OF value

    LFLowpn = 1                                     'start with PlayerNumber=1,
             if floor is flat, this serves as the base for convergence
    OFlow = PlayerOFValue(LFLowpn)
    For PlayerNumber = 2 To NumTeammates
        If PlayerOFValue(PlayerNumber) < OFlow Then             'Reassign if better
            LFLowpn = PlayerNumber
            OFlow = PlayerOFValue(LFLowpn)
        End If
    Next PlayerNumber
    For DVNumber = 1 To DVDimension
        LowPlayerPosition(DVNumber) = PlayerPosition(DVNumber, LFLowpn)
    Next DVNumber
                              'Display best of the initial set of players
    PlayerNumber = LFLowpn
    Call Assign
    Call Response_Model
    Call DataOut

End Sub
'
'   ************************************************
Sub Leapfrogging()
'   Relocate the player with the worst position to a random position to the other
                                                               side of the best.
'   If desired reevaluate the best to avoid finding a fortuitous best ever in
                                                          stochastic functions
                                   'relocate worst with the leapover the best
    constraint = "Unassessed"      'but must jump to an unconstrained area
    PlayerNumber = LFHighpn
    Do Until constraint = "PASS"
        For DVNumber = 1 To DVDimension
            PlayerLeapDelta(DVNumber) = LowPlayerPosition(DVNumber) - HighPlayer-
Position(DVNumber)          'difference between trial solutions with highest
                                                          and lowest OF values
           HighPlayerPosition(DVNumber) = LowPlayerPosition(DVNumber) + Rnd() *
PlayerLeapDelta(DVNumber) 'high (or infeasible) jumps to random position in
                                    window, repelled by recent vacated spots
        Next DVNumber
        For DVNumber = 1 To DVDimension
```

```
            PlayerPosition(DVNumber, LFHighpn) = HighPlayerPosition(DVNumber)
'reassign position of former high individual to its new feasible location
                                                               Next DVNumber
        Call ConstraintTest
    Loop

    Call Assign
    Call Response_Model
    Call OFCalculate
    PlayerOFValue(LFHighpn) = RMS
                                                    'find the individual
                                          with the lowest OF value presently
    If PlayerOFValue(LFHighpn) < OFlow Then         'If needed, reassign player with
                                                                  lowest OF value
        OFlow = PlayerOFValue(LFHighpn)
        For DVNumber = 1 To DVDimension
            LowPlayerPosition(DVNumber) = PlayerPosition(DVNumber, LFHighpn)
        Next DVNumber
        LFLowpn = LFHighpn
        Call DataOut
    End If
                                                    'find the individual with the
                                                      highest OF value presently
    If PlayerOFValue(LFHighpn) > OFhigh Then        'we know which is high
        OFhigh = PlayerOFValue(LFHighpn)
        For DVNumber = 1 To DVDimension
            HighPlayerPosition(DVNumber) = PlayerPosition(DVNumber, LFHighpn)
        Next DVNumber
    Else
        Call Find_High                              'need to search for the new high
    End If

    For DVNumber = 1 To DVDimension                 'display best player
        Cells(3 + DVNumber, 8) = PlayerPosition(DVNumber, LFLowpn)
    Next DVNumber

End Sub
'
'    ************************************************
Sub DeleteOldData()

    Range("D9:O2500").Select
    Selection.ClearContents
    Range("E1").Select

End Sub
'
'   ************************************************
Sub ClearOldResults()

    Sheets("DataRESULT").Select
    Range("I10:Y509").Select
    Selection.ClearContents
    Range("B9:E2009").Select
    Selection.ClearContents
    Range("A1").Select

    Sheets("DataIN").Select
    Range("D9:D1509").Select
```

```
    Selection.ClearContents
    Range("A1").Select

    Sheets("Main").Select
    Range("H4:H13").Select
    Selection.ClearContents
    Range("A16:X516").Select
    Selection.ClearContents
    Selection.Interior.ColorIndex = 0
    Range("F12").Select

End Sub
'
'   ***********************************************
Sub DisplayModel()

    Call InitializeData

    PlayerNumber = 1
    For DVNumber = 1 To DVDimension
        PlayerPosition(DVNumber, PlayerNumber) = Cells(3 + DVNumber, 8)
    Next DVNumber

    Call Assign
    Call ConstraintTest
    If constraint = "FAIL" Then End

    Call Response_Model

    Call DataOut

End Sub
'
'  ***********************************************
Sub DataOut()

    For DataNumber = ModelDelay + 1 To Ndata
            Sheets("DataIN").Cells(DataNumber + 8, 4) = ymodel(DataNumber)
    Next DataNumber

End Sub
'
'   ***********************************************
Sub DataInput()

    For DataNumber = 1 To 1500
        If Sheets("DataIN").Cells(DataNumber + 8, 5) = "" Then Exit For
    Next DataNumber
    Ndata = DataNumber - 1

    For DataNumber = 1 To Ndata
        ydata(DataNumber) = Sheets("DataIN").Cells(DataNumber + 8, 5)
'        ytime(DataNumber) = Sheets("DataIN").Cells(DataNumber + 8, 6)
        For InputNumber = 1 To NInputs
            udata(InputNumber, DataNumber) = Sheets("DataIN").Cells(DataNumber +
                                                       8, 5 + InputNumber)
        Next InputNumber
    Next DataNumber
```

```
End Sub
'
'   ************************************************
Sub Assign()
'   User needs to change this to fit the application

    ModelA = PlayerPosition(1, PlayerNumber)
    ModelB = PlayerPosition(2, PlayerNumber)
    ModelDelay = Int(PlayerPosition(3, PlayerNumber))
    ModelC = PlayerPosition(4, PlayerNumber)
    ModelD = PlayerPosition(5, PlayerNumber)
    ModelE = PlayerPosition(6, PlayerNumber)

End Sub
'
'   ************************************************
Sub Response_Model()
'   User needs to change this to fit the application

    NOFE = NOFE + 1           'Sum the Number Of Function Evaluations as a
                                            measure of computational work
    Cells(6, 6) = NOFE

    For DataNumber = ModelDelay + 1 To Ndata
        influence1 = udata(1, DataNumber - ModelDelay)
        influence2 = udata(2, DataNumber - ModelDelay)
'        ymodel(DataNumber) = ModelA * (influence1 ^ ModelB)       'good model
                                     for early time data (90000, -0.9, delay=4)
'        ymodel(DataNumber) = ModelA * (influence1 ^ ModelB) + ModelC *
(influence2 - ModelD)     'good model for all data rms=16 (1500, -0.4, delay=5,
                                                                   2.8, 1990)
         ymodel(DataNumber) = ModelA + ModelB * (influence2 - 1970) ^ ModelC -
ModelD * influence1 'best model for all data rms=10 (87, 3E-7, delay=4, 5.3, .008)
'        ymodel(DataNumber) = ModelA + ModelB * ((influence2 - 1970) ^ ModelC -
ModelD * influence1) / influence1 'good model for all data rms=12.4
                                                  (64, 0.8, delay=6, 3.6, 30)
'        ymodel(DataNumber) = ModelA * (influence1 ^ ModelB) * ((influence2 - Mod-
elC) ^ ModelD) 'good model for all data rms=13.2 (.0004, -0.5, Delay=5, 1916, 3.8)
'        ymodel(DataNumber) = ModelA + ModelB * (influence2 - 1970) ^ ModelC -
   ModelD * influence1 - ModelE * (influence1 / 6000) ^ 2 'best model for rms, but
                                                        one extra coefficient
    Next DataNumber

End Sub
'
'   ************************************************
Sub ConstraintTest()
'   User needs to change this to fit the application

    constraint = "PASS"

    ModelA = PlayerPosition(1, PlayerNumber)
    ModelB = PlayerPosition(2, PlayerNumber)
    ModelDelay = PlayerPosition(3, PlayerNumber)
    ModelC = PlayerPosition(4, PlayerNumber)
    ModelD = PlayerPosition(5, PlayerNumber)

    If ModelDelay < 0 Then constraint = "FAIL"
    If ModelDelay > 7 Then constraint = "FAIL"
```

```
    If ModelC > 1980 Then constraint = "FAIL"

        Cells(8, 6) = constraint
    If constraint = "FAIL" Then
        Cells(8, 6).Interior.ColorIndex = 3
    Else
        Cells(8, 6).Interior.ColorIndex = 4
    End If

End Sub
'
'   **********************************************
Sub RandomSubsetSS()

    If Iteration = 1 Then
        RRMSfilter = 0#
        varnum = 0#
        vardeno = 0#
        RRMSold = 0#
        NRSsets = Int(Ndata / 3 + 0.5)

        Sheets("DataRESULT").Select
        Range("B9:E1009").Select
        Selection.ClearContents
        Range("A1").Select
        Sheets("Main").Select
        Range("F12").Select
    End If

    PlayerNumber = LFLowpn
    Call Assign
    Call Response_Model

    For RSSIndex = 1 To Ndata
        RSdata(RSSIndex) = "open"
    Next RSSIndex

    SSDRSS = 0
    For SelectedSetNumber = 1 To NRSsets
        status = "taken"
        Do Until status = "open"
            RSSIndex = Int(Ndata * Rnd() + 0.5)
            status = RSdata(RSSIndex)
        Loop
        RSdata(RSSIndex) = "taken"
        SSDRSS = SSDRSS + (ydata(RSSIndex) - ymodel(RSSIndex)) ^ 2
    Next SelectedSetNumber
    RRMS = Sqr(SSDRSS / NRSsets)

    If Iteration = 1 Then RRMSfilter = RRMS

    varnum = 0.05 * (RRMS - RRMSfilter) ^ 2 + 0.95 * varnum
    RRMSfilter = 0.05 * RRMS + 0.95 * RRMSfilter
    vardeno = 0.05 * (RRMS - RRMSold) ^ 2 + 0.95 * vardeno
    RRMSold = RRMS
    If vardeno <> 0 Then RStatistic = (2 - 0.05) * varnum / vardeno
    Cells(13, 6) = RStatistic
    Sheets("DataRESULT").Cells(8 + Iteration, 2) = Iteration
    Sheets("DataRESULT").Cells(8 + Iteration, 3) = RRMS
```

```
    Sheets("DataRESULT").Cells(8 + Iteration, 4) = RRMSfilter
    Sheets("DataRESULT").Cells(8 + Iteration, 5) = LFLowpn

End Sub
'
'   ***********************************************
Sub InitializeData()

    If Cells(3, 15) > 0 And Cells(3, 15) < 1 And Cells(4, 15) > 0 And Cells(4, 15)
                                                                        < 1 Then
        EndNum = Int(0.5 + Log(1 - Cells(3, 15)) / Log(1 - Cells(4, 15)))
        If EndNum > Cells(6, 4) Then
            Cells(6, 13) = "End Number should be > " &   EndNum
            Range("M6:O6").Select
            Selection.Interior.ColorIndex = 3
            Range("F12").Select
        Else
            Cells(6, 13) = ""
            Range("M6:O6").Select
            Selection.Interior.ColorIndex = 0
            Range("F12").Select
        End If

    Else
        Cells(6, 13) = "c and f must both be between 0 and 1"
        Range("M6:O6").Select
        Selection.Interior.ColorIndex = 3
        Range("F12").Select
    End If

    Watch = Cells(8, 15)
    Include = Cells(9, 15)
    Show = Cells(10, 15)

    StartNum = Cells(5, 4)
    EndNum = Cells(6, 4)
    MaxIter = Cells(7, 4)

    RCritical = Cells(8, 4)

    NInputs = Cells(10, 4)

    DVDimension = Cells(4, 4)
    For DVNumber = 1 To 10
        CoefficientLabel(DVNumber) = Cells(3 + DVNumber, 7)
        Cells(15, DVNumber + 4) = CoefficientLabel(DVNumber)
        Sheets("DataRESULT").Cells(8, 13 + DVNumber) = CoefficientLabel(DVNumber)
    Next DVNumber

    Call Randomize

    Call DataInput

    If Show = "Y" Then
        Cells(27, 17) = "Player #"
        Cells(27, 18) = "RMS"
        For DVNumber = 1 To 10
            Cells(27, DVNumber + 18) = CoefficientLabel(DVNumber)
        Next DVNumber
```

```
    End If

End Sub
'
'   ************************************************
Sub OFCalculate()
'   This is based on vertical least squares.
'   User could make it based on Akaho's method, or other measures of closeness
                                                                        of fit

    SSD = 0
    For DataNumber = ModelDelay + 1 To Ndata
        SSD = SSD + (ydata(DataNumber) - ymodel(DataNumber)) ^ 2
        If SSD > 10000000000# Then Exit For
    Next DataNumber
    RMS = Sqr(SSD / Ndata)

End Sub
'
'   ************************************************
Sub Show_Players()

    For PlayerNumber = 1 To NumTeammates
        Cells(27 + PlayerNumber, 17) = PlayerNumber
        Cells(27 + PlayerNumber, 18) = PlayerOFValue(PlayerNumber)
        For DVNumber = 1 To DVDimension
            Cells(27 + PlayerNumber, 18 + DVNumber) = PlayerPosition(DVNumber,
                                                                PlayerNumber)
        Next DVNumber
    Next PlayerNumber

    Cells(LFHighpnold + 27, 17).Interior.ColorIndex = 0
    Cells(LFHighpn + 27, 17).Interior.ColorIndex = 3
    LFHighpnold = LFHighpn
    Cells(LFLowpnold + 27, 17).Interior.ColorIndex = 0
    Cells(LFLowpn + 27, 17).Interior.ColorIndex = 4
    LFLowpnold = LFLowpn

End Sub
'
'   ************************************************
Sub Sort_Trials()

    Sheets("DataRESULT").Select
    Range("J9:Z509").Select
    ActiveWorkbook.Worksheets("DataRESULT").Sort.SortFields.Clear
    ActiveWorkbook.Worksheets("DataRESULT").Sort.SortFields.Add Key:=Range( _
        "M9:M509"), SortOn:=xlSortOnValues, Order:=xlAscending, DataOption:= _
        xlSortNormal
    With ActiveWorkbook.Worksheets("DataRESULT").Sort
        .SetRange Range("J9:Z509")
        .Header = xlGuess
        .MatchCase = False
        .Orientation = xlTopToBottom
        .SortMethod = xlPinYin
        .Apply
    End With
    Range("F1").Select
    Sheets("Main").Select
```

```
    Range("F12").Select

End Sub
'
'   ************************************************
Sub DisplayBestModel()

    DVDimension = Cells(4, 4)
    For DVNumber = 1 To DVDimension
        Cells(3 + DVNumber, 8) = Sheets("DataRESULT").Cells(9, 13 + DVNumber)
    Next DVNumber

    Call DisplayModel

End Sub
```

Appendix C

Bootstrapping with Static Model

This is written in Excel VBA. Dots in the margin indicate the comment or code wraps around to the next line.

```
'   Bootstrapping with a steady state model
'       Scaled variables, Akaho's normal least squares objec-
tive function
'       use of either Solver or Heuris-
tic Cyclic Direct Search as the Optimizer
'
'   Copyright R. Russell Rhinehart
'       26 October 2015
'       School of Chemical Engineering, Oklahoma State University
'       rrr@okstate.edu
'   For a complementary copy of the Excel .xlsm file contact the
                                                          author.
'
'   This application is for a model of index of refraction as a
                                                    response to
'       a methanol-water mixture composition.
'
'   Computation of the residuals and SSD value is per-
formed in the Excel Cells
'   VBA Functions are used to calculate the model and the nor-
mal deviation squared

'   The dimensioning below limits data to 30 sets and realiza-
tions to 500
'   You can increase either if you wish.

Dim y(30, 500)
Dim DV(4)
Dim DDV(4)
```

Nonlinear Regression Modeling for Engineering Applications: Modeling, Model Validation, and Enabling Design of Experiments, First Edition. R. Russell Rhinehart.

```
'
'
Function model(x, a, b, c, d)
'   The commented out code includes a variety of optional four
                                                coefficient formula
'   The last is the simplest and best.
'   In this case scaling adds little value, but contributes to
                                                         complexity

    '  Scaled cubic has coefficient values of 0.0068, -0.0046,
                                                -0.0025, and 0.0008
    '  with Akaho SSD of 6.7 and r-lag-1 of 0.57 (reject)
'   xmax = Cells(1, 17)
'   ymax = Cells(2, 17)
'   xmin = Cells(1, 18)
'   ymin = Cells(2, 18)
'   xscaled = (x - xmin) / (xmax - xmin)
'   yscaled = a + b * xscaled + c * xscaled ^ 2 + d * xscaled ^ 3
                                    'traditional polynomial of order 3
'   model = ymin + yscaled * (ymax - ymin)

    '  Scaled power law has coefficient values of 0.026, 0.59,
                                                    -0.62, and 0.98
    '  with Akaho SSD of 0.22 and r-lag-1 of 0.18 (accept) but it is
                                      nearly a simple quadratic - reject)
'   xmax = Cells(1, 17)
'   ymax = Cells(2, 17)
'   xmin = Cells(1, 18)
'   ymin = Cells(2, 18)
'   xscaled = (x - xmin) / (xmax - xmin)
'   yscaled = a + b * xscaled + c * xscaled ^ d   'flexible power
                                                                law
'   model = ymin + yscaled * (ymax - ymin)

'   xmax = Cells(1, 17)
'   ymax = Cells(2, 17)
'   xmin = Cells(1, 18)
'   ymin = Cells(2, 18)
'   xscaled = (x - xmin) / (xmax - xmin)
'   yscaled = a + b * Sin(c * xscaled + d)
'   model = ymin + yscaled * (ymax - ymin)

    '  unscaled versions
'    model = a + b * x + c * x ^ 2 + d * x ^ 3
'    model = a + b * x + c * x ^ d
    model = a + b * Sin(c * (x + d))

End Function
'
'
Function akaho_d2(x, r, b, c, d)
'   Calculates the normal deviation squared
```

```
    sigmay = 0.001
    sigmax = 0.05

'    derivative = b + 2 * c * x + 3 * d * x ^ 2
'    derivative = b + c * d * x ^ (d - 1)
    derivative = b * c * Cos(c * (x + d))

    akaho_d2 = (r ^ 2) / (sigmay ^ 2 + (sigmax * derivative) ^ 2)

End Function
'
'
Sub bootstrap()
'   Routine to estimate the uncertainty on a Single-Input-Single
                                                        -Output
'       Steady-State model. x is input variable, y is response
                                                      variable.
'   Uses Excel Solver with GRG search for the optimizer for each
                                                 data realization.
'
'   R. R. Rhinehart - 30 August 2015
'   This reports the nominal model (based on all data) and the
'       extreme model values found in Bootstrapping with
'       user-defined Ntrials. It also reports the 95% probable
'       range on the model.
'   Since the extreme values are based on the extreme realization,
'       the values are dependent on the number of realizations.
'       Two for instance is not enough, and 1,000 might not randomly
'       select the data realization that reveals the extreme. My
'       experience is that 100 realizations is a good balance
                                                      of work
'       to finding the reasonable extremes.
'   By contrast, the 95% range is based on all of the realizations.
'       The 95% range is a more reproducable value, and more repre-
sentative.
'       But, it is based on the +/- 2-sigma limits, and predi-
cated on normal
'       (Gaussian) model variability. This is not checked.
    'Clear old data for sigma calculation
    Range("V8:Y510").Select
    Selection.ClearContents
    Range("Y2").Select
    Selection.ClearContents
    Range("G4").Select

    ' Count the number of data
    For i = 1 To 1000        'although the array will only hold up
                                                            to 30
        If Cells(7 + i, 17) = "" Then Exit For
    Next i
    Ndata = i - 1
```

```
' Transfer all data to realization set
For i = 1 To Ndata
    Cells(7 + i, 3) = Cells(7 + i, 17)
    Cells(7 + i, 4) = Cells(7 + i, 18)
Next i

' Run solver to get model based on all data - $G$6 is Akahko's
                                            normal least squares
' use $H$6 for standatd vertical least squares
SolverOk SetCell:="$G$6", MaxMinVal:=2, ValueOf:=0,
                                        ByChange:="$E$3:$E$6", _
    Engine:=1, EngineDesc:="GRG Nonlinear"
SolverSolve Userfinish:=True
solverfinish keepfinal:=1

' initialize nominal and high and low extremes with base case
                                                          model
For i = 1 To Ndata
    Cells(7 + i, 20) = Cells(7 + i, 19)
    Cells(7 + i, 21) = Cells(7 + i, 19)
Next i

'refresh graphs
DoEvents

' Start bootstrapping trials
Nrealizations = Cells(5, 18)
For bstrial = 1 To Nrealizations
    Cells(5, 21) = bstrial
    Application.ScreenUpdating = False 'turns off screen
updating
' Select data for realization
For i = 1 To Ndata
    j = 1 + Int(Ndata * Rnd())
    Cells(7 + i, 3) = Cells(7 + j, 17)
    Cells(7 + i, 4) = Cells(7 + j, 18)
Next i
' Sort selected data (just for human visualization, not needed
                                              for bootstrapping)
Range("C8:D58").Select
ActiveWorkbook.Worksheets("Sheet1").Sort.SortFields.Clear
ActiveWork-
book.Worksheets("Sheet1").Sort.SortFields.Add Key:=Range("C8:C58") _
, SortOn:=xlSortOnValues, Order:=xlAscending,
                                        DataOption:=xlSortNormal
With ActiveWorkbook.Worksheets("Sheet1").Sort
    .SetRange Range("C8:D58")
    .Header = xlGuess
    .MatchCase = False
    .Orientation = xlTopToBottom
```

```
      .SortMethod = xlPinYin
      .Apply
End With
' Return cursor to a convenient cell
Range("G4").Select
' Call Solver to optimize model based on this data realization
' $G$6 is Akaho's normal SSD, use $H$6 for vertical SSD
SolverOk SetCell:="$G$6", MaxMinVal:=2, ValueOf:=0,
                                        ByChange:="$E$3:$E$6", _
    Engine:=1, EngineDesc:="GRG Nonlinear"
SolverSolve Userfinish:=True
solverfinish keepfinal:=1
' Save in y-array and Update high and low extremes
For i = 1 To Ndata
    y(i, bstrial) = Cells(i + 7, 19)
    If Cells(i + 7, 19) > Cells(i + 7, 20) Then Cells(i + 7, 20)
                                                = Cells(i + 7, 19)
    If Cells(i + 7, 19) < Cells(i + 7, 21) Then Cells(i + 7, 21)
                                                = Cells(i + 7, 19)
 Next i
 ' Reveal progressive results to human observer
 Application.ScreenUpdating = True 'update the screen
 DoEvents 'pause to update all graphs in the worksheet
Next bstrial

' restore original data to realization set
For i = 1 To Ndata
    Cells(7 + i, 3) = Cells(7 + i, 17)
    Cells(7 + i, 4) = Cells(7 + i, 18)
Next i

'call Solver to reset the model coefficients for 100% of the
                                                         data
'   $G$6 is Akaho's normal SSD, use $H$6 for vertical SSD
SolverOk SetCell:="$G$6", MaxMinVal:=2, ValueOf:=0,
                                      ByChange:="$E$3:$E$6", _
    Engine:=1, EngineDesc:="GRG Nonlinear"
SolverSolve Userfinish:=True
solverfinish keepfinal:=1

'find 95% boundarys for model
For i = 1 To Ndata
    Cells(2, 25) = i
    For bstrial = 1 To Nrealizations     'list the y-values …
        Cells(bstrial + 7, 25) = y(i, bstrial)
    Next bstrial
    sigma = Cells(5, 25)                  '… so that Excel can
                                              calculate sigma
    Cells(i + 7, 22) = Cells(i + 7, 19) + 1.96 * sigma
    Cells(i + 7, 23) = Cells(i + 7, 19) - 1.96 * sigma
DoEvents
```

```
    Next i
    ' Return cursor to a convenient cell
    Range("G4").Select

End Sub
'
'
Sub bootstrap2()
'   Routine to estimate the uncertainty on a Single-Input-Single-
                                                          Output
'       Steady-State model. x is input variable, y is response vari-
able.
'   Uses Heuristic Cyclic Direct Search for the optimizer for each
                                                   data realization.
'
'   R. R. Rhinehart - 30 August 2015
'   This reports the nominal model (based on all data) and the
'          extreme model values found in Bootstrapping with
'          user-defined Ntrials. It also reports the 95% probable
'          range on the model.
'   Since the extreme values are based on the extreme realization,
'       the values are dependent on the number of realizations.
'       Two for instance is not enough, and 1,000 might not randomly
'       select the data realization that reveals the extreme. My
'       experience is that 100 realizations is a good balance of
                                                           work
'       to finding the reasonable extremes.
'   By contrast, the 95% range is based on all of the realizations.
'       The 95% range is a more reproducable value, and more repre-
sentative.
'       But, it is based on the +/- 2-sigma limits, and predi-
cated on normal
'       (Gaussian) model variability. This is not checked.

    'Clear old data for sigma calculation
    Range("V8:Y510").Select
    Selection.ClearContents
    Range("Y2").Select
    Selection.ClearContents
    Range("G4").Select

    ' Count the number of data
    For i = 1 To 1000     'although the array will only hold up to 30
        If Cells(7 + i, 17) = "" Then Exit For
    Next i
    Ndata = i - 1

    ' Transfer all data to realization set
    For i = 1 To Ndata
        Cells(7 + i, 3) = Cells(7 + i, 17)
        Cells(7 + i, 4) = Cells(7 + i, 18)
```

```
 Next i

' Run HCDS to get model based on all data
Call HCDS

' initialize nominal and high and low extremes with base case
                                                         model
For i = 1 To Ndata
 Cells(7 + i, 20) = Cells(7 + i, 19)
 Cells(7 + i, 21) = Cells(7 + i, 19)
Next i

'refresh graphs
DoEvents

' Start bootstrapping trials
Nrealizations = Cells(5, 18)
For bstrial = 1 To Nrealizations
    Cells(5, 21) = bstrial
    Application.ScreenUpdating = False 'turns off screen updating
    ' Select data for realization
    For i = 1 To Ndata
        j = 1 + Int(Ndata * Rnd())
        Cells(7 + i, 3) = Cells(7 + j, 17)
        Cells(7 + i, 4) = Cells(7 + j, 18)
    Next i
    ' Sort selected data (just for human visualization,
                                    not needed for bootstrapping)
    Range("C8:D58").Select
    ActiveWorkbook.Worksheets("Sheet1").Sort.SortFields.Clear
    ActiveWorkbook.Worksheets("Sheet1").Sort.SortFields
                                      .Add Key:=Range("C8:C58") _
    , SortOn:=xlSortOnValues, Order:=xlAscending,
                                          DataOption:=xlSortNormal
    With ActiveWorkbook.Worksheets("Sheet1").Sort
         .SetRange Range("C8:D58")
         .Header = xlGuess
         .MatchCase = False
         .Orientation = xlTopToBottom
         .SortMethod = xlPinYin
         .Apply
    End With
    ' Return cursor to a convenient cell
    Range("G4").Select
    ' Call HCDS to optimize model based on this data realization
    Call HCDS
    ' Save in y-array and Update high and low extremes
    For i = 1 To Ndata
        y(i, bstrial) = Cells(i + 7, 19)
        If Cells(i + 7, 19) > Cells(i + 7, 20) Then Cells(i + 7,
                                          20) = Cells(i + 7, 19)
```

```
            If Cells(i + 7, 19) < Cells(i + 7, 21) Then Cells(i + 7,
                                                    21) = Cells(i + 7, 19)
        Next i
        ' Reveal progressive results to human observer
        Application.ScreenUpdating = True 'update the screen
        DoEvents    'pause to update all graphs in the worksheet
    Next bstrial
    ' restore original data to realization set
    For i = 1 To Ndata
        Cells(7 + i, 3) = Cells(7 + i, 17)
        Cells(7 + i, 4) = Cells(7 + i, 18)
    Next i

    'call HCDS to reset the model coefficients for 100% of the data
    Call HCDS

    'find 95% boundarys for model
    For i = 1 To Ndata
        Cells(2, 25) = i
        For bstrial = 1 To Nrealizations    'list the y-values …
            Cells(bstrial + 7, 25) = y(i, bstrial)
        Next bstrial
        sigma = Cells(5, 25)                '… so that Excel can
                                               calculate sigma
        Cells(i + 7, 22) = Cells(i + 7, 19) + 1.96 * sigma
        Cells(i + 7, 23) = Cells(i + 7, 19) - 1.96 * sigma
        DoEvents
    Next i
    ' Return cursor to a convenient cell
    Range("G4").Select

End Sub
'
'
Sub HCDS()  'Heuristic Cyclic Direct Search optimization algorithm

    threshold = 0.0001        'convergence criterion for rela-
tive change in DV
    SSDColumn = 7            'to use Akaho's normal least squares
'    SSDColumn = 8            'to use standard vertical least squares

    For iDV = 1 To 4          'inialize the Trial Solution with
                                    values in Spread Sheet
        DV(iDV) = Cells(2 + iDV, 5)
        DDV(iDV) = 0.1 * DV(iDV)
    Next iDV

    OFbase = Cells(6, SSDColumn)
    For iCycle = 1 To 1000       'Limit of 1,000 iterations for the
                                                      optimizer
        Cells(1, 7) = iCycle
```

```
        For iDV = 1 To 4          'For each decision variable
            Cells(2 + iDV, 5) = DV(iDV) + DDV(iDV) 'increment the
                                                        DV value
            DoEvents                                 'use the
                                         worksheet to calculate results
            OF = Cells(6, SSDColumn)
            If OF < OFbase Then                      'optimizer logic
                                         to direct the next search step
                DV(iDV) = Cells(2 + iDV, 5)          'update base case
                                              and expand step if better
                DDV(iDV) = 1.2 * DDV(iDV)
                OFbase = OF
            Else
                Cells(2 + iDV, 5) = DV(iDV)          'replace base case
                                      adn reverse and contract if worse
                DDV(iDV) = -0.5 * DDV(iDV)
            End If
        Next iDV
        '    Test for convergence - either exit or return to the
                                                            search
        '    Test is based on relative changes so that a com-
mon threshold can be used for
        '        coefficients that are orders of magnitude
                                                   different.
        If Abs(DDV(1)) < thresh-
old * Abs(DV(1)) And Abs(DDV(2)) < threshold * Abs(DV(2)) And
Abs(DDV(3)) < threshold * Abs(DV(3)) And Abs(DDV(4)) < threshold
* Abs(DV(4)) Then Exit For
                                                           Next iCycle

End Sub
```

References and Further Reading

Akaho, S. (2008) Curve Fitting that Minimizes the Mean Square of Perpendicular Distances from Sample Points, citeseer.ist.psu.edu (accessed 15 September 2008).

Akaike, H. (1969) Fitting autoregressive models for prediction, *Annals of the Institute of Statistical Mathematics* **21**, 221–227.

Alekman, S. L. (1994) Significance tests can determine steady-state with confidence, *Control for the Process Industries*, **3**, 11, 62–64.

ASTM Standard E178-80 (1980) *Standard Practice for Dealing with Outlying Observations*, American Society for Testing and Materials, Philadelphia, PA, Annual Book of Standards.

ASTM Standard E29-88 (1989a) *Standard Practice for Using Significant Digits in Test Data to Determine Conformance with Specifications*, American Society for Testing and Materials Annual Book of Standards, Sect. 14, Vol. 2.

ASTM Standard E177-86 (1989b) *Standard Practice for Use of the Terms Precision and Bias in ASTM Test Methods*, American Society for Testing and Materials Annual Book of Standards, Sect. 14, Vol. 2.

Barolo, M., G. B. Guarise, C. Marchi, S. A. Rienzi, and A. Trotta (2003) Rapid and reliable tuning of dynamic distillation models on experimental data, *Chemical Engineering Communications*, **190**, 12, 1583–1600.

Bethea, R. M. and R. R. Rhinehart (1991) *Applied Engineering Statistics*, Marcel Dekker, Inc., New York.

Beyer, W. H. (ed.) (1998) *Handbook of Tables for Probability and Statistics*, CRC Press, Boca Raton, FL.

Bhat, S. A. and D. N. Saraf (2004) Steady-state identification, gross error detection, and data reconciliation for industrial process units, *Industrial and Engineering Chemistry Research*, **43**, 15, 4323–4336.

Bray, R. P. and R. R. Rhinehart (2001a) A simplified model for the etch rate of novolac-based photoresist, *Plasma Chemistry and Plasma Processing*, **21**, 1, 2149–162.

Bray, R. P. and R. R. Rhinehart (2001b) A method of mechanistic model validation with a case study in plasma etching, *Plasma Chemistry and Plasma Processing*, **21**, 1, 163–174.

Brown, P. R. and R. R. Rhinehart (2000) Demonstration of a method for automated steady-state identification in multivariable systems, *Hydrocarbon Processing*, **79**, 9, 79–83.

Cao, S. and R. R. Rhinehart (1995) An efficient method for on-line identification of steady-state, *Journal of Process Control*, **5**, 6, 363–374.

Cao, S. and R. R. Rhinehart (1997a) A self-tuning filter, *Journal of Process Control*, **7**, 2, 139–148.

Cao, S. and R. R. Rhinehart (1997b) Critical values for a steady-state identifier, *Journal of Process Control*, **7**, 2, 149–152.

Carlberg, P.J. and Feord, D.M. (1997) On-line model based optimization and control of a reactor system with heterogeneous catalyst, *Proceedings of the American Control Conference*, Albuquerque, NM, June 1997, pp. 2021–2022.

Castiglioni, P., G. Merati, and M. Di Rienzo (2004) Identification of steady states and quantification of transition periods from beat-by-beat cardiovascular time series: application to incremental exercise test, *Computers in Cardiology*, **31**, 2004, 269–272.

Chandak, C. (2009) *A utilitarian comparison of nonlinear regression methods*, MS thesis, Oklahoma State University.

Nonlinear Regression Modeling for Engineering Applications: Modeling, Model Validation, and Enabling Design of Experiments, First Edition. R. Russell Rhinehart.

Chandran, S. and R.R. Rhinehart (2002) Heuristic random optimization. *Proceedings of the 2002 American Control Conference*, Anchorage, AK, May 8–10, 2002, paper TM10-6.

Coleman, H. W. and W. G. Steele (1989) *Experimentation and Uncertainty Analysis for Engineers*, Wiley-Interscience, New York.

Crowe, E. L., F. A. Davis, and M. W. Maxfield (1955) *Statistics Manual*, Dover Publications, New York, **1955**.

Flehmig, F. and W. Marquardt (2006) Detection of multivariable trends in measured process quantities, *Journal of Process Control*, **16**, 947–957.

Forbes, J. F. and T. E. Marlin (1994) Model accuracy for economic optimizing controllers: the bias update case, *Industrial and Engineering Chemistry Research*, **33**, 1919–1929.

Fruehauf, P. S. and D. P. Mahoney (1993) Distillation column control design using steady state models: usefulness and limitations, *ISA Transactions*, **32**, 157–175.

Goncalves, C. M., M. Schwaab, and J. C. Pinto (2013) Comparison between linear and nonlinear regression, *Chemical Engineering Education*, **47**, 3, 161–169.

Hagan, M. T., H. B. Demuth, M. H. Beale, and O. De Jesus (2014) *Neural Network Design*, 2nd edn, Campus Publishing Service, University of Colorado, Boulder, CO.

Henley, W. J. (1993) Measurement and interpretation of photosynthetic light-response curves in algae in the context of photoinhibition and diel changes, *Journal of Phycology*, **29**, 6, 729–739.

Hooke, R. and T. A. Jeeves, (1961) Direct search solution of numerical and statistical problems, *Journal of the Association for Computing Machinery (ACM)*, **8**, 2, 212–229.

Huang, T (2013) *Steady state and transient state identification in an industrial process*, MS thesis, Oklahoma State University.

Huang, T. and R. R. Rhinehart (2013) Steady state and transient state identification for flow rate on a pilot-scale absorption column, *Proceedings of the 2013 American Control Conference*, Washington DC, June 2013.

Iyer, M. S. and R. R. Rhinehart (1999) A method to determine the required number of neural network training repetitions, *IEEE Transactions on Neural Networks*, **10**, 2, 427–432.

Iyer, M.S. and R. R. Rhinehart (2000) A novel method to stop neural network training. *Proceedings of the 2000 American Control Conference*, Chicago, IL, June 28–30, 2000, paper WM17-3, pp. 929–933.

Jayaraman, S. K. and R. R. Rhinehart (2015) Modeling and optimization of algae growth, *I&EC Research*, 2015, DOI: 10.1021/acs.iecr.5b01635, accepted 5 August 2015.

Jeison, D. and J. B. van Lier (2006) On-line cake layer management by trans membrane pressure steady state assessment in anaerobic membrane bioreactors, *Biochemical Engineering Journal*, **29**, 204–209.

Jiang, T., B. Chen, X. He, and P. Stuart (2003) Application of steady-state detection method based on wavelet transform, *Computers and Chemical Engineering*, **27**, 569–578.

Jubien, G. and G. Bihary (1994) Variation in standard deviation is best measure of steady state, *Control for the Process Industries*, **3**, 11, 64.

Katterhenry, P.R. and R. R. Rhinehart (2001) Use a virtual employee to trigger a sequence of conditions, *Control for the Process Industries*, **XIV**, **10**, 53–55.

Keeler, J. D., E. Hartman, and S. Piche (1998) Process modeling and optimization using focused attention neural networks, *ISA Transactions*, **37**, 1, 41–52.

Keller, J. Y., M. Darouach, and G. Krzakala (1994) Fault detection of multiple biases or process leaks in linear steady state systems, *Computers and Chemical Engineering* **18**, 10, 1001–1004.

Kennedy, J. and R. C. Eberhart (1995) Particle swarm optimization, *Proceedings of the 1995 IEEE International Conference on Neural Networks*, Perth, Australia, IEEE Service Center, Piscataway, NJ.

Kim, M., S. H. Yoon, P. A. Domanski, and W. V. Payne (2008) Design of a steady-state detector for fault detection and diagnosis of a residential air conditioner, *International Journal of Refrigeration*, **31**, 790–799.

Kuehl, P. and A. Horch (2005) Detection of sluggish control loops – experiences and improvements, *Control Engineering Practice*, **13**, 1019–1025.

Le Roux, G.A.C., B. F. Santoro, F. F. Sotelo, M. Teissier, and X. Joulia (2008) Improving steady-state identification, *Proceedings of the 18th European Symposium on Computer Aided Process Engineering – ESCAPE 18* (eds B. Braunschweig and X. Joulia), Elsevier B.V./Ltd, London.

Li, J. and R. R. Rhinehart (1998) Heuristic random optimization, *Computers and Chemical Engineering*, **22**, 3, 427–444.

Ljung, L. (1999) *System Identification: Theory for the User*, New York, Prentice Hall PTR.

Loar, J. (1994) Use SPC to compare process variables with control limits. *Control for the Process Industries*, **3**, 11, 62.

Mallows, C. L. (1973) Some comments on Cp, *Technometrics*, **15**, 661–675.

Mansour, M. and J. E. Ellis (2007) Methodology of on-line optimization applied to a chemical reactor, *Applied Mathematical Modelling*, DOI: 10.1016/j.apm.2006.11.014.

Meko, D. (2010) Notes on Autocorrelation, Notes_3, GEOS 585A, Spring 2009, from the course *Applied Time Series Analysis*, University of Arizona, http://www.ltrr.arizona.edu/~dmeko/notes_3.pdf (accessed 26 April 2010).

Miller, A. J. (1990) *Subset Selection in Regression*, Chapman & Hall, London.

Nachtwey, P. (2012) In LinkedIn Automation and Control Engineering Discussion Group, topic "Automated Steady State Identification", http://www.linkedin.com/groupAnswers?viewQuestionAndAnswers=&discussionID=144128653&gid=1967039&commentID=94720027&trk=view_disc&ut=3L2w_8e797U5o1.

Natarajan, S. and R. R. Rhinehart (1997) Automated stopping criteria for neural network training, *1997 American Control Conference, Proceedings*, Albuquerque, NM, June 4–6, 1997, paper TP09-4, pp. 2409–2413.

Nelder, J. A. and Mead, R. (1965) A Simplex method for function minimization, *The Computer Journal*, **7**, 308–313.

NIST/SEMATECH (2010) *e-Handbook of Statistical Methods*, http://www.itl.nist.gov/div898/handbook/ (accessed 13 September 2010).

Ochiai, S. (1997) Calculating process control parameters from steady state operating data, *ISA Transactions*, **36**, 4, 313–320.

Padmanabhan, V. and R. R. Rhinehart (2005) A novel termination criterion for optimization, *Proceedings of the 2005 American Control Conference*, Portland, OR, June 8–10, 2005, paper ThA18.3, pp. 2281–2286.

Padmanabhan, V. and R. R. Rhinehart, R.R (2008) Evaluation of a novel stopping criterion for optimization, *Proceedings, IASTED (International Association of Science and Technology for Development) Conference*, Quebec City, May 26–28, 2008, paper 622-045.

Prata, D. M., M. Schwaab, E. L. Lima, and J. C. Pinto (2009) Nonlinear dynamic data reconciliation and parameter estimation through particle swarm optimization: application for an industrial polypropylene reactor, *Chemical Engineering Science*, **64**, 3953–3967.

Ratakonda, S., U. M. Manimegalai-Sridhar, R. R. Rhinehart, and S. V. Madihally (2012) Assessing viscoelastic properties of chitosan scaffolds and validating with cyclical tests, *Acta Biomaterialia*, **8**, 4, 1566–1575.

Rhinehart, R. R. (1995) A novel method for automated identification of steady-state, *Proceedings of the 1995 American Control Conference*, Seattle, WA, June 21–23, 1995, paper FP05-3.

Rhinehart, R. R. (2002) Automated steady-state identification: experimental demonstrations, *Journal of Process Analytical Chemistry*, **7**, 2, 1–4.

Rhinehart, R. R. (2011a) Neural networks in processes and automation, Chapter 13, *Instrument Engineers' Handbook, Process Software and Digital Networks*, Vol. **3**, 4th edn (eds B. Liptak and H. Eren), Taylor & Francis Group, CRC Press, Boca Raton, FL.

Rhinehart, R. R. (2011b) Fuzzy logic control in processes and automation, Chapter 14, *Instrument Engineers' Handbook, Process Software and Digital Networks*, Vol. **3**, 4th edn (eds B. Liptak and H. Eren), Taylor & Francis Group, CRC Press, Boca Raton, FL.

Rhinehart, R.R. (2013) Tutorial: automated steady and transient state identification in noisy processes, *Proceedings of the 2013 American Control Conference*, Washington, DC, June, 2013.

Rhinehart, R. R. (2014) Convergence criterion in optimization of stochastic processes, *Computers and Chemical Engineering*, **68**, 1–6.

Rhinehart, R. R. (2016) Uncertainty – estimation, propagation, and reporting, Section 1.14, *Instrument Engineers' Handbook, Vol. I, Process Measurement and Analysis*, 5th edn (eds B. Liptak and K. Venczel), Taylor & Francis, CRC Press, Boca Raton, FL (in press).

Rhinehart, R. R., S. Gebreyohannes, U. Manimegalai-Sridhar, A. Patrachari, and Md S. Rahaman (2011) A power-law approach to orifice flow rate calibration, *ISA Transactions: The Journal of Automation*, **50**, 2, 329–341.

Rhinehart, R. R., M. Su, and U. Manimegalai-Sridhar (2012) Leapfrogging and synoptic leapfrogging: a new optimization approach, *Computers and Chemical Engineering*, **40**, 11, 67–81.

Salsbury, T.I. and Singhal, A. (2006) Method of and apparatus for evaluating the performance of a control system, US Patent 7,024,336, April 4, 2006.

Schladt, M., and B. Hu (2007) Soft sensors on nonlinear steady-state data reconciliation in the process industry, *Chemical Engineering and Processing*, **46**, 1107–1115.

Shrowti, N., K. Vilankar, and R. R. Rhinehart (2010) Type-II critical values for a steady-state identifier, *Journal of Process Control*, **20**, 7, 885–890.

Snyman, J. A. and L. P. Fatti (1987) A multi-start global minimization algorithm with dynamic search trajectories, *Journal of Optimization Theory and Applications* **54**, 121–141.

Snyman, J. A., and S. Kok (2008) A reassessment of the Snyman–Fatti dynamic search trajectory method for unconstrained global minimization, *Journal of Global Optimization*, DOI: 10.1007/s10898-008-9293-y.

Spendly, W., G. R. Hext, and F. R. Himsworth (1962) Sequential application of Simplex designs in optimization and evolutionary operation. *Technometrics*, **4**, 441–461.

Stanley, G. (2012) In LinkedIn Automation and Control Engineering Discussion Group, topic "Automated Steady State Identification", http://www.linkedin.com/groupAnswers?viewQuestionAndAnswers=&discussionID=144128653&gid=1967039&commentID=94720027&trk=view_disc&ut=3L2w_8e797U5o1.

Stephens, S. S. (1946) On the theory of scales of measurement, *Science*, **103**, 2684, 677–680, American Association for the Advancement of Science, June 7, 1946.

Su, M. and R. R. Rhinehart (2010) A generalized TSK model with a novel rule antecedent structure: structure and parameter identification, *Computers and Chemical Engineering*, **34**, 8, 1199–1219.

Subawalla, H., A. J. Zehnder, M. Turcotte, and M. J. Bonin (2004) Multivariable optimization of a multiproduct continuous chemicals facility, *ISA Transactions*, **43**, 153–168.

Svensson, C. (2012) In LinkedIn Automation and Control Engineering Discussion Group, topic "Automated Steady State Identification", http://www.linkedin.com/groupAnswers?viewQuestionAndAnswers=&discussionID=144128653&gid=1967039&commentID=94720027&trk=view_disc&ut=3L2w_8e797U5o1.

Szela, J. T. and R. R. Rhinehart (2003) A virtual employee to trigger experimental conditions, *Journal of Process Analytical Chemistry*, **8**, 1.

Vasyutynskyy, V. (2005) Passive Monitoring of Control Loops in Building Automation, Technical Report, Department of Computer Science, Technische Universitat Dresden.

Vennavelli, A. N. and M. R. Resetarits (2013) Demonstration of the SS and TS identifier at the Fractionation Research Inc. (FRI) Distillation Unit, *Proceedings of the 2013 American Control Conference*, Washington DC, June, 2013.

Wikipedia, Correlogram, http://en.wikipedia.org.wiki/autocorrelation_plot (accessed 13 September 2010).

Yao, Y., C. Zhao, and F. Gao (2009) Batch-to-batch steady state identification based on variable correlation and Mahalanobis distance, *Industrial and Engineering Chemistry Research*, **48**, 24, 11060–11070.

Ye, L., Y. Liu, Z. Fei, and J. Liang (2009) Online probabilistic assessment of operating performance based on safety and optimality indices for multimode industrial processes, *Industrial and Engineering Chemistry Research*, **48**, 24, 10912–10923.

Zhang, J. (2001) Developing robust neural network models by using both dynamic and static process operating data, *Industrial and Engineering Chemistry Research*, **40**, 1, 234–241.

Index

Nonlinear Regression Modeling for Engineering Applications: Modeling, Model Validation, and Enabling Design of Experiments, First Edition. R. Russell Rhinehart.